P2H4 table has problems

Mechanical Engineering Series

Frederick F. Ling
Editor-in-Chief

For other volumes in this series, go to
http://www.springer.com/series/1161

Sara McAllister · Jyh-Yuan Chen
A. Carlos Fernandez-Pello

Fundamentals of
Combustion Processes

 Springer

Sara McAllister
University of California, Berkeley
Department of Mechanical Engineering
Berkeley, CA
USA
Currently:
Research Mechanical Engineer
USDA Forest Service RMRS
Missoula Fire Sciences Laboratory
Missoula, MT
smcallister@fs.fed.us

Jyh-Yuan Chen
University of California, Berkeley
Department of Mechanical Engineering
Berkeley, CA
USA
jychen@me.berkeley.edu

A. Carlos Fernandez-Pello
University of California, Berkeley
Department of Mechanical Engineering
Berkeley, CA
USA
ferpello@me.berkeley.edu

Please note that additional material for this book can be downloaded from
http://extras.springer.com

ISBN 978-1-4419-7942-1   e-ISBN 978-1-4419-7943-8
DOI 10.1007/978-1-4419-7943-8
Springer New York Dordrecht Heidelberg London

Library of Congress Control Number: 2011925371

Printed on acid-free paper

Springer is part of Springer Science+Business Media (www.springer.com)

Mechanical Engineering Series

Frederick F. Ling
Editor-in-Chief

The Mechanical Engineering Series features graduate texts and research monographs to address the need for information in contemporary mechanical engineering, including areas of concentration of applied mechanics, biomechanics, computational mechanics, dynamical systems and control, energetics, mechanics of materials, processing, production systems, thermal science, and tribology.

Series Preface

Mechanical engineering, an engineering discipline forged and shaped by the needs of the industrial revolution, is once again asked to do its substantial share in the call for industrial renewal. The general call is urgent as we face profound issues of productivity and competitiveness that require engineering solutions, among others. The Mechanical Engineering Series features graduate texts and research monographs intended to address the need for information in contemporary areas of mechanical engineering.

The series is conceived as a comprehensive one that covers a broad range of concentrations important to mechanical engineering graduate education and research. We are fortunate to have a distinguished roster of consulting editors on the advisory board, each an expert in one of the areas of concentration. The names of the consulting editors are listed on the facing page of this volume. The areas of concentration are applied mechanics, biomechanics, computational mechanics, dynamic systems and control, energetics, mechanics of materials, processing, production systems, thermal science, and tribology.

Austin, Texas   Frederick F. Ling

Preface

Combustion is present continuously in our lives. It is a major source of energy conversion for power generation, transportation, manufacturing, indoor heating and air conditioning, cooking, etc. It is also a source of destructive events such as explosions and building and wildland fires. Its uncontrolled use may have damaging health effects through contamination of air and water. While combustion has helped humanity to prosper greatly, particularly with the use of fossil fuels, its indiscriminate use is altering the current global ecological balance through contamination and global warming. Thus, it is natural that combustion concerns people of all education levels, and it is important that the subject of combustion is taught at several levels of technical depth in schools and colleges.

Combustion is an interdisciplinary field with the interaction of thermodynamics, chemistry, fluid mechanics, and heat transfer, and, consequently, difficult to describe in simple terms and in a balanced manner between the different basic sciences. Many of the books currently available in combustion are geared to researchers in the field or to students conducting graduate studies. There are few books that are planned for teaching students that are not advanced in their technical studies. It is for this reason we have written this book aiming at readers that have not been previously exposed to combustion science, and that is at the undergraduate college level. We have often traded accuracy in our description and explanation of combustion processes for simplicity and easiness of understanding. Our readers should have knowledge of basic sciences, but are not necessarily advanced in their studies.

The book is based on lectures given by the authors through the years in a senior elective undergraduate combustion class in the Department of Mechanical Engineering at the University of California, Berkeley. The organization of the book chapters follows more or less those of other combustion textbooks, starting with a review of thermodynamics, chemical kinetics and the transport conservation equations. This is followed with chapters on the basic concepts of ignition, premixed and non-premixed combustion, and a chapter on emissions from combustion. The application of these basic concepts in practical combustion systems is implemented in a chapter devoted to internal combustion engines. Examples of problem solutions of different combustion processes are given through the book to help the student understand the material. A few problems are also given at the end of the different chapters.

In addition to the traditional class lectures, the course has a weekly demonstration laboratory where the students are exposed to the actual combustion processes presented in class.[1] We feel that these demonstration laboratories are very valuable to the students since they help them visualize the somewhat abstract concepts presented in class. For this reason, we have included as an appendix a description of several of the laboratories used in the class together with videos of some of the lab experiments to help a potential user of the book implement the laboratories.[2]

Finally, we would like to thank the graduate students that through the years have helped us as Teaching Assistants of the course and have helped us refine our class notes, and the Mechanical Engineering technical staff for the invaluable help running the demonstration laboratories. Our special thanks goes to Anthony DeFilippo for his unconditional help in commenting about the content of the book and revising and editing each chapter.

---

[1]Labs are located on Springer Extras at http://extras.springer.com/2011/978-1-4419-7942-1

[2]Links to laboratory video demonstrations are located in each lab. Readers can also find them at http://www.youtube.com/user/FndmtlsofCombustion

Contents

Nomenclature

a   exponent of Arrhenius reaction rate; crankshaft radius
A   area
A_o pre-exponential factor
[A]   molar concentration of species A
AFR   air-fuel ratio by mass ($1/f$)
AKI   anti-knock index

b   exponent of Arrhenius reaction rate
B   bore (engine cylinder diameter)
BMEP   brake mean effective pressure (atm)
BSFC   brake specific fuel consumption (g/kW-h)
BTDC   before top dead center

c   specific heat
c_p specific heat at constant pressure
c_v specific heat at constant volume
CAD   crank angle degree (θ)
CFD   computational fluid dynamics
CFR   cooperative fuel research
CHF   critical heat flux
CI   compression ignited
CN   cetane number
CNF   cumulative number function
CR   compression ratio, max cylinder volume/min cylinder volume
CVF   cumulative volume function

d   diameter
D_i diffusivity of species i
DI   direct injection
DPF   diesel particulate filter

E   total system energy
E_a activation energy

EA	excess air
EI	emission index
EGR	exhaust gas recirculation

f	fuel-air ratio by mass
f_s	stoichiometric fuel-to-air ratio by mass
F	radiation geometrical factor
FAR	fuel-air ratio (same as f)

g	Gibbs free energy per unit mass; acceleration due to gravity
G	Gibbs free energy
GDI	gasoline direct injection

h	enthalpy per unit mass
H	total enthalpy, kJ
$\hat{h}$	enthalpy per mole
$\tilde{h}$	convective heat transfer coefficient
h_{fg}	latent heat of vaporization
HCCI	homogeneous charge compression ignition
HHV	higher heating value per mass of fuel
HRR	heat release rate, btu/kW-h
$\Delta h°$	enthalpy of formation

IC	internal combustion
IDI	indirect injection
IMEP	indicated mean effective pressure

$\tilde{k}, k$	thermal conductivity
k_B	Boltzmann constant
k_i	Arrhenius kinetic rate constant
K	thermodynamic equilibrium constant

l, L	length
L_p	spray penetration distance
LFL	lean flammability limit
LHV	lower heating value per mass of fuel
LPG	liquified petroleum gas

m	mass
$\dot{m}$	mass flow rate
$\dot{m}''$	mass flux
M	molecular mass; third body species
MBT	max brake torque
MIE	minimum ignition energy
MON	motor octane number
MSE	mass species emission

n	moles, mol
$\dot{n}$	molar flow rate
OFR	oxygen/fuel ratio

P	pressure
PFI	port fuel injection
PM	particulate matter
PRF	primary reference fuels

$\dot{q}$	heat transfer rate
$\dot{q}''$	heat transfer rate per unit area
$\dot{q}'''$	rate of heat release per unit volume
$\dot{q}_{RxT}$	rate of reaction progress
Q_{12}	total heat input for process from state 1 to state 2
Q_c	heat of combustion
$Q_{rxn,p}$	heat of reaction at constant pressure
$Q_{rxn,v}$	heat of reaction at constant volume

r	radius
$\hat{r}$	reaction rate (rate of production or destruction of a chemical species per unit volume)
r_c	cut-off ratio
$\hat{R}_u$	universal gas constant
R_i	specific gas constant
RFL	rich flammability limit
RON	research octane number
RPM	revolutions per minute

s	entropy per unit mass
S	total entropy; surface area; molar stoichiometric air/fuel ratio
S_L	laminar flame speed
S_T	turbulent flame speed
SI	spark ignited
SMD	Sauter mean diameter
STP	standard conditions (25°C and 1 atm)

t	time
T	temperature
T_a	activation temperature
TDC	top dead center

u	internal energy per unit mass; velocity in x-direction
u'	characteristic turbulence velocity
U	total internal energy

v	specific volume
$V, \mathcal{V}$	volume
$\dot{V}, \dot{\mathcal{V}}$	volumetric flow rate
V	velocity

W	work
$\dot{W}$	power

x	distance
x_i	mole fraction of species i
X	body force
y_i	mass fraction of species i
α	thermal diffusivity; number of carbon atoms in fuel
β	droplet constant; number of hydrogen atoms in fuel
γ	ratio of specific heats; number of oxygen atoms in fuel
λ	normalized air-fuel ratio ($AFR/AFR_{\text{stoichiometric}}$)
δ	laminar flame thickness; boundary layer thickness
ε	emissivity; eddy diffusivity
η	thermal efficiency
η_c	combustion efficiency
η_v	volumetric efficiency
θ	crank angle, degrees; degrees of angle
μ	absolute viscosity
ν	kinematic viscosity
ρ	density
σ	surface tension
σ_s	Stefan-Boltzmann constant $= 5.67 \cdot 10^{-8}$ W/m^2-K^4
ϕ	equivalence ratio, f/f_s
Φ	spray cone angle
τ	characteristic time
ω_c, ω_p	net consumption/production rate

Subscripts

a	air
b	background (temperature); backward
c	characteristic; clearance
e	effective
eq	equilibrium
f	fuel; forward
g	gas
i	species, initial
l	liquid
L	losses; laminar
m	mean
o	outside; reference condition; orifice
P	product; constant pressure
R	reactant
s	solid; surface; stoichiometric
sat	saturation
st	stoichiometric

T   turbulent
v   vapor; constant volume
w   water

Superscripts

^{0} standard conditions (STP)

Overbars

^   quantity per mole
-    average value; nondimensional variable

Dimensionless numbers

Bi   Biot number $= \tilde{h}L/\tilde{k}_s$
Da   Damköhler number
Le   Lewis number $= \alpha/D_{AB}$
Nu   Nusselt number $= \tilde{h}L/\tilde{k}_a$
Pe   Peclet number $= lu/\alpha$
Pr   Prandtl number $= \nu = c_p\mu/\tilde{k}$
Re   Reynolds number $= \nu L/\nu$
Sc   Schmidt number $= \nu/D_{AB}$
We   Weber number $= \rho\nu^2 L/\sigma$

Physical Constants

Standard atmosphere (atm)	101.325 kPa
Universal gas constant ($\hat{R}_u$)	8.31447 kJ/kmol-K^3
	8.31447 kPa · m^3/kmol-K
	1.98591 kcal/kmol-K
	0.0831447 bar · m^3/kmol-K
	83.1447 bar·cm^3/mol-K
	82.0574 atm·cm^3/mol-K
Acceleration of gravity	9.807 m/s^2
Planck's constant	$6.625 \cdot 10^{-34}$ J-s
Stefan-Boltzmann constant	$5.67 \cdot 10^{-8}$ W/m^2-K^4

---

[3]The notation kJ/kmol-K means kJ divided by the product of kmol and K; equivalent to kJ/(kmol·K).

Conversion Factors

British units to SI units	SI units to British units
Density	
$1\ lb/ft^3 = 16.02\ kg/m^3$	$1\ kg/m^3 = 0.0624\ lb/ft^3$
Energy	
$1\ Btu = 1.054\ kJ$	$1\ kJ = 0.949\ Btu$
$1\ kcal = 4.184\ kJ$	$1\ kJ = 0.239\ kcal$
$1\ therm = 10^5\ Btu = 105.4\ MJ$	$1\ MJ = 9.49 \cdot 10^{-3}\ therm$
$1\ quad = 10^{15}\ Btu = 1.05 \cdot 10^{15}\ kJ$	$1\ kJ = 9.52 \cdot 10^{-16}\ quad$
Energy per unit mass	
$1\ Btu/lb = 2.324\ kJ/kg$	$1\ kJ/kg = 0.430\ Btu/lb$
$1\ cal/g = 4.184\ kJ/kg$	$1\ kJ/kg = 0.239\ cal/g$
Energy flux	
$1\ Btu/(h\text{-}ft^2) = 3.152\ W/m^2$	$1\ W/m^2 = 0.3173\ Btu/(h\text{-}ft^2)$
Force	
$1\ lb = 4.448\ N$	$1\ N = 0.2248\ lb$
Heat transfer coefficient	
$1\ Btu/ft^2\text{-}h\text{-}^\circ R = 5.678\ W/m^2\text{-}K$	$1\ W/m^2\text{-}K = 0.1761\ Btu/ft^2\text{-}h\text{-}^\circ R$
Kinematic Viscosity	
$1\ stokes = 10^{-4}\ m^2/s$	$1\ m^2/s = 10^4\ stokes$
Length	
$1\ ft = 0.3048\ m$	$1\ m = 3.281\ ft$
Mass	
$1\ lb = 0.4536\ kg$	$1\ kg = 2.2\ lb$
Power	
$1\ hp = 0.7458\ kW$	$1\ kW = 1.341\ hp$
Pressure	
$1\ atm = 101.3\ kPa = 1.013\ bar$	$1\ bar = 0.9871\ atm$
$1\ in.\ Hg = 3.376\ kPa$	$1\ kPa = 0.2962\ in.\ Hg$
$1\ in.\ H_2O = 0.2488\ kPa$	$1\ kPa = 4.019\ in.\ H_2O$
Specific heat	
$1\ Btu/lb\text{-}^\circ R = 4.188\ kJ/kg\text{-}K$	$1\ kJ/kg\text{-}K = 0.2388\ Btu/lb\text{-}^\circ R$
Surface tension	
$1\ lb/ft = 14.59\ N/m$	$1\ N/m = 0.06854\ lb/ft$
Temperature	
$1^\circ R = 0.5556\ K$	$1\ K = 1.8^\circ R$
Thermal conductivity	
$1\ Btu/h\text{-}ft\text{-}^\circ R = 1.73\ W/m\text{-}K$	$1\ W/m\text{-}K = 0.5780\ Btu/h\text{-}ft\text{-}^\circ R$

(continued)

British units to SI units	SI units to British units
Torque	
1 ft-lb = 1.356 N · m	1 N · m = 0.7375 ft-lb
Viscosity	
1 poise = 0.1 kg/m-s	1 kg/m-s = 10 poise
Volume	
1 ft^3 = 0.02832 m^3	1 m^3 = 35.31 ft^3
1 gal = 0.003785 m^3 = 3.785 Liter	1 Liter = 0.2640 gal
1 barrel = 42 gal = 0.15897 m^3= 158.97 Liter	1 Liter = 6.291 · 10^{-3} barrel
1 cord = 128 ft^3 = 3.625 m^3	1 m^3 = 0.2759 cord

Chapter 1
Fuels

Fuel and oxidizer are the two essential ingredients of a combustion process. Fuels can be classified as substances that liberate heat when reacted chemically with an oxidizer. Practical application of a fuel requires that it be abundant and inexpensive, and its use must comply with environmental regulations. Most fuels currently used in combustion systems are derived from non-renewable fossil sources. Use of these "fossil fuels" contributes to global warming effects because of the net-positive amount of carbon dioxide emissions inherent to their utilization. Fuels derived from biomass or from other renewable means represent potentially attractive alternatives to fossil fuels and are currently the subject of intensive research and development. This chapter provides a short introduction to the terminology for describing fuels commonly used in combustion.

1.1 Types of Fuel

Fuels for transportation and power generation can come in all phases: solid, liquid, or gas. Naturally occurring solid fuels include wood and other forms of biomass, peat, lignite, and coal. Liquid fuels are derived primarily from crude oil. The refining processes of fractional distillation, cracking, reforming, and impurity removal are used to produce many products including gasoline, diesel fuels, jet fuels, and fuel oils. Figure 1.1 shows typical end products from crude oil, with the lighter, more volatile components at the top.

The most widely used gaseous fuels for power generation and home heating are natural gas and liquid petroleum gas. In nature, natural gas is found compressed in porous rock and shale formations sealed in rock strata below ground. Natural gas frequently exists near or above oil deposits. Raw natural gas from northern America contains methane (~87.0–96.0% by volume) and lesser amounts of ethane, propane, butane, and pentane. Liquefied petroleum gas (LPG) consists of ethane, propane, and butane produced at natural gas processing plants. LPG also includes liquefied refinery gases such as ethylene, propylene, and butylene. Gaseous fuels can also be produced from coal and wood but are more expensive. Gasoline is used primarily in lightweight vehicles. As seen in Fig. 1.1, gasoline is a mixture of light

S. McAllister et al., *Fundamentals of Combustion Processes*,
Mechanical Engineering Series, DOI 10.1007/978-1-4419-7943-8_1,
© Springer Science+Business Media, LLC 2011

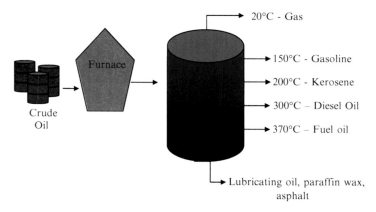

20°C - Gas

150°C - Gasoline

200°C - Kerosene

300°C – Diesel Oil

370°C – Fuel oil

Lubricating oil, paraffin wax, asphalt

Fig. 1.1 Typical end products from refining and distilling crude oil

distillate hydrocarbons from refined crude oil. The precise composition of gasoline varies seasonally and geographically and depends on the producer of the fuel. Diesel fuel is used in medium and heavy vehicles, as well as in rail and marine engines. Typical diesel fuel is also a mixture of hydrocarbons from refined crude oil, but it is composed of a blend of fuels with a higher boiling point range than that of gasoline. Fuel oil (commonly called "bunker" fuel) is widely used in large marine vessels.

Hydrocarbon fuels can come from sources other than fossil fuels as well. For example, biofuels are any kind of solid, liquid, or gaseous fuel derived from biomass, or recently living organisms. There are several types of biofuels: vegetable oil, biodiesel, bioalcohols, biogas, solid biofuels (wood, charcoal, etc.), and syngas. Notably, all of these forms of biofuels still require combustion of the fuel for power production, highlighting the continuous future dependence on combustion-related technology for transportation and power generation. Straight vegetable oil can be used in some diesel engines (those with indirect injection in warm climates), but typically it is first converted into biodiesel. Biodiesel is a liquid fuel that can be used in any diesel engine and is made from oils and fats through a process called transesterification. Figure 1.2 shows this process in detail. Compared to traditional diesel fuel, biodiesel can substantially reduce emissions of unburnt hydrocarbons, carbon monoxide (CO), sulfates, and particulate matter. Unfortunately, emissions of nitrogen oxides (NO_x) are not reduced.

Bioalcohols, such as ethanol, propanol, and butanol, are produced by microorganisms and enzymes that ferment sugars, starches, or cellulose. Ethanol from corn or sugar cane is perhaps the most common, but any sugar or starch that can be used to make alcoholic beverages will work. Currently in the U.S., ethanol is often blended with normal gasoline by about 5% by volume to increase efficiency and reduce emissions. With some modifications, many vehicles can operate on pure ethanol. The production of ethanol is a multi-stage process that involves enzyme digestion to release the sugar from the starch (hydrolysis), fermentation, distillation, and drying. Some opponents argue that the move toward an ethanol economy will have a

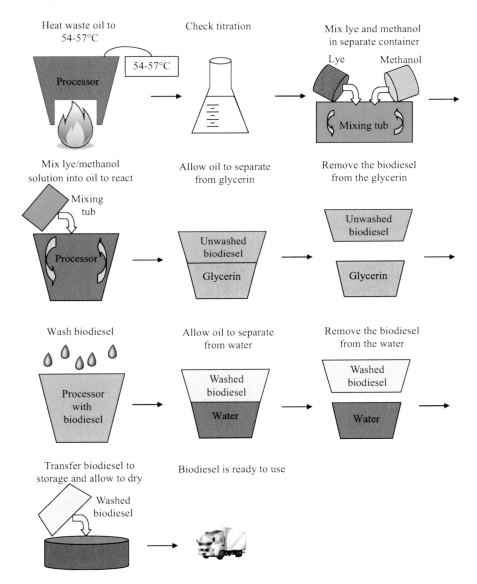

Fig. 1.2 Biodiesel production process

negative impact on global food production, impacting the poorest countries the most. Using cellulose from nonfood crops or inedible waste products would help alleviate this potential problem. However, cellulose is much more difficult to break down with standard enzymes and therefore requires a longer, more expensive process. Figure 1.3 details the additional steps required to isolate the sugar from the cellulose before the fermentation process can begin. An alternative approach is

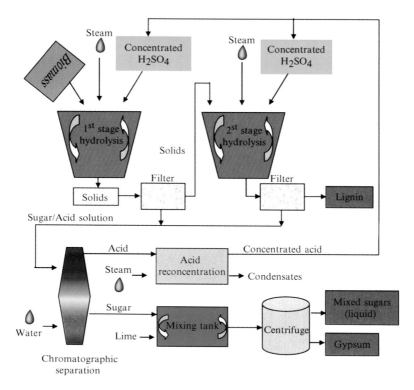

Fig. 1.3 Simplified flow diagram of the conversion of cellulose/hemicellulose to mixed sugars using Arkenol's concentrated acid hydrolysis

the thermal pyrolysis (degradation) of wood to produce biofuel. Because the heat of combustion of the pyrolysis products is larger than that of the heat of pyrolysis of wood, the overall energy balance may be positive, making this method viable. One problem with this approach is that the pyrolysis products are gaseous, thus they are not easily condensed and often have low energetic value.

Biogas is generated from the anaerobic digestion of organic material, such as municipal (landfills) and animal waste. When these materials decompose, they release methane. If this gas is collected and used for power generation, greenhouse gas emissions are reduced both directly and indirectly by reducing the amount of methane released into the atmosphere and by displacing the use of non-renewable fuels. Several biogas power plants are currently in operation, such as the Short Mountain Power Plant in Eugene, Oregon that produces 2.5 MW annually and provides electricity for about 1,000 homes [2].

Syngas (from synthesis gas) is a mixture of combustible gases produced by the gasification of a carbon-containing fuel such as coal or municipal waste. Another method of producing syngas is through steam reforming of natural gas. Typically, syngas is a combination of carbon monoxide, carbon dioxide, and hydrogen.

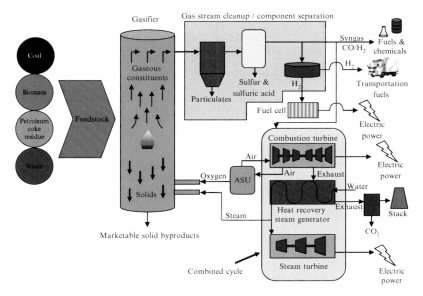

Fig. 1.4 Gasification process for syngas production [3]

The benefit of syngas is that it converts solid feedstock into a gaseous form that can be more easily used for power generation. Figure 1.4 details this process.

An alternative to hydrocarbon fuels is hydrogen. The use of hydrogen in the transportation and power generation industries is receiving increased attention, primarily because hydrogen provides a means for energy storage and subsequent conversion into power with reduced pollutant emissions. When hydrogen combusts in air, the products are water and nitrogen, but there is potential to form nitrogen oxides (NO_x). The main advantages of hydrogen are that it burns easily, it can be used almost directly in systems that are well developed and reliable, and it can significantly reduce fossil fuel consumption. However, because hydrogen burns so easily, safety is a major concern. Hydrogen can be produced two ways: by the decomposition of water through electrolysis or by the reformation of fossil fuels. Electrolysis is attractive because it can generate hydrogen from carbonless energy sources such as solar, wind, or nuclear, without emissions of CO_2. In this way, hydrogen production provides a means to store the energy generated from sources normally limited by their variability (i.e. solar and wind).

1.2 Fuel Usage

Figure 1.5 shows the energy consumption of the United States from 1949 through 2008 [1]. Energy consumption has steadily increased during this period. The primary source of energy by far has been from petroleum products. The only

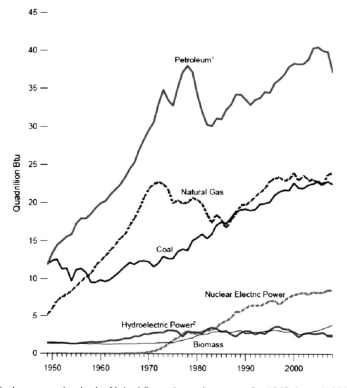

Fig. 1.5 Fuel consumption in the United States by major source for 1949 through 2008 [1]

major declines in petroleum consumption occurred during the energy crises in 1973 and in 1977. By 2000, U.S. petroleum imports had reached an annual record of 11 million barrels per day. Despite an increase in alternative energy sources, nearly 40% of the energy consumption in 2008 was from petroleum (see Fig. 1.6).

From Fig. 1.6, 89% of the energy consumption in the United States in 2008 was from technologies that require combustion. This figure is not expected to change dramatically in the near future, so there is a clear need for ongoing research and development on combustion systems so that the consumption of fossil fuels and the resulting emissions can be reduced.

1.3   Basic Considerations of the Choice of Fuels

Fuel and oxidizer are the primary components in combustion. For most combustion processes, air is used as the oxidizer because air is free and available almost everywhere on earth. The choice of fuel will depend on the purpose of the

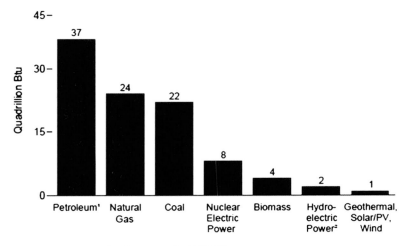

Fig. 1.6 Energy consumption by source for 2008 [1]

combustion process and is subject to local safety and emission regulations. Several factors listed below affect the choice of fuel.

1. *Energy content per volume or per mass.* When space (or weight) is limited, the energy content of a fuel per unit volume plays an important role in determining the amount of volume needed. Normal liquid hydrocarbon fuels contain about 33 MJ/L. Due to oxygen content, alcohol fuels such as ethanol contain a slightly less energy, about 29 MJ/L. Gaseous fuels often contain much less energy per unit volume due to the large volume occupied by the gaseous molecules. Hydrogen at standard conditions (STP) contains only 12 kJ/L (note though that hydrogen has higher energy content per mass). Therefore gaseous hydrogen needs to be compressed to about 2,500 atm to get the equivalent energy per volume as hydrocarbon fuels. This obviously raises safety issues and also weight issues since the hydrogen must be stored in heavy bottles. For the purpose of heating a home or providing hot water, fuels with low heat content may be adequate. If pipelines are available for delivery of gaseous fuels, the heating content may be less important in the selection of fuel. For transportation applications, liquid fuels are preferred due to their high energy content. Most cars are currently operated with liquid fuels. Liquid hydrogen and oxygen are used in the Space Shuttles. Due to its very low boiling point ($-252.76°C$ or $-422.93°F$), liquid hydrogen can be stored in the tank for only a few hours before it starts boiling due to heat transfer from its surroundings. When converted from liquid to gas, hydrogen expands approximately 840 times. Its low boiling point and low density result in rapid dispersion of liquid hydrogen spills. For applications in vehicles, the liquid hydrogen would start boiling within a couple of days even with the current best insulation technologies. For fuel cell vehicles using hydrogen, the low energy density of gaseous hydrogen presents a technical problem. Therefore, storage of hydrogen is

currently a research topic being pursued worldwide. Potential options include high-pressure tanks, metal hydrates to absorb hydrogen, and ammonia as a hydrogen carrier.

2. *Safety.* Safety is an important factor in selecting fuels, especially for transportation applications. The fuel must be safe to handle yet easy to burn under the designed engine conditions. Many properties of the fuel, such as vapor pressure, minimum ignition energy, flammability, toxicity, and heat release rate, can influence safety in different ways. Although volatile liquid fuels such as gasoline present safety issues if spilled because they ignite easily, they are quite safe in a fuel container. Similarly, heavy hydrocarbons, such as naphthalene (used to make moth balls), are solid at room temperature and are easy to handle, but they may melt if exposed to heat and burn releasing high amounts of heat. The ease of ignition and the rate of heat release are important factors in the rapid development of fire. Plastics, for example, ignite relatively easily and release large amounts of heat when burning. Consequently they are more dangerous from a fire safety point of view than wood, which is difficult to ignite and burn. The products of combustion from plastics are also more toxic than those from wood.

3. *Combustion and fuel properties.* Different applications of combustion processes pose different requirements on combustion characteristics. For instance, spark ignition engines require the fuel to meet certain anti-knock criterion. Octane number is a commonly used parameter in gauging such a fuel property. In diesel engines, the requirements are different due to the different combustion process used. The ease of autoignition is important because diesel engines rely on compression ignition. Such a property is quantified by the cetane number. In gas turbine engines, the tendency of the fuel to form soot is an important characteristic and is quantified by the smoke point. Liquid fuel properties such as viscosity and cloudiness can affect both the storage/handling of fuels and their combustion processes. For instance, high viscosity may prevent economic transport of some fuels through pipelines. High viscosity can also cause problems in the fuel injection process of internal combustion engines.

4. *Cost.* From an economic viewpoint, the cheapest fuel that meets the purpose of combustion while maintaining compliance with local safety and emissions laws will be chosen. Fuel cost and availability has determined the selection of fuels to use in the transportation and power industry from the beginning. The relatively low cost of fossil fuels has enhanced the dependence on these fuels and deterred the development of alternative fuels or energy sources.

1.4   Classification of Fuels by Phase at Ambient Conditions

Distribution methods and combustion processes vary based on a fuel's state of matter, making the phase of a fuel at standard conditions a logical basis for classification.

- *Solid fuels* (wood, coal, biomass) – $C_\alpha H_\beta O_\gamma$ with $\alpha > \beta$ – produce more CO_2 when burned.
- *Liquid fuels* (oil, gasoline, diesel fuel) – $C_\alpha H_\beta O_\gamma$ with $\alpha < \beta$
- *Gaseous fuels* (natural gas, hydrogen gas, syngas) – $C_\alpha H_\beta O_\gamma$ with $n\alpha \ll \beta$ – have the lowest C/H ratio, thus producing the least green house gas (CO_2) per unit energy output.

1.5 Identification of Fuel by Molecular Structure: International Union of Pure and Applied Chemistry (IUPAC)

The identification of a fuel can be best defined by its molecular structure. For organic chemistry, the convention adopted by International Union of Pure and Applied Chemistry (IUPAC) is well established and should be used. Most hydrocarbon fuels can be classified by their types of carbon-to-carbon (C—C) bonds as listed in Table 1.1. When a fuel contains all single C—C bonds, it is classified as an alkane. The chemical composition is $C_\alpha H_{2\alpha+2}$ where α denotes the total number of C atoms in the molecule. The names of hydrocarbon fuels are assigned by

Table 1.1 Naming conventions for hydrocarbon fuels commonly used in combustion

Family Name	Formula	C-C	Structure	Example
Alkanes (saturated, Paraffins)	$C_\alpha H_{2\alpha+2}$	Single	Straight or branched	Ethane $CH_3\text{-}CH_3$
Alkenes (olefins)	$C_\alpha H_{2\alpha}$	One double bond remaining single	Straight or branched	Ethene $CH_2{=}CH_2$
Alkynes (Acetylenes)	$C_\alpha H_{2\alpha-2}$	One triple bond remaining single	Straight or branched	Ethyne $HC{\equiv}CH$
Cyclanes (cycloalkanes)	$C_\alpha H_{2\alpha}$	Single bond	Closed rings	Cyclopropane $H_2C \overset{}{\underset{CH_2}{\diagdown\diagup}} CH_2$
Aromatics (benzene family)	$C_\alpha H_{2\alpha-6}$	Aromatic bond	Closed ring	Benzene

Table 1.2 Naming conventions – prefixes for hydrocarbon fuels	Number of carbon atoms (α)	Prefix
	1	Meth-
	2	Eth-
	3	Prop-
	4	But-
	5	Pent-
	6	Hex-
	7	Hept-
	8	Oct-
	9	Non-
	10	Dec-

Fig. 1.7 Molecular structure of n-octane

combining the prefix based on the number of carbon atoms (α) (see Table 1.2) with the suffix based on the type of bonds between the carbon atoms (Table 1.1). Examples of small alkanes are methane (CH_4), ethane (C_2H_6), propane (C_3H_8), and n-butane (C_4H_{10}). Alkanes with $\alpha \geq 4$ can have branches, and such alternative fuel structures are called *isomers*. By definition, isomers are molecules with the same chemical formula and often the same kinds of chemical bonds between atoms, but in which the atoms are arranged differently (analogous to a chemical anagram). Larger molecules tend to have more isomers. Many isomers share similar if not identical properties in most chemical contexts. However, combustion characteristics of isomers, particularly their ignition properties, may be quite different. Fuel structures can contain branches; the naming of such molecules is defined by IUPAC. For example, n-octane is an isomer of octane with a straight chain structure as sketched in Fig. 1.7.

Due the long straight chain, n-octane has a high tendency to knock in a spark ignition engine. Isooctane is another isomer of octane with a branched structure and an IUPAC name

$$
\underbrace{2,2,4}_{\text{positions of three } CH_3} - \underbrace{\text{trimethyl}}_{\text{branch species}} - \underbrace{\text{pentane}}_{\text{base}-\text{longest sraight chain}} \quad .
$$

It has a relatively low tendency to knock in a spark ignition engine. The structures of these isomers are compared in Fig. 1.8.

Fig. 1.8 Straight n-octane, *left*, has a higher tendency to autoignite than isooctane, *right*

The three branched CH_3 radicals attached to the pentane leads to a relatively low tendency to knock in a spark ignition engine. In general, a straight chain molecule becomes easier to break and burn when the size of molecule increases. In total, octane has 18 isomers: (1) Octane (n-octane) (2) 2-Methylheptane (3) 3-Methylheptane, (4) 4-Methylheptane, (5) 3-Ethylhexane, (6) 2,2-Dimethylhexane, (7) 2,3-Dimethylhexane, (8) 2,4-Dimethylhexane (9) 2,5-Dimethylhexane, (10) 3,3-Dimethylhexane, (11) 3,4-Dimethylhexane, (12) 2-Methyl-3-ethylpentane, (13) 3-Methyl-3-ethylpentane, (14) 2,2,3-Trimethylpentane, (15) 2,2,4-Trimethyl-pentane (isooctane), (16) 2,3,3-Trimethylpentane, (17) 2,3,4-Trimethylpentane, (18) 2,2,3,3-Tetramethylbutane. Table 1.1 summarizes the conventions used in identifying hydrocarbon fuels commonly used in combustion.

Because both gasoline and diesel fuel are composed of an unknown blend of various hydrocarbons, most analysis is performed assuming a surrogate fuel. Gasoline is often assumed to consist primarily of isooctane, whereas diesel fuel is often represented by n-heptane. However, there are limitations to using these model fuels to represent real fuels. For example, autoignition characteristics of 87 octane gasoline are not perfectly predicted by isooctane, which has an octane number of 100.

Example 1.1 Write the structural formula for the following species:

(a) 2-2-dimethylpropane
(b) 2-4-5-trimethyl-3-ethyloctane

Solution:

(a) 2-2-dimethylpropane

$$CH_3 \ \overset{\displaystyle CH_3}{\underset{\displaystyle CH_3}{\overset{|}{\underset{|}{C}}}} \ CH_3$$

(b) 2-4-5-trimethyl-3-ethyloctane

Example 1.2 Write all structures of isomers for pentane.

Solution:
Pentane is C_5H_{12}, so:

(a) n-pentane:

$$CH_3 \text{ --- } CH_2 \text{ --- } CH_2 \text{ --- } CH_2 \text{ --- } CH_3$$

(b) iso-pentane:

$$
\begin{array}{c}
CH_3 \\
| \\
CH_3 \text{ --- } CH \text{ --- } CH_2 \text{ ---} CH_3
\end{array}
$$

(c) neo-pentane: (also called 2-2-dimethylpropane)

$$
\begin{array}{c}
CH_3 \\
| \\
CH_3 \text{ --- } C \text{ --- } CH_3 \\
| \\
CH_3
\end{array}
$$

1.6   Some Related Properties of Liquid Fuels

1. *Flash point of liquid fuels* Flash point is the lowest temperature at which a fuel will liberate vapor at a sufficient rate such that the vapor will form a mixture with ambient air that will ignite in the presence of an ignition source. When the fuel reaches the flash point, the fuel is ready to combust when there is ignition source. If a spill of fuel occurs, the possibility of fire is very high if the air/fuel temperature reaches the flash point. Table 1.3 lists flash points of some common

Table 1.3 Flash points of commonly used fuels

Fuel	Flash point (°F)	Flash point (°C)
Gasoline	−45	−43
Iso-octane	10	−12.2
Kerosene	100	38
Diesel[a]	125	51.7
n-Heptane	25	−3.9
Toluene	40	4.4
Biodiesel	266	130
Jet fuel	100	38
Ethanol	55	12.8
n-Butane	−90	−68
Iso-butane	−117	−82.8
Xylene	63	17.2

[a] There are three classes of diesel fuel #1, #2 and #4. The values here are referring to #2 diesel commonly used in transportation

fuels showing that gasoline is a dangerous transportation fuel with low flash point of $-40°C$.

2. **Pour point** Pour point is defined as the lowest temperature (in $°F$ or $°C$) at which a liquid will flow easily (meaning it still behaves as a fluid). Hence, pour point is a rough indication of the lowest temperature at which oil is readily pumped.

3. **Cloud point** The cloud point is the temperature at which wax crystals begin to form in a petroleum product as it is cooled. Wax crystals depend on nucleation sites to initiate growth. The difference in the cloud points between two samples can sometimes be explained by the fact that any fuel additive will increase the number of nucleation sites, which initiates clouding. A change in temperature at which clouding starts to occur is therefore expected upon addition of any additive.

Exercise

1.1 Ethanol and dimethyl ether (DME), which happen to be chemical isomers, have been considered as potential fuels for the future. At ambient conditions, determine the phase of these two fuels.

References

1. Department of Energy/Energy Information Administration (2008), *Annual Energy Review*, Report Number DOE/EIA-0384(2008).
2. http://www.epud.org/shmtn.aspx
3. http://www.fossil.energy.gov/programs/powersystems/gasification/howgasificationworks

Chapter 2
Thermodynamics of Combustion

2.1 Properties of Mixtures

The thermal properties of a pure substance are described by quantities including internal energy, u, enthalpy, h, specific heat, c_p, etc. Combustion systems consist of many different gases, so the thermodynamic properties of a mixture result from a combination of the properties of all of the individual gas species. The ideal gas law is assumed for gaseous mixtures, allowing the ideal gas relations to be applied to each gas component. Starting with a mixture of K different gases, the total mass, m, of the system is

$$m = \sum_{i=1}^{K} m_i, \qquad (2.1)$$

where m_i is the mass of species i. The total number of moles in the system, N, is

$$N = \sum_{i=1}^{K} N_i, \qquad (2.2)$$

where N_i is the number of moles of species i in the system. Mass fraction, y_i, and mole fraction, x_i, describe the relative amount of a given species. Their definitions are given by

$$y_i \equiv \frac{m_i}{m} \quad \text{and} \quad x_i \equiv \frac{N_i}{N}, \qquad (2.3)$$

where $i = 1, 2, \ldots, K$. By definition,

$$\sum_{i=1}^{K} y_i = 1 \quad \text{and} \quad \sum_{i=1}^{K} x_i = 1.$$

S. McAllister et al., *Fundamentals of Combustion Processes*,
Mechanical Engineering Series, DOI 10.1007/978-1-4419-7943-8_2,
© Springer Science+Business Media, LLC 2011

With M_i denoting the molecular mass of species i, the average molecular mass, M, of the mixture is determined by

$$M = \frac{m}{N} = \frac{\sum_i N_i M_i}{N} = \sum_i x_i M_i. \tag{2.4}$$

From Dalton's law of additive pressures and Amagat's law of additive volumes along with the ideal gas law, the mole fraction of a species in a mixture can be found from the partial pressure of that species as

$$\frac{P_i}{P} = \frac{N_i}{N} = \frac{V_i}{V} = x_i, \tag{2.5}$$

where P_i is the partial pressure of species i, P is the total pressure of the gaseous mixture, V_i the partial volume of species i, and V is the total volume of the mixture. The average intrinsic properties of a mixture can be classified using either a molar base or a mass base. For instance, the internal energy per unit mass of a mixture, u, is determined by summing the internal energy per unit mass for each species weighted by the mass fraction of the species.

$$u = \frac{U}{m} = \frac{\sum_i m_i u_i}{m} = \sum_i y_i u_i, \tag{2.6}$$

where U is the total internal energy of the mixture and u_i is the internal energy per mass of species i. Similarly, enthalpy per unit mass of mixture is

$$h = \sum_i y_i h_i$$

and specific heat at constant pressure per unit mass of mixture is

$$c_p = \sum_i y_i c_{p,i}.$$

A molar base property, often denoted with a $^\wedge$ over bar, is determined by the sum of the species property per mole for each species weighted by the species mole fraction, such as internal energy per mole of mixture

$$\hat{u} = \sum_i x_i \hat{u}_i,$$

enthalpy per mole of mixture

$$\hat{h} = \sum_i x_i \hat{h}_i,$$

and entropy per mole of mixture

$$\hat{s} = \sum_i x_i \hat{s}_i.$$

Assuming constant specific heats during a thermodynamic process, changes of energy, enthalpy, and entropy of an individual species per unit mass are described as follows:

$$\Delta u_i = c_{v,i}(T_2 - T_1) \tag{2.7}$$

$$\Delta h_i = c_{p,i}(T_2 - T_1) \tag{2.8}$$

$$\Delta s_i = c_{p,i} \ln \frac{T_2}{T_1} - R_i \ln \frac{P_{i,2}}{P_{i,1}} \tag{2.9}$$

$P_{i,1}$ and $P_{i,2}$ denote the partial pressures of species i at state 1 and state 2, respectively. R_i is the gas constant for species i ($R_i = \hat{R}_u/M_i =$ universal gas constant/molecular mass of species i). The overall change of entropy for a combustion system is

$$\Delta S = \sum_i m_i \Delta s_i.$$

A summary of the thermodynamic properties of mixtures is provided at the end of the chapter.

2.2   Combustion Stoichiometry

For a given combustion device, say a piston engine, how much fuel and air should be injected in order to completely burn both? This question can be answered by balancing the combustion reaction equation for a particular fuel. A stoichiometric mixture contains the exact amount of fuel and oxidizer such that after combustion is completed, all the fuel and oxidizer are consumed to form products. This ideal mixture approximately yields the maximum flame temperature, as all the energy released from combustion is used to heat the products. For example, the following reaction equation can be written for balancing methane-air combustion

$$CH_4 + \underline{?}\left(O_2 + \frac{79}{21}N_2\right) \rightarrow \underline{?}CO_2 + \underline{?}H_2O + \underline{?}N_2, \tag{2.10}$$

where air consisting of 21% O_2 and 79% N_2 is assumed.[1] The coefficients associated with each species in the above equation are unknown. By balancing the atomic abundance on both the reactant and product sides, one can find the coefficient for each species. For instance, let's determine the coefficient for CO_2: on the reactant side, we have 1 mol of C atoms; hence the product side should also have 1 mol of C atoms. The coefficient of CO_2 is therefore unity. Using this procedure we can determine all the coefficients. These coefficients are called the reaction stoichiometric coefficients. For stoichiometric methane combustion with air, the balanced reaction equation reads:

$$CH_4 + \underline{2}(O_2 + 3.76N_2) \rightarrow \underline{1}CO_2 + \underline{2}H_2O + \underline{7.52}N_2. \tag{2.11}$$

Note that on the reactant side there are $2 \cdot (1 + 3.76)$ or 9.52 mol of air and its molecular mass is 28.96 kg/kmol. In this text, the reactions are balanced using 1 mol of fuel. This is done here to simplify the calculations of the heat of reaction and flame temperature later in the chapter. Combustion stoichiometry for a general hydrocarbon fuel, $C_\alpha H_\beta O_\gamma$, with air can be expressed as

$$C_\alpha H_\beta O_\gamma + \left(\alpha + \frac{\beta}{4} - \frac{\gamma}{2}\right)(O_2 + 3.76N_2) \rightarrow \alpha CO_2 + \frac{\beta}{2}H_2O + 3.76\left(\alpha + \frac{\beta}{4} - \frac{\gamma}{2}\right)N_2. \tag{2.12}$$

The amount of air required for combusting a stoichiometric mixture is called *stoichiometric* or *theoretical* air. The above formula is for a single-component fuel and cannot be applied to a fuel consisting of multiple components. There are two typical approaches for systems with multiple fuels. Examples are given here for a fuel mixture containing 95% methane and 5% hydrogen. The first method develops the stoichiometry of combustion using the general principle of atomic balance, making sure that the total number of each type of atom (C, H, N, O) is the same in the products and the reactants.

$$0.95CH_4 + 0.05H_2 + 1.925(O_2 + 3.76N_2) \rightarrow$$
$$0.95CO_2 + 1.95H_2O + 7.238N_2.$$

The other method of balancing a fuel mixture is to first develop stoichiometry relations for CH_4 and H_2 individually:

$$CH_4 + 2(O_2 + 3.76N_2) \rightarrow CO_2 + 2H_2O + 2 \cdot 3.76N_2$$

$$H_2 + 0.5(O_2 + 3.76N_2) \rightarrow H_2O + 0.5 \cdot 3.76N_2$$

Then, multiply the individual stoichiometry equations by the mole fractions of the fuel components and add them:

$$0.95 \cdot \{CH_4 + 2(O_2 + 3.76N_2) \rightarrow CO_2 + 2H_2O + 2 \cdot 3.76N_2\}$$
$$0.05 \cdot \{H_2 + 0.5(O_2 + 3.76N_2) \rightarrow H_2O + 0.5 \cdot 3.76N_2\}$$
$$\Rightarrow 0.95CH_4 + 0.05H_2 + 1.925(O_2 + 3.76N_2) \rightarrow$$
$$0.95CO_2 + 1.95H_2O + 7.238N_2$$

2.2.1 Methods of Quantifying Fuel and Air Content of Combustible Mixtures

In practice, fuels are often combusted with an amount of air different from the stoichiometric ratio. If less air than the stoichiometric amount is used, the mixture is described as fuel rich. If excess air is used, the mixture is described as fuel lean. For this reason, it is convenient to quantify the combustible mixture using one of the following commonly used methods:

Fuel-Air Ratio (FAR): The fuel-air ratio, f, is given by

$$f = \frac{m_f}{m_a}, \tag{2.13}$$

where m_f and m_a are the respective masses of the fuel and the air. For a stoichiometric mixture, Eq. 2.13 becomes

$$f_s = \left. \frac{m_f}{m_a} \right|_{stoichiometric} = \frac{M_f}{(\alpha + \frac{\beta}{4} - \frac{\gamma}{2}) \cdot 4.76 \cdot M_{air}}, \tag{2.14}$$

where M_f and M_{air} (~28.84 kg/kmol) are the average masses per mole of fuel and air, respectively. The range of f is bounded by zero and ∞. Most hydrocarbon fuels have a stoichiometric fuel-air ratio, f_s, in the range of 0.05–0.07. The air-fuel ratio (AFR) is also used to describe a combustible mixture and is simply the reciprocal of FAR, as AFR $= 1/f$. For instance, the stoichiometric AFR of gasoline is about 14.7. For most hydrocarbon fuels, 14–20 kg of air is needed for complete combustion of 1 kg of fuel.

Equivalence Ratio: Normalizing the actual fuel-air ratio by the stoichiometric fuel-air ratio gives the equivalence ratio, ϕ.

$$\phi = \frac{f}{f_s} = \frac{m_{as}}{m_a} = \frac{N_{as}}{N_a} = \frac{N_{O2s}}{N_{O2,a}} \tag{2.15}$$

The subscript s indicates a value at the stoichiometric condition. $\phi < 1$ is a *lean* mixture, $\phi = 1$ is a *stoichiometric* mixture, and $\phi > 1$ is a *rich* mixture. Similar to f, the range of ϕ is bounded by zero and ∞ corresponding to the limits of pure air and fuel respectively. Note that equivalence ratio is a normalized quantity that provides the information regarding the content of the combustion mixture. An alternative variable based on AFR is frequently used by combustion engineers and is called lambda (λ). Lambda is the ratio of the actual air-fuel ratio to the stoichiometric air-fuel ratio defined as

$$\lambda = \frac{AFR}{AFR_s} = \frac{1/f}{1/f_s} = \frac{1}{f/f_s} = \frac{1}{\phi} \tag{2.16}$$

Lambda of stoichiometric mixtures is 1.0. For rich mixtures, lambda is less than 1.0; for lean mixtures, lambda is greater than 1.0.

Percent Excess Air: The amount of air in excess of the stoichiometric amount is called excess air. The percent excess air, %*EA*, is defined as

$$\%EA = 100\frac{m_a - m_{as}}{m_{as}} = 100\left(\frac{m_a}{m_{as}} - 1\right) \tag{2.17}$$

For example, a mixture with %EA = 50 contains 150% of the theoretical (stoichiometric) amount of air.

Converting between quantification methods: Given one of the three variables (f, ϕ, and %EA), the other two can be deduced as summarized in Table 2.1 with their graphic relations. In general, the products of combustion include many different

Table 2.1 Relations among the three variables for describing reacting mixtures

f (fuel air ratio by mass)	ϕ (equivalence ratio)	%EA (% of excess air)
$f = f_s \cdot \phi$	$\phi = \dfrac{f}{f_s}$	$\%EA = 100\dfrac{1-\phi}{\phi}$
$f = \dfrac{100 \cdot f_s}{\%EA + 100}$	$\phi = \dfrac{100}{\%EA + 100}$	$\%EA = 100\dfrac{1 - f/f_s}{f/f_s}$

species in addition to the major species (CO_2, H_2O, N_2, O_2), and the balance of the stoichiometric equation requires the use of thermodynamic equilibrium relations. However, assuming that the products contain major species only (complete combustion) and excess air, the global equation for lean combustion $\phi \leqslant 1$ is

$$C_\alpha H_\beta O_\gamma + \frac{1}{\phi}\left(\alpha + \frac{\beta}{4} - \frac{\gamma}{2}\right)(O_2 + 3.76N_2) \rightarrow$$
$$\alpha CO_2 + \frac{\beta}{2}H_2O + \frac{3.76}{\phi}\left(\alpha + \frac{\beta}{4} - \frac{\gamma}{2}\right)N_2 + \left(\alpha + \frac{\beta}{4} - \frac{\gamma}{2}\right)\left(\frac{1}{\phi} - 1\right)O_2 \tag{2.18}$$

In terms of %EA, we replace ϕ by $\dfrac{100}{\%EA + 100}$ and the result is

$$C_\alpha H_\beta O_\gamma + \left(\frac{\%EA}{100} + 1\right)\left(\alpha + \frac{\beta}{4} - \frac{\gamma}{2}\right)(O_2 + 3.76N_2) \rightarrow$$
$$\alpha CO_2 + \frac{\beta}{2}H_2O + 3.76\left(\frac{\%EA}{100} + 1\right)\left(\alpha + \frac{\beta}{4} - \frac{\gamma}{2}\right)N_2 + \left(\alpha + \frac{\beta}{4} - \frac{\gamma}{2}\right)\frac{\%EA}{100}O_2 \tag{2.19}$$

The amount of excess air can be deduced from measurements of exhaust gases. The ratio of mole fractions between CO_2 and O_2 is

$$\frac{x_{CO2}}{x_{O2}} = \frac{\alpha}{\left(\alpha + \frac{\beta}{4} - \frac{\gamma}{2}\right)\frac{\%EA}{100}} \rightarrow \frac{\%EA}{100} = \frac{\alpha}{\left(\alpha + \frac{\beta}{4} - \frac{\gamma}{2}\right)\frac{x_{CO_2}}{x_{O_2}}}$$

or using Table 2.1

$$\phi = \frac{100}{100 + \%EA} \rightarrow \phi = \frac{1}{1 + \dfrac{\alpha}{\left(\alpha + \frac{\beta}{4} - \frac{\gamma}{2}\right)\frac{x_{CO_2}}{x_{O_2}}}} \tag{2.20}$$

For rich combustion ($\phi > 1$), the products may contain CO, unburned fuels, and other species formed by the degradation of the fuel. Often additional information on the products is needed for complete balance of the chemical reaction. If the products are assumed to contain only unburned fuel and major combustion products, the corresponding global equation can be written as

$$C_\alpha H_\beta O_\gamma + \frac{1}{\phi}\left(\alpha + \frac{\beta}{4} - \frac{\gamma}{2}\right)(O_2 + 3.76N_2) \rightarrow$$
$$\frac{\alpha}{\phi}CO_2 + \frac{\beta}{2\phi}H_2O + \frac{3.76}{\phi}\left(\alpha + \frac{\beta}{4} - \frac{\gamma}{2}\right)N_2 + \left(1 - \frac{1}{\phi}\right)C_\alpha H_\beta O_\gamma \tag{2.21}$$

Example 2.1 Considering a stoichiometric mixture of isooctane and air, determine:

(a) the mole fraction of fuel
(b) the fuel-air ratio
(c) the mole fraction of H_2O in the products
(d) the temperature of products below which H_2O starts to condense into liquid at 101.3 kPa

Solution:
The first step is writing and balancing the stoichiometric reaction equation. Using Eq. 2.12,

$$C_8H_{18} + \left(8 + \frac{18}{4} - 0\right)(O_2 + 3.76N_2) \rightarrow 8CO_2 + 9H_2O + 3.76\left(8 + \frac{18}{4} - 0\right)N_2$$

$$C_8H_{18} + 12.5(O_2 + 3.76N_2) \rightarrow 8CO_2 + 9H_2O + 3.76 \cdot 12.5 \cdot N_2$$

From here:

(a) $x_{C_8H_{18}} = \dfrac{N_{C_8H_{18}}}{N_{C_8H_{18}} + N_{air}} = \dfrac{1}{1 + 12.5 \cdot 4.76} = 0.0165$

(b) $f_s = \dfrac{M_f}{\left(\alpha + \frac{\beta}{4} - \frac{\gamma}{2}\right) \cdot 4.76 \cdot M_{air}} = \dfrac{114}{12.5 \cdot 4.76 \cdot 28.96} = 0.066$

(c) $x_{H_2O} = \dfrac{N_{H_2O}}{N_{CO_2} + N_{H_2O} + N_{N_2}} = \dfrac{9}{8 + 9 + 3.76 \cdot 12.5} = 0.141$

(d) The partial pressure of water is 101 kPa $\cdot$ 0.141 = 14.2 kPa. A saturation table for steam gives the saturation temperature at this water pressure $\cong 53°C$.

Example 2.2 How many kg (lb) of air are used to combust 55.5 L (~14.7 US gallons) of gasoline?

Solution:
We will use isooctane C_8H_{18} to represent gasoline. The stoichiometric fuel-air ratio is

$$f_s = \frac{M_f}{\left(\alpha + \frac{\beta}{4} - \frac{\gamma}{2}\right) \cdot 4.76 \cdot M_{air}}$$

$$= \frac{114 \text{ kg/kmol}}{(8 + 18/4 - 0) \cdot 4.76 \cdot 28.84 \text{ kg/kmol}}$$

$$= 0.066$$

One gallon of gasoline weighs about 2.7 kg (6 lb). The total fuel thus weighs about 40 kg (88 lb). The required air weighs about $40/f_s \approx 610$ kg $\approx 1,300$ lb. This is a lot of weight if it must be carried. Hence, for transportation applications, free ambient air is preferred.

Example 2.3 In a model "can-combustor" combustion chamber, n-heptane (C_7H_{16}) is burned under an overall lean condition. Measurements of dry exhaust give mole fractions of CO_2 and O_2 as $x_{CO_2} = 0.084$ and $x_{O_2} = 0.088$. Determine the %EA, equivalence ratio ϕ, and λ.

Solution:
To avoid condensation of water inside the instruments, measurements of exhaust gases are taken on a 'dry' mixture that is obtained by passing the exhaust gases through an ice bath so that most water is condensed. Further removal of water can be done with desiccants. The mole fractions measured under dry conditions will be larger than at real conditions since water is removed. However, this will not impact the relation deduced above, as both x_{CO_2} and x_{O_2} are increased by the same factor.

$$\frac{\%EA}{100} = \frac{\alpha}{\left(\alpha + \frac{\beta}{4} - \frac{\gamma}{2}\right)\frac{x_{CO_2}}{x_{O_2}}} = \frac{7}{(7 + 16/4 - 0)(0.084/0.088)} = 0.667 \rightarrow \%EA$$

$$= 66.7$$

Next we use the relations given in Table 2.1 to convert %EA to ϕ and λ

$$\phi = \frac{100}{\%EA + 100} = \frac{100}{66.7 + 100} = 0.6$$

$$\lambda = \frac{1}{\phi} = 1.67$$

2.3 Heating Values

Heating values of a fuel (units of kJ/kg or MJ/kg) are traditionally used to quantify the maximum amount of heat that can be generated by combustion with air at standard conditions (STP) (25°C and 101.3 kPa). The amount of heat release from combustion of the fuel will depend on the phase of water in the products. If water is in the gas phase in the products, the value of total heat release is denoted as the lower heating value (LHV). When the water vapor is condensed to liquid, additional energy (equal to the latent heat of vaporization) can be extracted and the total energy release is called the higher heating value (HHV). The value of the LHV can be calculated from the HHV by subtracting the amount of energy released during the phase change of water from vapor to liquid as

$$LHV = HHV - \frac{N_{H2O,P} M_{H2O} h_{fg}}{N_{fuel} M_{fuel}} \quad (\text{MJ/kg}), \qquad (2.22)$$

where $N_{H2O,P}$ is the number of moles of water in the products. Latent heat for water at STP is $h_{fg} = 2.44$ MJ/kg $= 43.92$ MJ/kmol. In combustion literature, the LHV is normally called the enthalpy or heat of combustion (Q_C) and is a positive quantity.

2.3.1 Determination of HHV for Combustion Processes at Constant Pressure

A control volume analysis at constant pressure with no work exchanged can be used to theoretically determine the heating values of a particular fuel. Suppose reactants with 1 kmol of fuel enter the inlet of a control volume at standard conditions and products leave at the exit. A maximum amount of heat is extracted when the products are cooled to the inlet temperature and the water is condensed. Conservation of energy for a constant pressure reactor, with H_P and H_R denoting the respective total enthalpies of products and reactants, gives

$$- Q_{rxn,p} = H_R - H_p. \qquad (2.23)$$

The negative value of $Q_{rxn,p}$ indicates heat transfer out of the system to the surroundings. It follows from above that the heating value of the fuel is the difference in the enthalpies of the reactants and the products. However, in combustion systems, the evaluation of the enthalpies is not straightforward because the species entering the system are different than those coming out due to chemical reactions. $Q_{rxn,p}$ is often referred to as the enthalpy of reaction or heat of reaction, with the subscript p indicating that the value was calculated at constant pressure. The enthalpy of reaction is related to the enthalpy of combustion by $Q_{rxn,p} = -Q_C$.

2.3.1.1 Enthalpy of Formation

In combustion processes, reactants are consumed to form products and energy is released. This energy comes from a rearrangement of chemical bonds in the reactants to form the products. The standard enthalpy of formation, $\Delta \hat{h}_f^o$, quantifies the chemical bond energy of a chemical species at standard conditions. The enthalpy of formation of a substance is the energy needed for the formation of that substance from its constituent elements at STP conditions (25°C and 1 atm). The molar base enthalpy of formation, $\Delta \hat{h}_f^o$, has units of MJ/kmol, and the mass base enthalpy of formation, $\Delta \hat{h}_f^o$, has units of MJ/kg. Elements in their most stable forms, such as $C_{(graphite)}$, H_2, O_2, and N_2, have enthalpies of formation of zero. Enthalpies of formation of commonly encountered chemical species are tabulated in Table 2.2.

A departure from standard conditions is accompanied by an enthalpy change. For thermodynamic systems without chemical reactions, the change of enthalpy of an ideal gas is described by the sensible enthalpy,

Table 2.2 Enthalpy of formation of common combustion species

Species	$\Delta \hat{h}^o$ (MJ/kmol)	Species	$\Delta \hat{h}^o$ (MJ/kmol)
H_2O (g)	−241.83	H	+217.99
CO_2	−393.52	N	+472.79
CO	−110.53	NO	+90.29
CH_4	−74.87	NO_2	+33.10
C_3H_8	−104.71	O	+249.19
C_7H_{16} (g) (n-heptane)	−224.23	OH	+39.46
C_8H_{18} (g) (isooctane)	−259.25	C (g)	+715.00
CH_3OH (g) (methanol)	−201.54	C_2H_2 (acetylene)	+226.73
CH_3OH (l) (methanol)	−238.43	C_2H_4 (ethylene)	+52.28
C_2H_6O (g) (ethanol)	−235.12	C_2H_6 (ethane)	−84.68
C_2H_6O (l) (ethanol)	−277.02	C_4H_{10} (n-butane)	−126.15

Fig. 2.1 Constant-pressure flow reactor for determining enthalpy of formation

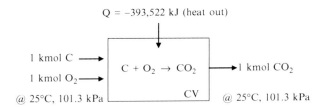

$$\hat{h}_{si} = \int_{T_o}^{T} \hat{c}_p(T)dT,$$

where the subscript i refers to species i, T_0 denotes the standard temperature (25°C), and $^\wedge$ indicates that a quantity is per mole. Note that the sensible enthalpy of any species is zero at standard conditions. The 'absolute' or 'total' enthalpy, h_i, is thus the sum of the sensible enthalpy and the enthalpy of formation:[2]

$$\hat{h}_i = \Delta \hat{h}_i^o + \hat{h}_{si} \tag{2.24}$$

One way to determine the enthalpy of formation of a species is to use a constant-pressure flow reactor. For instance, the enthalpy of formation of CO_2 is determined by reacting 1 kmol of $C_{(graphite)}$ with 1 kmol of O_2 at 25°C at a constant pressure of 101.3 kPa. The product, 1 kmol of CO_2, flows out of this reactor at 25°C as sketched in Fig. 2.1. An amount of heat produced in the reaction is transferred

---

[2] When phase change is encountered, the total enthalpy needs to include the latent heat, $\hat{h}_i = \Delta \hat{h}_i^o + \hat{h}_{si} + \hat{h}_{latent}$.

out of this system, therefore the enthalpy formation of CO_2 is negative $\Delta \hat{h}^o_{CO2} = -393.52$ MJ/kmol. This means that CO_2 at 25°C contains less energy than its constituent elements $C_{(graphite)}$ and O_2, which have zero enthalpy of formation. The enthalpy of formation is not negative for all chemical species. For instance, the enthalpy formation of NO is $\Delta \hat{h}^o_{NO} = +90.29$ MJ/kmol, meaning that energy is needed to form NO from its elements, O_2 and N_2. For most unstable or "radical" species, such as O, H, N, and CH_3, the enthalpy of formation is positive.

2.3.1.2 Evaluation of the Heat of Combustion from a Constant-Pressure Reactor

Using the conservation of energy equation (2.23), we can now evaluate the enthalpies of the reactants and products. Inserting the expression for the total enthalpy,

$$
\begin{aligned}
-Q_{rxn,p} = H_R - H_p &= \sum_i N_{i,R}\left(\Delta \hat{h}^o_{i,R} + \hat{h}_{si,R}\right) - \sum_i N_{i,P}\left(\Delta \hat{h}^o_{i,P} + \hat{h}_{si,P}\right) \\
&= \left[\sum_i N_{i,R}\Delta \hat{h}^o_{i,R} - \sum_i N_{i,P}\Delta \hat{h}^o_{i,P}\right] + \sum_i N_{i,R}\hat{h}_{si,R} - \sum_i N_{i,P}\hat{h}_{si,P},
\end{aligned}
\tag{2.25}
$$

where N_i is the number of moles of species i. The sensible enthalpies of common reactants and products can be found in Appendix 1. When the products are cooled to the same conditions as the reactants, the amount of heat transfer from the constant-pressure reactor to the surroundings is defined as the heating value. At STP the sensible enthalpy terms drop out for both reactants and products and the heat release is

$$
- Q^0_{rxn,p} = \sum_i N_{i,R}\Delta \hat{h}^o_{i,R} - \sum_i N_{i,P}\Delta \hat{h}^o_{i,P}
\tag{2.26}
$$

Usually excess air is used in such a test to ensure complete combustion. The amount of excess air used will not affect $-Q^0_{rxn,p}$ at STP. Unless the reactant mixtures are heavily diluted, the water in the products at STP normally will be liquid.[3] Assuming that water in the products is liquid, HHV is determined:

$$
HHV = \frac{-Q^0_{rxn,p}}{N_{fuel}M_{fuel}}.
\tag{2.27}
$$

The negative sign in front of $Q^0_{rxn,p}$ ensures that HHV is positive.

2.3.2 Determination of HHV for Combustion Processes from a Constant-Volume Reactor

A constant-volume reactor is more convenient than the constant-pressure reactor to experimentally determine the HHV of a particular fuel. For a closed system, conservation of energy yields

$$- Q_{rxn,v} = U_R - U_p \tag{2.28}$$

Because of the combustion process, the same type of accounting must be used to include the change in chemical bond energies. The internal energy will be evaluated by using its relation to enthalpy. Note that relation $h = u + pv$ is mass based and the corresponding molar base relation is $\hat{h} = \hat{u} + \hat{R}_u T$. At STP (T $= T_0 = 25°$C), the total internal energy of the reactants, U_R, inside the closed system is

$$
\begin{aligned}
U_R &= H_R - PV \\
&= H_R - \sum_i N_{i,R} \hat{R}_u T_0 \\
&= \sum_i N_{i,R} \Delta \hat{h}^o_{i,R} - \sum_i N_{i,R} \hat{R}_u T_0
\end{aligned}
\tag{2.29}
$$

The total internal energy of products is evaluated in a similar manner:

$$U_P = \sum_i N_{i,P} \Delta \hat{h}^o_{i,P} - \sum_i N_{i,P} \hat{R}_u T_0 \tag{2.30}$$

Using the internal energy relations, we can re-express the heat release at constant volume in terms of enthalpies as

$$
\begin{aligned}
-Q^0_{rxn,v} &= U_R - U_P \\
&= \sum_i N_{i,R} \Delta \hat{h}^o_{i,R} - \sum_i N_{i,R} \hat{R}_u T_0 - \left[\sum_i N_{i,P} \Delta \hat{h}^o_{i,P} - \sum_i N_{i,P} \hat{R}_u T_0 \right] \\
&= \sum_i N_{i,R} \Delta \hat{h}^o_{i,R} - \sum_i N_{i,P} \Delta \hat{h}^o_{i,P} + \left(\sum_i N_{i,P} - \sum_i N_{i,R} \right) \hat{R}_u T_0
\end{aligned}
\tag{2.31}
$$

Therefore, HHV for combustion processes is calculated as

$$HHV = \frac{-Q^0_{rxn,v} - \left(\sum_i N_{i,P} - \sum_i N_{i,R} \right) \hat{R}_u T_0}{N_{fuel} M_{fuel}}, \tag{2.32}$$

where N_{fuel} is the number of moles of fuel burned and M_{fuel} is the molecular mass of the fuel. The negative sign in front of $Q^0_{rxn,v}$ is to make sure that HHV is positive. For a general fuel, $C_\alpha H_\beta O_\gamma$, the difference between $-Q_{rxn,v}$ and $-Q_{rxn,p}$ is

$$\left(\sum_i N_{i,P} - \sum_i N_{i,R} \right) \hat{R}_u T_0 = \Delta N \hat{R}_u T_0 = \left(\frac{\beta}{4} + \frac{\gamma}{2} - 1 \right) \hat{R}_u T_0 \qquad (2.33)$$

and is usually small in comparison to HHV; therefore normally no distinction is made between the heat of reaction at constant pressure or constant volume.

2.3.2.1 Experimental Determination of HHV: The Bomb Calorimeter

To experimentally measure the heating value of a fuel, the fuel and air are often enclosed in an explosive-proof steel container (see Fig. 2.2), whose volume does not change during a reaction. The vessel is then submerged in water or another liquid that absorbs the heat of combustion. The heat capacitance of the vessel plus the liquid is then measured using the same technique as other calorimeters. Such an instrument is called a bomb calorimeter.

A constant-volume analysis of the bomb calorimeter data is used to determine the heating value of a particular fuel. The fuel is burned with sufficient oxidizer in a closed system. The closed system is cooled by heat transfer to the surroundings such that the final temperature is the same as the initial temperature. The standard conditions are set for evaluation of heating values. Conservation of energy gives

$$U_P - U_R = Q^0_{rxn,v} \qquad (2.34)$$

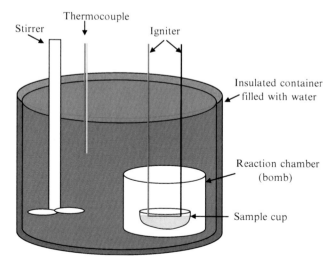

Fig. 2.2 Bomb calorimeter

Because the final water temperature is close to room temperature, the water in the combustion products is usually in liquid phase. Therefore the measurement leads to the HHV from a constant-volume combustion process as described by Eq. 2.32:

$$HHV = \left\{ -Q^0_{rxn,v} - \left(\sum_i N_{i,P} - \sum_i N_{i,R} \right) \hat{R}_u T_0 \right\} / \left[N_{fuel} M_{fuel} \right],$$

where N_{fuel} is the number of moles of fuel burned and M_{fuel} is the molecular weight of the fuel. The negative sign in front of $Q^0_{rxn,v}$ ensures that HHV is positive. In a bomb calorimeter, if the final temperature of the combustion products is higher than the reactants by only a few degrees ($<10°C$), the error is negligible. The amount of heat transfer is estimated by

$$- Q^0_{rxn,v} = (m_{steel} \cdot c_{p,steel} + m_{water} \cdot c_{p,water}) \Delta T, \tag{2.35}$$

where ΔT is the temperature change of the water and the steel container.

The bomb calorimeter can also measure the enthalpy of formation of a chemical species. For instance, to determine enthalpy of formation of H_2O, we start out with 1 mol of H_2 and 0.5 mol of O_2. These element species have zero enthalpy of formation; therefore

$$\sum_i N_{i,R} \Delta \hat{h}^0_{i,R} = 0.$$

The only product is the species of interest, namely H_2O. We therefore can write the enthalpy of formation of H_2O, $\Delta \hat{h}^0_{i,P}$, as

$$\Delta \hat{h}^0_{i,P} = \frac{Q^0_{rxn,v} + \left(\sum_i N_{i,P} - \sum_i N_{i,R} \right) \hat{R}_u T_0}{N_{i,P}} = \frac{Q^0_{rxn,v} + \Delta N \hat{R}_u T_0}{N_{i,P}} \tag{2.36}$$

where

$$\Delta N = \sum_i N_{i,P} - \sum_i N_{i,R}.$$

2.3.3 Representative HHV Values

Listed in Table 2.3 are higher heating values of some common and less common fuels.

Example 2.4 A table of thermodynamic data gives the enthalpy of formation for liquid water as $\Delta \hat{h}^0_{H_2O(l)} = -285.8$ kJ/mol. A bomb calorimeter burning 1 mol of H_2 with O_2 measures 282.0 kJ of heat transfer out of the reacted mixture. Estimate the error of the enthalpy measurement.

Solution:

We start out with the combustion stoichiometry

$$H_2(g) + 0.5O_2(g) = H_2O(liq),$$

$$\Delta N = -1.5 \quad \text{(Change in moles of } gas \text{ in the mixture)}$$

Applying the ideal gas approximation to the energy balance with $Q^0_{rxn,v} = -282.0$ kJ,

$$\Delta\hat{h}^0_{H_2O(l)} = Q^0_{rxn,v} + \Delta N\hat{R}_u T_0,$$
$$= -282.0 \text{ kJ/mol} \cdot 1 \text{ mol} + (-1.5 \text{ mol} \cdot 8.314 \text{ J/mol} - \text{K} \cdot 298\,\text{K} \cdot 0.001 \text{ kJ/J})$$
$$= (-282.0 - 3.72)\text{kJ} = -285.7 \text{ kJ}$$

The error is $(285.8 - 285.7)/285.8 = 0.03\%$. In this case, more heat is given off if the reaction is carried out at constant pressure, since the P-V work $(1.5\hat{R}_u T_0)$ due to the compression of 1.5 mol of gases in the reactants would contribute to $\Delta\hat{h}^0_{H_2O(l)}$. However, this difference is only about 1–2% of the enthalpy of formation. The enthalpy of formation for gaseous H_2O is obtained by adding the latent heat to $\Delta\hat{h}^0_{H_2O(l)}$:

$$\Delta\hat{h}^0_{H_2O(g)} = \Delta\hat{h}^0_{H_2O(l)} + \hat{h}_{fg} = -241.88 \text{ kJ/mol},$$

Table 2.3 Heat values of various fuels

Fuel	Heating value		
	MJ/kg	BTU/lb	kJ/mol
Hydrogen	141.8	61,100	286
Methane	55.5	23,900	890
Ethane	51.9	22,400	1,560
Propane	50.35	21,700	2,220
Butane	49.5	20,900	2,877
Gasoline	47.3	20,400	~5,400
Paraffin	46	19,900	16,300
Diesel	44.8	19,300	~4,480
Coal	15–27	8,000–14,000	200–350
Wood	15	6,500	300
Peat	6–15	2,500–6,500	
Methanol	22.7	9,800	726
Ethanol	29.7	12,800	1,368
Propanol	33.6	14,500	2,020
Acetylene	49.9	21,500	1,300
Benzene	41.8	18,000	3,270
Ammonia	22.5	9,690	382
Hydrazine	19.4	8,370	622
Hexamine	30.0	12,900	4,200
Carbon	32.8	14,100	393.5

where

$$\hat{h}_{fg} = 43.92 \text{ kJ/mol}.$$

Example 2.5 The heat released by 1 mol of sugar in a bomb calorimeter experiment is 5,648 kJ/mol. Calculate the enthalpy of combustion per mole of sugar.

Solution:
The balanced chemical reaction equation is

$$C_{12}H_{22}O_{11}(s) + 12O_2(g) = 12CO_2(g) + 11H_2O(liq)$$

Since the total number of moles of gas is constant (12) in the products and reactants, $\Delta N = 0$. Therefore, work is zero and the enthalpy of combustion equals the heat transfer: $-5,648$ kJ/mol.

2.4 Adiabatic Flame Temperature

One of the most important features of a combustion process is the highest temperature of the combustion products that can be achieved. The temperature of the products will be greatest when there are no heat losses to the surrounding environment and all of the energy released from combustion is used to heat the products. In the next two sections, the methodology used to calculate the maximum temperature, or *adiabatic flame temperature*, will be presented.

2.4.1 Constant-Pressure Combustion Processes

An adiabatic constant-pressure analysis is used here to calculate the adiabatic flame temperature. Under this idealized condition, conservation of energy is:

$$H_P(T_P) = H_R(T_R), \tag{2.37}$$

where

$$H_P(T_P) = \sum_i N_{i,P}\hat{h}_{i,P} = \sum_i N_{i,P}[\Delta\hat{h}^o_{i,P} + \hat{h}_{si,P}(T_P)]$$

and

$$H_R(T_R) = \sum_i N_{i,R}\hat{h}_{i,R} = \sum_i N_{i,R}[\Delta\hat{h}^o_{i,R} + \hat{h}_{si,R}(T_R)].$$

Figure 2.3 is a graphic explanation of how the adiabatic flame temperature is determined. At the initial reactant temperature, the enthalpy of the product mixture

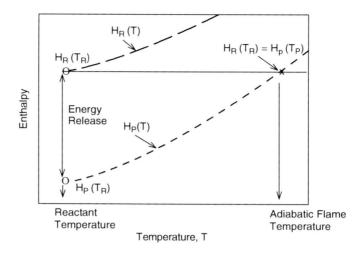

Fig. 2.3 Graphical interpretation of adiabatic flame temperature

is lower than that of the reactant mixture. The energy released from combustion is used to heat up the products such that the condition $H_P(T_P) = H_R(T_R)$ is met.

The task is finding the product temperature given the enthalpy of reactants. Three different methods can be used to obtain T_P:

1. Using an average c_p value,
2. An iterative enthalpy balance,
3. Finding the equilibrium state using computer software (such as Cantera).

The first two methods can be performed manually if complete combustion is considered and provide only quick estimates. An equilibrium state solver takes into account dissociation of products at high temperature, making it more accurate than the first two methods.

Method 1: Constant, average c_p
From conservation of energy, $H_p(T_p) = H_R(T_R)$, which can be expressed as

$$\sum_i N_{i,P}[\Delta \hat{h}_{i,P}^o + \hat{h}_{si,P}(T_P)] = \sum_i N_{i,R}[\Delta \hat{h}_{i,R}^o + \hat{h}_{si,R}(T_R)]$$

Rearranging yields

$$\sum_i N_{i,P} \hat{h}_{si,P}(T_P) = -\left\{ \sum_i N_{i,P} \Delta \hat{h}_{i,P}^o - \sum_i N_{i,R} \Delta \hat{h}_{i,R}^o \right\} + \sum_i N_{i,R} \hat{h}_{si,R}(T_R)$$

$$= -Q_{rxn,p}^0 + \sum_i N_{i,R} \hat{h}_{si,R}(T_R) \qquad (2.38)$$

with

$$- Q^0_{rxn,p} = \sum_i N_{i,R} \Delta \hat{h}^o_{i,R} - \sum_i N_{i,P} \Delta \hat{h}^o_{i,P}. \tag{2.39}$$

Note that water in the products is likely in gas phase due to the high combustion temperature; therefore $- Q^0_{rxn,p} = \text{LHV} \cdot N_{fuel} \cdot M_{fuel} = \text{LHV} \cdot m_f$ when the fuel is completely consumed. The second term, $\sum_i N_{i,R} \hat{h}_{si,R}(T_R)$, in Eq. 2.38 represents the difference of sensible enthalpy between T_R and T_0 (25°C) for the reactant mixture. With the assumption that the sensible enthalpy can be approximated by $\hat{h}_{si,P}(T_P) \approx \hat{c}_{pi} (T_P - T_0)$ with $\hat{c}_{pi} \approx$ constant, we have

$$(T_P - T_0) \sum_i N_{i,P} \hat{c}_{pi} \equiv \hat{c}_p (T_P - T_0) \sum_i N_{i,P} = -Q^0_{rxn,p} + \sum_i N_{i,R} \hat{h}_{si,R}(T_R) \tag{2.40}$$

Rearranging the equation one finds T_P as

$$T_P = T_0 + \frac{-Q^0_{rxn,p} + \sum_i N_{i,R} \hat{h}_{si,R}(T_R)}{\sum_i N_{i,P} \hat{c}_{pi}}$$

$$\approx T_R + \frac{-Q^0_{rxn,p}}{\sum_i N_{i,P} \hat{c}_{pi}} \tag{2.41}$$

$$= T_R + \frac{\text{LHV} \cdot N_{fuel} \cdot M_{fuel}}{\sum_i N_{i,P} \hat{c}_{pi}},$$

where the following approximation has been applied[4]

$$\frac{\sum_i N_{i,R} \hat{h}_{si,R}(T_R)}{\sum_i N_{i,P} \hat{c}_{pi}} = \frac{\sum_i N_{i,R} \hat{c}_{p_{i,R}}(T_R - T_0)}{\sum_i N_{i,P} \hat{c}_{pi}} \approx T_R - T_0$$

When reactants enter the combustor at the standard conditions, the above equation reduces to (as sensible enthalpies of reactants are zero at T_0)

$$T_P = T_0 + \frac{\text{LHV} \cdot N_{fuel} \cdot M_{fuel}}{\sum_i N_{i,P} \hat{c}_{pi}}. \tag{2.42}$$

---

[4] $\sum_i N_{i,R} \hat{c}_{p_{i,R}}$ and $\sum_i N_{i,p} \hat{c}_{pi}$ are assumed to be approximately equal.

The above procedure is general and can be applied to any mixture. Note that the specific heat is a function of temperature, so the accuracy of this approach depends on the value selected for the specific heat $\hat{c}_p$.

If the heating value of a fuel is given, a mass-based analysis for the same control volume can be conducted. The initial mixture consists of fuel and air with m_f and m_a, respectively. By mass conservation, the products have a total mass of $m_f + m_a$. The sensible enthalpy of the products is approximated by $H_{s,P} = (m_a + m_f) \cdot \bar{c}_{p,P} \cdot (T_P - T_0)$, where $\bar{c}_{p,P}$ is an average value of specific heat evaluated at the average temperature of the reactants and products, i.e., $\bar{c}_{p,P} = c_p(\bar{T})$, where $\bar{T} = (T_p + T_R)/2$. Similarly, the sensible enthalpy of the reactants is estimated by $H_{s,R} = (m_a + m_f) \cdot \bar{c}_{p,R} \cdot (T_R - T_0)$, where $\bar{c}_{p,R}$ is an average value of specific heat evaluated at the average temperature of reactants and the standard temperature, i.e., $\bar{c}_{p,R} = c_p(\bar{T})$, where $\bar{T} = (T_R + T_0)/2$. From conservation of energy, $H_{s,P}$ equals the amount of heat released from combustion plus the sensible enthalpy of the reactants, $H_{s,P} = -Q^0_{rxn,p} + H_{s,R} = m_{fb}$ $\cdot LHV + H_{s,R}$, where m_{fb} is the amount of fuel burned. For $\phi \leqslant 1$, $m_{fb} = m_f$ since there is enough air to consume all the fuel in a lean mixture. For rich combustion ($\phi > 1$), the limiting factor is the amount of air available, m_a. Therefore, for $\phi > 1$, the amount of fuel burned (with air, m_a) is $m_{fb} = m_a f_s$, where f_s is the stoichiometric fuel/air ratio by mass. Then the adiabatic flame temperature is calculated for a lean mixture as $\phi \leqslant 1$

$$
\begin{aligned}
T_P &\cong T_0 + \frac{m_f \cdot LHV + (m_a + m_f)\bar{c}_{p,R}(T_R - T_0)}{(m_a + m_f)\bar{c}_{p,P}} \\
&\approx T_R + \frac{m_f \cdot LHV}{(m_a + m_f)\bar{c}_{p,P}} = T_R + \frac{m_f/m_a \cdot LHV}{(1 + m_f/m_a)\bar{c}_{p,P}} \\
&= T_R + \frac{f \cdot LHV}{(1 + f)\bar{c}_{p,P}} = T_R + \frac{\phi \cdot f_s \cdot LHV}{(1 + \phi \cdot f_s)\bar{c}_{p,P}}
\end{aligned} \tag{2.43}
$$

where $\bar{c}_{p,R} \approx \bar{c}_{p,P}$ is used in deriving the second line. Similarly, for the rich mixtures one gets

$$
\phi \geqslant 1 \quad T_p = T_R + \frac{f_s \cdot LHV}{(1 + f)\bar{c}_{p,P}} = T_R + \frac{f_s \cdot LHV}{(1 + \phi \cdot f_s)\bar{c}_{p,P}} \tag{2.44}
$$

Note that f_s is very small for hydrocarbon fuels (e.g., $f_s = 0.058$ for methane). As such, the product (flame) temperature increases almost linearly with equivalence ratio, ϕ, for lean combustion as shown in Fig. 2.4. As expected, the flame temperature peaks at the stoichiometric ratio. In rich combustion, the flame temperature decreases with ϕ.

Method 2: Iterative enthalpy balance

A more accurate approach is to find the flame temperature by iteratively assigning the flame temperature T_p until $H_p(T_p) \approx H_R(T_R)$. The enthalpy of reactants is assumed given. The enthalpy of products can be expressed in the following form

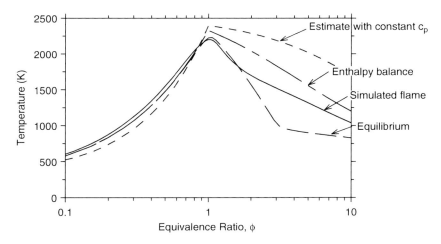

Fig. 2.4 Comparison of flame temperatures with different approaches

$$H_P(T_P) = \sum_i N_{i,P}\hat{h}_{i,P} = \sum_i N_{i,P}[\Delta\hat{h}^o_{i,P} + \hat{h}_{si,P}(T_P)] = H_R(T_R) = \sum_i N_{i,R}\hat{h}_{i,R}$$

Next, we rearrange the above equation to find an expression for the sensible enthalpy of the products as

$$\sum_i N_{i,P}\Delta\hat{h}^o_{i,P} + \sum_i N_{i,P}\hat{h}_{si,P}(T_P) = \sum_i N_{i,R}\Delta\hat{h}^o_{i,R} + \sum_i N_{i,R}\hat{h}_{si,R}(T_R)$$

$$\sum_i N_{i,P}\hat{h}_{si,P}(T_P) = \sum_i N_{i,R}\Delta\hat{h}^o_{i,R} - \sum_i N_{i,P}\Delta\hat{h}^o_{i,P} + \sum_i N_{i,R}\hat{h}_{si,R}(T_R) \qquad (2.45)$$

$$\sum_i N_{i,P}\hat{h}_{si,P}(T_P) = -Q^0_{rxn,p} + \sum_i N_{i,P}\hat{h}_{si,R}(T_R).$$

With an initial guess of flame temperature, T_{p1}, one evaluates $H_p(T_{p1})$ from tables such as those in Appendix 3. If $H_p(T_{p1}) < H_R(T_R)$, we guess a higher flame temperature, T_{p2}. One repeats this process until the two closest temperatures are found such that $H_p(T_{f1}) < H_R(T_R) < H_p(T_{f2})$. The product temperature can be estimated by linear interpolation. This method, although more accurate, still assumes complete combustion to the major products.

Method 3: Equilibrium State (Free software: Cantera; Commercial software: Chemkin)
Dissociation[5] of products at high temperature ($T > 1,500$ K at ambient pressure) can take a significant portion of energy from combustion and hence the product

---

[5] Dissociation is the separation of larger molecules into smaller molecules. For example, $2H_2O \leftrightarrow 2H_2 + O_2$.

temperature is lower than that calculated with only major components as products. The equilibrium state determines the species concentrations and temperature under certain constraints such as constant enthalpy, pressure, or temperature. The equilibrium flame temperature is expected to be lower than the temperatures estimated with Method 1 or Method 2. In addition, the chemical equilibrium state is often used in combustion engineering as a reference point for chemical kinetics (the subject of Chap. 3) if infinite time is available for chemical reactions. At this ideal state, forward and backward reaction rates of any chemical reaction steps are balanced. By constraining certain variables such as constant pressure and enthalpy, the chemical equilibrium state can be determined by minimizing the Gibbs free energy, even without knowledge of the chemical kinetics. Computer programs (such as STANJAN, Chemkin, Cantera) are preferred for this task, as hand calculations are time consuming.

2.4.2 Comparison of Adiabatic Flame Temperature Calculation Methods

The presented methods of estimating adiabatic flame temperature will produce different values from each other. Predicted adiabatic flame temperatures of a methane/air mixture at ambient pressure using these methods are compared in Fig. 2.4 for a range of equivalence ratios. Also included are the results from a flame calculation using a detailed, non-equilibrium flame model. On the lean side, the results agree reasonably well among all methods, as the major products are CO_2, H_2O, unburned O_2, and N_2. Visible deviations arise near stoichiometric conditions and become larger in richer mixtures. One reason for the deviation is the assumptions made about product species in the rich mixtures. For rich mixtures at the equilibrium state, CO is preferred over CO_2 due to the deficiency in O_2. Because the conversion of CO into CO_2 releases a large amount of energy, the rich mixture equilibrium temperatures are lower than those from the flame calculation, which has a residence time of less than 1 s. Among the methods, the results from the detailed flame model calculations are closest to reality, as real flames have finite residence times and generally do not reach equilibrium.

Example 2.6. Estimate the adiabatic flame temperature of a constant-pressure reactor burning a stoichiometric mixture of H_2 and air at 101.3 kPa and 25°C at the inlet.

Solution:
The combustion stoichiometry is $H_{2(g)} + 0.5 \ (O_{2(g)} \ +3.76 \ N_{2(g)}) \rightarrow H_2O_{(g)} + 1.88 \ N_{2(g)}$

$$-Q^0_{rxn,p} = \sum_i N_{i,R}\Delta\hat{h}^o_{i,R} - \sum_i N_{i,P}\Delta\hat{h}^o_{i,P}$$
$$= \Delta\hat{h}^o_{H2} + 0.5\Delta\hat{h}^o_{O2} + 1.88\Delta\hat{h}^o_{N2} - 1\cdot\Delta\hat{h}^o_{H2O}$$
$$= 0 + 0 + 0 - 1\,\text{mol}\cdot(-241.88\,\text{kJ/mol}) = 241.88\,\text{kJ}$$

Method 1: Assuming a constant (average) $\hat{c}_p$ at 1,500 K, $\hat{c}_{p,H_2O}(1,500\,\text{K}) = 0.0467\,\text{kJ/mol} - \text{K}$
and

$$\hat{c}_{p,N_2}(1,500\,\text{K}) = 0.0350\,\text{kJ/mol} - \text{K}:$$

$$T_p = T_0 + \frac{-Q^0_{rxn,p} + \sum_i N_{i,R}\hat{h}_{si,R}(T_R)}{\sum_i N_{i,p}\hat{c}_{p,i}}$$

$$= 300 + \frac{(241.88 + 0)\text{kJ/mol}}{(0.047 + 1.88 \cdot 0.035)\ \text{kJ/mol} - \text{K}}$$

$$\sim 2,148\,\text{K}$$

The average temperature of the products and reactants is now (2,148 K + 298 K)/ 2 ~ 1,223 K, indicating that the initial assumption of $T_{ave} = 1,500$ K was too high. Using the new average temperature of 1,223 K to evaluate the specific heats, the calculated flame temperature becomes $T_p \sim 2,253$ K. The average temperature is now $T_{ave} = 1,275$ K. This new average temperature can be used to calculate the specific heats and the process should be continued until the change in the average temperature is on the order of 20 K. By doing this procedure, we obtain $T_P \sim 2,230$ K.

Method 2: Iterative enthalpy balance:

$$H_P(T_P) = H_R(T_R)$$

$$\sum_i N_{i,p}\Delta\hat{h}^o_{i,p} + \sum_i N_{i,p}\hat{h}_{si,p}(T_p) = \sum_i N_{i,R}\Delta\hat{h}^o_{i,R} + \sum_i N_{i,R}\hat{h}_{si,R}(T_R)$$

$$N_{H_2O}\Delta\hat{h}^o_{H_2O} + N_{H_2O}\hat{h}_{s,H_2O}(T_P) + N_{N_2}\Delta\hat{h}^o_{N_2} + N_{N_2}\hat{h}_{s,N_2}(T_P)$$
$$= N_{H_2}\Delta\hat{h}^o_{H_2} + N_{H_2}\hat{h}_{s,H_2}(T_R) + N_{O_2}\Delta\hat{h}^o_{O_2} + N_{O_2}\hat{h}_{s,O_2}(T_R)$$
$$+ N_{N_2}\Delta\hat{h}^o_{N_2} + N_{N_2}\hat{h}_{s,N_2}(T_R)$$

$$1 \cdot \Delta\hat{h}^0_{H_2O} + \hat{h}_{s,H_2O}(T_P) + 0 + 1.88 \cdot \hat{h}_{s,N_2}(T_P) = 0 + 0 + 0 + 0 + 0 + 0$$

$$\Delta\hat{h}^0_{H_2O} + \hat{h}_{s,H_2O}(T_P) + 1.88 \cdot \hat{h}_{s,N_2}(T_P) = 0.$$

The first step is to guess the product temperature. For this case, let's pick $T_P = 2,000$ K. We now plug in the value for the heat of formation of water and use thermodynamic property tables to evaluate the sensible enthalpy terms.

T_P (K)	$H_P(T_P)$ (MJ)
2,000 K	$-241.83 + 72.69 + 1.88 \cdot 56.14 = -63.6$ MJ
2,500 K	$-241.83 + 98.96 + 1.88 \cdot 74.31 = -3.1$ MJ

Our initial guess of $T_P = 2,000$ K was too low. The process was repeated with a higher guess of $T_P = 2,500$ K which resulted in a much smaller remainder, implying that $T_P \sim 2,500$ K. For more accuracy, we can use linear extrapolation (or interpolation if we bracketed the real value):

$$\frac{T_P - 2,500}{2,500 - 2,000} = \frac{0 + 3.1}{-3.1 + 63.6}$$

$$T_P = 2,526K$$

Method 3: Cantera. Assume H_2, O_2, and H_2O are the only species in the system; equilibrium temperature is 2,425.1 K. The equilibrium mole fractions are listed below

Mole fractions		
Species	$x_{reactant}$	$x_{product}$
H_2	0.2958	0.0153
O_2	0.1479	0.0079
N_2	0.5563	0.6478
H_2O	0	0.3286

Note that there is a small amount ($\sim 1.5\%$) of H_2 existing in the products due to the dissociation of H_2O at high temperature. Results of the above three methods agree with each other within 100–200 K which is less than 12% of the flame temperature. If radicals, such as H, OH, and O, are also included in the products, the equilibrium temperature drops to 2,384 K because additional dissociation occurs. This 41 K difference is about 1.7% of the flame temperature.

Example 2.7 The space shuttle burns liquid hydrogen and oxygen in the main engine. To estimate the maximum flame temperature, consider combustion of 1 mol of gaseous hydrogen with 0.5 mol of gaseous O_2 at 101.3 kPa. Determine the adiabatic flame temperatures using the average c_p method.

Solution:
The combustion stoichiometry is

$$H_{2(g)} + 0.5O_{2(g)} \rightarrow H_2O_{(g)}$$

$$-Q^0_{rxn,p} = \text{LHV of } H_2 \text{ at constant pressure}$$

$$-Q^0_{rxn,p} = \sum_i N_{i,R}\Delta\hat{h}^o_{i,R} - \sum_i N_{i,P}\Delta\hat{h}^o_{i,P} = \Delta\hat{h}^o_{H2} + 0.5\Delta\hat{h}^o_{O2} - 1\Delta\hat{h}^o_{H2O}$$

$$= 0 + 0 - 1 \text{ mol}(-241.88 \text{ kJ/mol}) = 241.88 \text{ kJ}$$

Guessing a final temperature of about 3,000 K, we use average specific heats evaluated at 1,500 K

$$T_P = T_0 + \frac{-Q^0_{rxn,p} + \sum_i N_{i,R} \hat{h}_{si,R}(T_R)}{\sum_i N_{i,P} \hat{c}_{pi}}$$

$$= 300\,K + \frac{241.88\,kJ/mol}{0.047\,kJ/mol - K}$$

$$\sim 5,822\,K$$

Discussion:

This temperature is evidently much higher than the NASA reported value of ~3,600 K. What is the main reason for such a **BIG** discrepancy? The estimated temperature is well above 2,000 K and one expects a substantial dissociation of H_2O back to H_2 and O_2. That is, $H_2(g) + 0.5\,O_2\,(g) \leftrightarrow H_2O\,(g)$. Now we use Cantera or a commercial software program, such as Chemkin, to compute the equilibrium temperature with only three species H_2, O_2, and H_2O. The predicted adiabatic flame temperature drops to 3508.7 K. The mole fractions of these three before reaction and after combustion are listed below.

Species	Reactant	Product
H_2	0.6667	0.2915
O_2	0.3333	0.1457
H_2O	0	0.5628

As seen in the table, the dissociation is *very significant*; about 30% of the products is H_2. Let's find out how much fuel is not burned by considering the following stoichiometric reaction:

$$H_2(g) + 0.5O_2(g) \rightarrow X \cdot H_2 + 0.5X \cdot O_2 + (1 - X) \cdot H_2O(g)$$

The mole fraction of H_2 in the products is

$$x_{H_2} = \frac{X}{X + 0.5X + 1 - X} = \frac{X}{0.5X + 1}.$$

With $x_{H_2} = 0.2915$, we get $X = 0.3412$. If we assume 66% of fuel is burned, a new estimate based on $\hat{c}_p$ at 1,500 K leads to

$$T_p = 300\,K + \frac{0.66 \cdot 241.88\,kJ/mol}{0.047\,kJ/mol - K} \sim 3,700\,K$$

that is in much better agreement with the equilibrium result. If we estimate $\hat{c}_p$ at 1,800 K we get

$$T_p = 300\,K + \frac{0.66 \cdot 241.88\,kJ/mole}{0.04966\,kJ/mole - K} \sim 3,514.7\,K.$$

If we include additional species, H, OH, and O in the products, the predicted equilibrium temperature drops to 3,076 K. The table below shows the mole fractions of each species in this case.

Species	Reactant	Product
H_2	0.6667	0.1503
O_2	0.3333	0.0510
H_2O	0	0.5809
OH	0	0.1077
O	0	0.0330
H	0	0.0771

Evidently, the radicals OH, H, and O take some energy to form; note that their values for enthalpy of formation are positive. Because the space shuttle engine operates at 18.94 MPa (2,747 psi, ~186 atm) at 100% power, the pressure needs to be taken into consideration as the combination of radicals occurs faster at higher pressures. The predicted equilibrium temperature at 18.94 MPa is 3,832.4 K and the mole fractions are listed below.

Species	Reactant	Product
H_2	0.6667	0.1169
O_2	0.3333	0.0336
H_2O	0	0.7051
OH	0	0.1005
O	0	0.0143
H	0	0.0296

The energy needed to vaporize liquid H_2 and O_2 and heat them from their boiling temperatures to 25°C are estimated to be 8.84 kJ/mol and 12.92 kJ/mol (energy = latent heat + sensible energy from boiling point to STP). With $H_2 + 0.5O_2$, the total energy required is then $8.84 + 0.5 \cdot 12.92$ or about 15.3 kJ/mol. The temperature drop due to this process is about ~15.3 kJ/(0.049 kJ/mol-K) = 148 K. With this, we estimate the space shuttle main engine temperature is $3,832 - 148$ K or ~3,675 K. The following information is used for estimating energy to vaporize H_2 and O_2: (1) for H_2, latent heat of vaporization 445.7 kJ/kg, boiling temperature $= -252.8$°C, $c_p \sim 4.12$ kJ/kg-K; (2) for O_2, latent heat of vaporization 212.7 kJ/kg, boiling temperature $= -183$°C, $c_p \sim 0.26$ kJ/kg-K.

2.5   Chapter Summary

The following shows the relations among different thermodynamics properties expressed in terms of mass fractions and mole fractions.

Property	Mass fraction, y_i	Mole fraction x_i
Species density ρ_i (kg/m^3)	ρy_i	$\rho \dfrac{x_i M_i}{\sum_{j=1}^{K} x_j M_j}$
Mole fraction, x_i $[-]$	$\dfrac{y_i/M_i}{\sum_{j=1}^{K} y_j/M_j}$	$-$
Mass fraction, y_i.	$-$	$\dfrac{x_i M_i}{\sum_{j=1}^{K} x_j M_j}$
Mixture molecular mass, M (kg/kmol)	$\dfrac{1}{\sum_{j=1}^{K} y_j/M_j}$	$\sum_{j=1}^{K} x_j M_j$
Internal energy of mixture, u (kJ/kg)	$\sum_{j=1}^{K} y_j \cdot u_j$	$\dfrac{1}{M}\sum_{j=1}^{K} x_j \cdot \hat{u}_j$
Enthalpy of mixture, h (kJ/kg)	$\sum_{j=1}^{K} y_j \cdot h_j$	$\dfrac{1}{M}\sum_{j=1}^{K} x_j \cdot \hat{h}_j$
Entropy of mixture, s (kJ/kg-K)	$\sum_{j=1}^{K} s_j \cdot h_j$	$\dfrac{1}{M}\sum_{j=1}^{K} x_j \cdot \hat{s}_j$
Specific heat at constant pressure c_p (kJ/kg-K)	$\sum_{j=1}^{K} y_j \cdot c_{pj}$	$\dfrac{1}{M}\sum_{j=1}^{K} x_j \cdot \hat{c}_{pj}$
Specific heat at constant volume c_v (kJ/kg-K)	$\sum_{j=1}^{K} y_j \cdot c_{vj}$	$\dfrac{1}{M}\sum_{j=1}^{K} x_j \cdot \hat{c}_{vj}$
Internal energy of mixture, $\hat{u}$ (kJ/kmol)	$M\sum_{j=1}^{K} y_j \cdot u_j$	$\sum_{j=1}^{K} x_j \cdot \hat{u}_j$
Enthalpy of mixture, $\hat{h}$ (kJ/kmol)	$M\sum_{j=1}^{K} y_j \cdot h_j$	$\sum_{j=1}^{K} x_j \cdot \hat{h}_j$
Entropy of mixture, $\hat{s}$ (kJ/kmol-K)	$M\sum_{j=1}^{K} y_j \cdot s_j$	$\sum_{j=1}^{K} x_j \cdot \hat{s}_j$
Specific heat at constant pressure $\hat{c}_p$ (kJ/kmol-K)	$M\sum_{j=1}^{K} y_j \cdot c_{pj}$	$\sum_{j=1}^{K} x_j \cdot \hat{c}_{pj}$
Specific heat at constant volume $\hat{c}_v$ (kJ/kmol-K)	$M\sum_{j=1}^{K} y_j \cdot c_{vj}$	$\sum_{j=1}^{K} x_j \cdot \hat{c}_{vj}$

Definitions

Enthalpy of combustion or heat of combustion: Ideal amount of energy that can be released by burning a unit amount of fuel.

Enthalpy of reaction or heat of reaction: Energy that must be supplied in the form of heat to keep a system at constant temperature and pressure during a reaction.

Enthalpy of formation or heat of formation: Heat of reaction per unit of product needed to form a species by reaction from the elements at the most stable conditions.

Combustion stoichiometry for a general hydrocarbon fuel, $C_\alpha H_\beta O_\gamma$

$$C_\alpha H_\beta O_\gamma + \left(\alpha + \frac{\beta}{4} - \frac{\gamma}{2}\right)(O_2 + 3.76N_2) \rightarrow \alpha CO_2 + \frac{\beta}{2}H_2O + 3.76\left(\alpha + \frac{\beta}{4} - \frac{\gamma}{2}\right)N_2$$

Variables to quantify combustible mixtures

Fuel/air ratio by weight: $f = \frac{m_f}{m_a}$

For stoichiometric mixture: $f_s = \frac{m_f}{m_{as}}$

Equivalence ratio: $\phi = \frac{f}{f_s} = \frac{m_{as}}{m_a}$

Normalized air/fuel ratio $\lambda = \frac{AFR}{AFR_s} = \frac{1/f}{1/f_s} = \frac{1}{f/f_s} = \frac{1}{\phi}$

Percent of excess air

$$\%EA = 100\frac{(m_a - m_{as})}{m_{as}} = 100\left(\frac{m_a}{m_{as}} - 1\right) = 100\left(\frac{1}{\phi} - 1\right)$$

Global equation for lean combustion $\phi \leqslant 1$

$$C_\alpha H_\beta O_\gamma + \frac{1}{\phi}\left(\alpha + \frac{\beta}{4} - \frac{\gamma}{2}\right)(O_2 + 3.76N_2)$$

$$\rightarrow \alpha CO_2 + \frac{\beta}{2}H_2O + \frac{3.76}{\phi}\left(\alpha + \frac{\beta}{4} - \frac{\gamma}{2}\right)N_2 + \left(\alpha + \frac{\beta}{4} - \frac{\gamma}{2}\right)\left(\frac{1}{\phi} - 1\right)O_2$$

in terms of λ

$$C_\alpha H_\beta O_\gamma + \lambda\left(\alpha + \frac{\beta}{4} - \frac{\gamma}{2}\right)(O_2 + 3.76N_2)$$

$$\rightarrow \alpha CO_2 + \frac{\beta}{2}H_2O + 3.76 \cdot \lambda \cdot \left(\alpha + \frac{\beta}{4} - \frac{\gamma}{2}\right)N_2 + (\lambda - 1)\left(\alpha + \frac{\beta}{4} - \frac{\gamma}{2}\right)O_2$$

Global equation for rich combustion $\phi > 1$ with the assumption that products contain unburned fuel

$$C_\alpha H_\beta O_\gamma + \frac{1}{\phi}\left(\alpha + \frac{\beta}{4} - \frac{\gamma}{2}\right)(O_2 + 3.76N_2)$$

$$\rightarrow \frac{\alpha}{\phi}CO_2 + \frac{\beta}{2\phi}H_2O + \frac{3.76}{\phi}\left(\alpha + \frac{\beta}{4} - \frac{\gamma}{2}\right)N_2 + \left(1 - \frac{1}{\phi}\right)C_\alpha H_\beta O_\gamma$$

Enthalpy of formation (heat of formation) determined by bomb calorimeter

$$\Delta\hat{h}_i^o = \frac{Q_{rxn,v}^0 + \Delta N \cdot \hat{R}_u T_0}{N_{i,P}}$$

$$\Delta N = \sum_i N_{i,P} - \sum_i N_{i,R} = \frac{\beta}{4} + \frac{\gamma}{2} - 1$$

where $Q^0_{rxn,v}$ is the heat released from a constant-volume reactor where the products and reactants are at STP.

Heating values at STP (T_0) from a constant-volume reactor

$$HHV = \frac{\sum_i N_{i,R}\Delta\hat{h}^o_{i,R} - \sum_i N_{i,P}\Delta\hat{h}^o_{i,P} + \left(\sum_i N_{i,p} - \sum_i N_{i,R}\right)\hat{R}_u T_0}{N_{fuel}M_{fuel}} \quad (MJ/kg)$$

$$LHV = HHV - \frac{N_{H2O,P}M_{H2O}h_{fg}}{N_{fuel}M_{fuel}} , h_{fg} = 2,440 kJ/kg$$

Heating values at STP (T_0) determined from a constant-pressure reactor

$$HHV = \frac{\sum_i N_{i,R}\Delta\hat{h}^o_{i,R} - \sum_i N_{i,P}\Delta\hat{h}^o_{i,P}}{N_{fuel}M_{fuel}}$$

Adiabatic flame temperature for reactants at standard conditions
Method 1: Estimate based on average $\hat{c}_p$ values

$$T_P = T_0 + \frac{N_{fuel}M_{fuel}LHV + \sum_i N_{i,R}\hat{h}_{si,R}(T_R)}{\sum_i N_{i,P}\hat{c}_{pi}}$$

$$T_P \approx T_R + \frac{N_{fuel}M_{fuel}LHV}{\sum_i N_{i,P}\hat{c}_{pi}}$$

or if mixture is not stoichiometric: mass-base analysis using LHV and ϕ

$$\phi \leq 1 \; T_P = T_R + \frac{f \cdot LHV}{(1+f)\bar{c}_p} = T_R + \frac{\phi \cdot f_s \cdot LHV}{(1+\phi \cdot f_s)\bar{c}_p}$$

$$\phi > 1 \; T_P = T_R + \frac{f_s \cdot LHV}{(1+f)\bar{c}_p} = T_R + \frac{f_s \cdot LHV}{(1+\phi \cdot f_s)\bar{c}_p}$$

Method 2: Enthalpy Balance

$$H_P(T_P) = H_R(T_R)$$
$$H_P(T_P) = \sum_i N_{i,P}\hat{h}_{i,P} = \sum_i N_{i,P}[\Delta\hat{h}^o_{i,P} + \hat{h}_{si,P}(T_P)]$$

Trial and error of T_P such that $H_P(T_P)$ matches $H_R(T_R)$

Exercises

2.1 Consider an isentropic combustion system with a total of K species. Assuming constant specific heats, show that the mixture temperature and pressure at two different states are related to the respective pressures as

$$\frac{T_2}{T_1} = \left(\frac{P_2}{P_1}\right)^{(\gamma-1)/\gamma}$$

where

$$\gamma = \frac{\displaystyle\sum_{i=1}^{K} m_i c_{p,i}}{\displaystyle\sum_{i=1}^{K} m_i c_{v,i}}.$$

2.2 Measurements of exhaust gases from a methane-air combustion system show 3% of oxygen by volume (dry base) in the exhaust. Assuming complete combustion, determine the excess percentage of air, equivalence ratio, and fuel/air ratio.

2.3 There has been a lot of interest about replacing gasoline with ethanol, but is this really a good idea? We're going to compare a blend of ethanol (70% ethanol and 30% gasoline by volume) to gasoline. Calculate the lower heating value (LHV) of a 70% ethanol/30% isooctane mixture in terms of kJ/mol of fuel. Assume complete combustion. How does this compare to the tabulated value for gasoline (isooctane)? Assuming a 20% thermal efficiency, if you need to get 100 kW of power from an engine, how much of each fuel (in mol/s) do you need? If you have a stoichiometric mixture of the ethanol/gasoline blend and air in your 100 kW engine, how much CO_2 are you emitting in g/s? How does this compare to the same engine running a stoichiometric mixture of 100% gasoline and air?

2.4 Gasoline is assumed to have a chemical composition of $C_{8.26} H_{15.5}$.

   (a) Determine the mole fractions of CO_2 and O_2 in the exhaust for an engine with normalized air/fuel ratio $\lambda = 1.2$ with the assumption of complete combustion.

   (b) The enthalpy of formation of $C_{8.26} H_{15.5}$ is -250 MJ/kmol. Determine the LHV of gasoline in terms of MJ/kg. The molecular mass of $C_{8.26} H_{15.5}$ is 114.62 kg/kmol.

   (c) Using an average c_p for the products at 1,200 K, estimate the adiabatic flame temperature at constant pressure of 1 atm for the lean ($\lambda = 1.2$) mixture.

2.5 A mixture of methane gas and air at 25°C and 1 atm is burned in a water heater at 150% theoretical air. The mass flow rate of methane is 1.15 kg/h. The exhaust gas temperature was measured to be 500°C and approximately

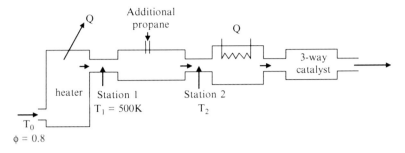

Fig. 2.5 Exercise 2.7

1 atm. The volumetric flow rate of cold water (at 22°C) to the heater is 4 L/min.

(a) Draw a schematic of the water heater and name its most important elements.
(b) Using Cantera, determine the amount of heat generated from burning of 1 kg of methane.
(c) Calculate the temperature of the hot water if the heat exchanger were to have an efficiency of 1.0, i.e., perfect heat transfer.

2.6 An acetylene-oxygen torch is used in industry for cutting metals.

(a) Estimate the maximum flame temperature using average specific heat c_p.
(b) Measurements indicate a maximum flame temperature of about 3,300 K. Compare with the result from (a) and discuss the main reasons for the discrepancy.

2.7 A space heater burns propane and air with intake temperature at $T_0 = 25°C$ and pressure at 1 atm (see Fig. 2.5). The combustible mixture enters the heater at an equivalence ratio $\phi = 0.8$. The exhaust gases exit at temperature $T_1 = 500$ K and contain CO_2, H_2O, O_2, and N_2 only at station 1. In order to use a 3-way catalyst for exhaust treatment, additional propane is injected into the exhaust to consume all the remaining oxygen in the exhaust such that the gases entering the catalyst contain only CO_2, H_2O, and N_2 at station 2. Assume that the entire system is at $P = 1$ atm and complete combustion occurs in both the heater and in the exhaust section.

(a) The volumetric flow rate of propane entering the heater is 1 L/min. Determine the injection rate of propane into the exhaust between station 1 and station 2 (see Fig. 2.5). Note that the propane at the injection station is at the same conditions as heater inlet, i.e., $T = 25°C$ and $P = 1$ atm.
(b) With the assumption of constant specific heats for the gases, estimate the temperature at station 2, T_2. The specific heat can be approximated by that of N_2 at 700 K as $\hat{c}_p = 30.68 \, kJ/kmol - K$,

Fuel:
$T_{fuel} = 25°C$
$P_{fuel} = 1\ atm$

Air:
$T_{air} = 427°C$
$P_{air} = 1\ atm$

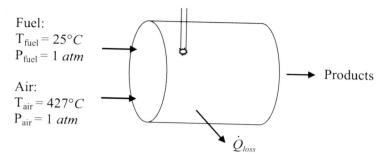

Products

$\dot{Q}_{loss}$

Fig. 2.6 Exercise 2.9

2.8 Two grams of solid carbon, C(s), are combusted with pure oxygen in a 500 cm³ bomb calorimeter initially at 300 K. After the carbon is placed inside the bomb, the chamber is evacuated and then filled with gaseous oxygen from a pressurized tank.

(a) Determine the minimum O_2 pressure inside the bomb necessary to allow complete combustion of the solid carbon.
(b) When the bomb is cooled back to its initial temperature of 300 K, determine the pressure inside the bomb.

2.9 Consider the combustion chamber in a jet engine at cruising altitude. For simplicity, the combustor is operated at 1 atm of pressure and burns a stoichiometric ($\phi = 1$) mixture of n-heptane (C_7H_{16}) and air. The intake conditions are as indicated in Fig. 2.6.

(a) Write the stoichiometric chemical reaction for the fuel with air.
(b) If the mass flow rate of fuel is 1 kg/s, what is the mass flow rate of air?
(c) What is the rate of heat loss from the combustion chamber if 10% of the LHV (heat of combustion) of the fuel is lost to surroundings?
(d) What is the temperature of the products?
(e) How does the temperature change if we burn fuel rich ($\phi > 1$)? How about fuel lean ($\phi < 1$)? (Hint: Easiest to show with a plot)

2.10 An afterburner is a device used by jet planes to increase thrust by injecting fuel after the main combustor. A schematic of this system is shown in Fig. 2.7. In the main combustor, hexane is burned with air at an equivalence ratio of $\phi = 0.75$. The products of the main combustor are CO_2, H_2O, O_2 and N_2, all of which enter the afterburner. In the afterburner, additional hexane is injected such that the equivalence ratio is $\phi = 1.25$. In the afterburner the hexane reacts with the excess O_2 from the main combustor to form CO, H_2O, and CO_2 only. Combined with the products of the main combustor, the gases exiting the afterburner are CO, CO_2, H_2O, O_2 and N_2. The entire system is

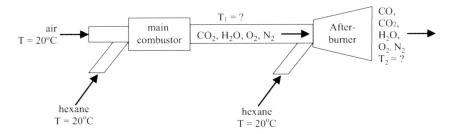

Fig. 2.7 Exercise 2.10

insulated, and the pressure everywhere is atmospheric. The inlet temperature of the hexane and air is 20°C. Determine the temperature of the exhaust gases at each stage (Fig. 2.7). **Note:** An approximate answer is sufficient and it can be assumed that the specific heats for the gases are constant and approximately equal to that of N_2 at 1,000 K.

Chapter 3
Chemical Kinetics

While thermodynamics provides steady state information of the combustion process, chemical kinetics describes the transient states of the system during the combustion process. Particularly important is information related to the rate at which species are consumed and produced, and the rate at which the heat of reaction is released. Combustion chemistry has two important characteristics not commonly observed in other chemical systems. First, combustion reaction rates are highly sensitive to temperature. Second, a large amount of heat is released during a chemical reaction. The heat release provides the positive feedback that sustains combustion: heat transfer from products to reactants raises the reactant temperature so that the chemical reaction proceeds at a high rate. The rate at which fuel and oxidizer are consumed is of great importance to combustion engineering, as one needs to ensure sufficient time for chemical reactions when designing a combustion system. Chemical kinetics is the science of chemical reaction rates. When chemical kinetics is coupled with fluid dynamics and heat transfer, a combustion system can be characterized. For instance, when air is blown onto a burning candle, the flame can respond by burning more vigorously because extra oxygen is present. If the feeding rate of air is too large and the chemical reaction rate cannot keep up to consume the combustible mixture, the flame will be extinguished. Another example is the combustion of a torch, such as a propane torch for soldering a copper pipe. If the fuel flow rate is increased to a certain point, the flame detaches from the nozzle. A further increase in fuel flow rate results in the flame blowing out. Another important area related to combustion chemistry is emissions. The formation of pollutants is controlled primarily by chemical kinetics. Pollutants are present in small amounts in the products, yet their impact on the environment and human health can be significant. The issues of pollutant formation will be addressed in a later chapter.

S. McAllister et al., *Fundamentals of Combustion Processes*,
Mechanical Engineering Series, DOI 10.1007/978-1-4419-7943-8_3,
© Springer Science+Business Media, LLC 2011

3.1   The Nature of Combustion Chemistry

A chemical reaction can be described by an overall stoichiometric relation as

$$
C_\alpha H_\beta O_\gamma + \left(\alpha + \frac{\beta}{4} - \frac{\gamma}{2}\right)(O_2 + 3.76N_2) \rightarrow
$$
$$
\alpha CO_2 + \frac{\beta}{2}H_2O + 3.76\left(\alpha + \frac{\beta}{4} - \frac{\gamma}{2}\right)N_2,
\tag{3.1}
$$

but the actual chemical kinetics in combustion rarely proceed in such a simple manner. For one of the simplest combustion systems, hydrogen with oxygen, the overall stoichiometric relation is

$$
H_2 + 0.5O_2 \rightarrow H_2O.
\tag{3.2}
$$

The chemical reaction does not start with H_2 and O_2 directly. In fact, H_2 and O_2 do not directly react with each other at all; breaking both H–H and O–O bonds simultaneously during a single molecular collision is less probable than other chemical pathways. The initiation of the chemical reaction is either through $H_2 + M \rightarrow H + H + M$ or $O_2 + M \rightarrow O + O + M$ to generate unstable, highly reactive molecules called 'radicals' which then react with H_2 and O_2 to produce more radicals leading to the build-up of a radical pool. The notation 'M' denotes all molecules that collide with H_2 or O_2, and are referred to as the third body molecules. The third body molecules serve as energy carriers. The above relation in Eq. 3.2 is only a "global" reaction; the combustion of hydrogen involves many "elementary reactions," each containing only two or three species.

The collection of elementary reactions that describe the overall, global reaction is referred to as a reaction or combustion mechanism. Depending on the amount of detail, a combustion mechanism can consist of only a couple of steps, themselves semi-global reactions, or thousands of elementary reactions. For instance, a detailed hydrogen-oxygen combustion mechanism contains about 9 species and 21 elementary reaction steps as shown in Table A in Appendix 4. For hydrocarbon fuels, due to the large number of isomers and many possible intermediate species, the number of species and steps in a detailed mechanism can grow substantially with the size of the fuel molecule. For CH_4/air combustion, the chemical kinetics can be reasonably described by 53 species and 400 steps (using the so-called GRI3.0 combustion mechanism). A recent detailed mechanism for isooctane contains 860 species and 3,606 steps [1]. Computing of chemical kinetics with such a large mechanism requires a significant amount of computer resources even for one-dimensional flames. Figure 3.1 presents the number of species in typical detailed combustion chemistry and its relation to the carbon content of fuels.

In general, there are four main types of elementary reactions that are important in combustion: chain initiation, chain branching, chain terminating or recombination, and chain propagating.

Fuel	Species
CH_4	53
C_2H_4	75
C_3H_8	176
$n\text{-}C_7H_{16}$	561
$i\text{-}C_8H_{18}$	857
$n\text{-}C_7H_{16}+i\text{-}C_8H_{18}$	1,033

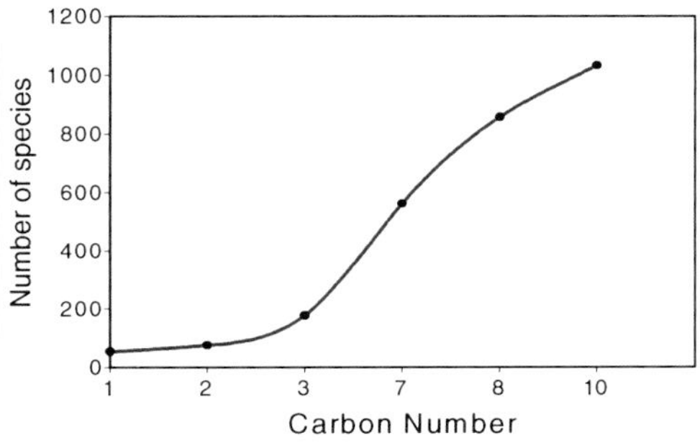

Fig. 3.1 *Left*: typical numbers of species in detailed reaction mechanisms. *Right*: number of species increases rapidly with the total number of carbon elements in fuels

3.1.1 Elementary Reactions: Chain Initiation

The initiation of the combustion reaction is through reactions such as

$$H_2 + M \rightarrow H + H + M$$

$$O_2 + M \rightarrow O + O + M$$

where M is a third body with enough energy to break the H_2 or O_2 bonds.

3.1.2 Elementary Reactions: Chain Branching

Chain branching reactions, such as

$$H + O_2 \rightarrow OH + O \tag{3.3}$$

$$O + H_2 \rightarrow H + OH, \tag{3.4}$$

produce two radicals on the product side (OH and O in Eq. 3.3, H and OH in Eq. 3.4) and consume one on the reactant side (H in Eq. 3.3, O in Eq. 3.4). The net gain of one radical is significant because these reactions increase the pool of radicals rapidly, leading to the explosive nature of combustion. If each collision leads to the products, the radical growth rate is 2^{Nc}, where N_c is the number of collisions. For instance, ten collisions would increase the radical population by about 1,000 times. Because the number of collisions among molecules at standard conditions (STP) is of the order of 10^9/s, the number of radicals can grow enormously in a short period of time.

3.1.3 Elementary Reactions: Chain Terminating or Recombination

When sufficient radicals or third bodies are present, radicals can react among themselves to recombine or react to form stable species. Recombination steps (also called termination steps) are depicted by

$$H + O_2 + M \rightarrow HO_2 + M \tag{3.5}$$

$$O + H + M \rightarrow OH + M \tag{3.6}$$

$$H + OH + M \rightarrow H_2O + M \tag{3.7}$$

and they decrease the radical pool by half.

3.1.4 Elementary Reactions: Chain Propagating

Chain propagating steps are reactions involving radicals where the total number of radicals remains unchanged. Different radicals can appear on both the reactant and product sides, but the total number of radicals in the reactant and product sides stays the same. For instance, the reaction step

$$H_2 + OH \rightarrow H_2O + H \tag{3.8}$$

consumes 1 mol of OH radicals and produces 1 mol of H radicals so that the net change in the number of radicals is zero. This reaction is still very important, as it produces most of the H_2O formed in hydrogen-oxygen combustion.

3.2 Elementary Reaction Rate

3.2.1 Forward Reaction Rate and Rate Constants

The chemical expression of an elementary reaction can be described by the following general expression

$$aA + bB \rightarrow cC + dD, \tag{3.9}$$

where a, b, c, d are the respective stoichiometric coefficients. Usually the values of a, b, c, d are one or two as not more than two molecules are likely involved in elementary reactions. The corresponding rate of reaction progress is often expressed by the following empirical form (often referred to as the law of mass action)

$$\text{Rate of reaction progress}: \dot{q}_{RxT} = k[A]^a[B]^b, \tag{3.10}$$

which states that the reaction rate is proportional to the concentration of reactants. The constant of proportionality is called the Arrhenius rate constant k and is of the form

$$k = A_o \exp\left(-\frac{E_a}{\hat{R}_u T}\right) = A_o \exp\left(-\frac{T_a}{T}\right), \tag{3.11a}$$

where A_0 is the pre-exponential factor, E_a is the activation energy, and $\hat{R}_u$ is the universal gas constant (1.987 cal/mol-K, 1 cal $= 4.184$ J)[1]. The ratio $E_a/\hat{R}_u$ has the unit of temperature and is referred to as the activation temperature (T_a). The pre-exponential factor (A_0) expresses the frequency of the reactants molecules colliding with each other and the activation energy (E_a) can be viewed as the energy barrier required for breaking the chemical bonds of the molecules during a collision. The exponential term, $\exp(-T_a/T)$, can be interpreted as the probability of a successful collision leading to products. Combustion chemistry often has reaction steps with high activation temperatures such that rates are very sensitive to temperature. On the other hand, recombination reactions, such as those in Eqs. 3.5–3.7, usually have very low or no activation energies so that the forward rate constants are insensitive to temperature. Because recombination reactions require three molecules to occur, the overall forward rate scales with P^3. As the pressure increases, the molecules are forced closer together so that the likelihood of three molecules colliding at the same time increases. Therefore the forward rate of a recombination step increases more rapidly with pressure than two body reaction steps that scale with P^2. The values of A_0 and E_a are determined experimentally using shock tubes or flow reactors. An example of the data obtained by such an experiment is shown in Fig. 3.2. The Arrhenius rate constant k is calculated from the rate of progress of the experimental data and the values of A_0 and E_a are found by plotting $\ln k = \ln A_o - \frac{E_a}{\hat{R}_u T}$ versus $1/T$ as shown in Fig. 3.3. The rate of the reaction is then expressed as

$$\dot{q}_{RxT} = A_o[A]^a[B]^b \exp\left(-\frac{E_a}{\hat{R}_u T}\right) \tag{3.11b}$$

The consumption rate of reactant A is then expressed by

$$\frac{d[A]}{dt} = \hat{r}_A = -a \cdot \dot{q}_{RxT}, \tag{3.11c}$$

and similar formulas can be used for products.

---

[1]Collision theory gives $k = A_o T^{1/2} \exp\left(-\frac{E_a}{R_u T}\right)$ and in general $k = A_o T^b \exp\left(-\frac{E_a}{R_u T}\right)$.

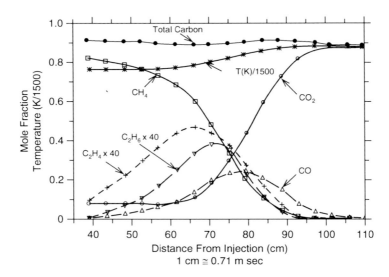

Fig. 3.2 Experimental measurements of the reaction rate of methane/air (Reprinted with permission from Dryer and Glassman [2])

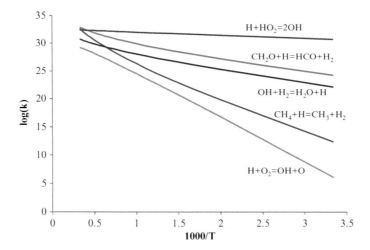

Fig. 3.3 Rate constant $k = k(T)$ for reactions in main pathway of methane-air combustion (Appendix 4 Table B)

3.2.2 Equilibrium Constants

The above procedure can be generalized to reversible reactions

$$aA + bB \leftrightarrow cC + dD \tag{3.12}$$

Designating the forward and backward reaction rate constants as k_f and k_b, respectively, the net rate of reaction progress becomes $\dot{q}_{RxT} = k_f[A]^a[B]^b - k_b[C]^c[D]^d$. At the chemical equilibrium state, forward and reverse reaction rates are equal, and $\dot{q}_{RxT} = k_f[A]_{eq}^a[B]_{eq}^b - k_b[C]_{eq}^c[D]_{eq}^d = 0$. The ratio $K_c = k_f/k_b$, is the *equilibrium constant based on concentrations*. K_c can be determined by thermodynamics properties of the reaction.

$$K_c = \frac{k_f}{k_b} = \frac{[C]_{eq}^c[D]_{eq}^d}{[A]_{eq}^a[B]_{eq}^b} = K_p(T)\left(\frac{\hat{R}_u T}{101.3 \text{ kPa}}\right)^{a+b-c-d} \tag{3.13}$$

where

$$K_p(T) = \exp\left\{\frac{a\hat{g}_A^0 + b\hat{g}_B^0 - c\hat{g}_C^0 - d\hat{g}_D^0}{\hat{R}_u T}\right\}$$

is the *equilibrium constant based on partial pressures*. The Gibbs free energy at the reference pressure (101.3 kPa) $\hat{g}_i^0(T) = \hat{h}_i(T) - T\hat{s}_i^0(T)$, is found in the thermodynamics tables in Appendix 3. K_p is dimensionless and depends on temperature only.

3.3 Simplified Model of Combustion Chemistry

As mentioned earlier, the complex chemical kinetics of practical, higher hydrocarbon fuels are described by chemical mechanisms with many hundreds or thousands of chemical species. The number of species and reaction steps grows nearly exponentially with the number of carbon atoms in the fuel; it becomes impractical for a human to comprehend physical significance from such large mechanisms. Computers can model detailed chemical kinetics in simplified reactors, but often engineers want to know the behavior of practical, multi-dimensional systems. Large-scale computational fluid dynamics simulations of practical systems can be coupled with chemical kinetics calculations, but processor and memory demands are intense when hundreds of chemical species and the corresponding reactions must be tracked at every point in the domain. A simplified description of chemical kinetics is thus extremely useful for practical applications of combustion sciences to engineering problems. For single component fuels, a one-step global reaction is often used in practical simulations due to its simplicity.

3.3.1 Global One-Step Reaction

For a general hydrocarbon fuel with an overall combustion stoichiometry as shown in Eq. 3.1, the corresponding global rate of progress can be expressed as

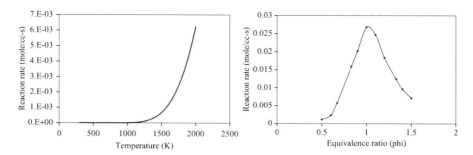

Fig. 3.4 Reaction rate for methane as a function of temperature and equivalence ratio

$$\dot{q}_{RxT} = A_o \exp\left(-\frac{E_a}{\bar{R}_u T}\right) [Fuel]^a [O_2]^b. \tag{3.14}$$

The pre-exponential factor, activation energy and exponents a and b are obtained experimentally in flow reactors (see Table 3.1). Typical units for the fuel and oxidizer concentrations are expressed in terms of mol/cm^3 so that the rate of progress has units of mol/cm^3-s. Note that A_0 has the unit of $(mol/cm^3)^{1-(a+b)}\cdot s^{-1}$. Because of the high activation energy in the exponential term, one can expect that the rate of progress is highly dependent on temperature as presented in Fig. 3.4. Because of this strong temperature dependence, the reaction rate can be quite sensitive to the equivalence ratio of the mixture due to the change in flame temperature as exemplified in Fig. 3.4. The consumption rates of fuel and oxygen are

$$\hat{r}_{fuel} = \frac{d[Fuel]}{dt} = -\dot{q}_{RxT}, \quad \text{and} \quad \hat{r}_{O_2} = \frac{d[O_2]}{dt} = -\left(\alpha + \frac{\beta}{4} - \frac{\gamma}{2}\right)\dot{q}_{RxT}. \tag{3.15}$$

The production rates of CO_2 and H_2O are

$$\hat{r}_{CO_2} = \frac{d[CO_2]}{dt} = \alpha \dot{q}_{RxT} \text{ and } \hat{r}_{H_2O} = \frac{d[H_2O]}{dt} = \frac{\beta}{2} \dot{q}_{RxT} \tag{3.16}$$

Table 3.1 gives empirically determined values of the pre-exponential factor (A_0), the activation energy (E_a), and the exponents a and b. Note that the exponents a and b in the global reaction rate equations are *not* the stoichiometric coefficients of the reaction as they would be if the reaction were elementary.

Example 3.1 Consider combustion of stoichiometric methane-air at a constant temperature of 1,800 K and 101.3 kPa. Using a one-step reaction formulation for the rate constant, estimate the amount of time required to completely consume the fuel.

Solution:
Stoichiometric methane-air combustion is

$$CH_4 + 2(O_2 + 3.76N_2) \rightarrow CO_2 + 2H_2O + 7.52N_2$$

The global rate of reaction progress is

$$\dot{q}_{RxT} = A_o \exp\left(-\frac{E_a}{\hat{R}_u T}\right)[Fuel]^a[O_2]^b.$$

From Table 3.1, $A_0 = 1.3 \cdot 10^9$, $E_a = 48.4$ kcal/mol, $a = -0.3$, $b = 1.3$. $E_a/\hat{R}_u = 24,358$ K. Note that the exponent of fuel concentration is negative meaning that if more fuel is present, the rate of chemical kinetics is slower. This peculiar behavior is due to the role of methane in the oxidation process as a strong radical consumer. That is, methane is competing for radicals leading to a negative effect on the build-up of radical pool. The global consumption rate for methane is

$$\frac{d[CH_4]}{dt} = \hat{r}_{CH_4} = -\dot{q}_{RxT} = -1.3 \cdot 10^9 \cdot \exp\left(-\frac{24,358}{T(K)}\right)[CH_4]^{-0.3}[O_2]^{1.3}$$

Next the concentrations of methane and oxygen are evaluated at $T = 1,800$ K using the ideal gas law

$$P_i V = N_i \hat{R}_u T$$

$$[C_i] = \frac{N_i}{V} = \frac{P_i}{\hat{R}_u T} = \frac{Px_i}{\hat{R}_u T}$$

Table 3.1 Global reaction rate constants for hydrocarbon fuels (Data reprinted with permission from Westbrook and Dryer [3])[a]

Fuel	A_0	E_a (kcal/mol)	a	b
CH_4^*	$1.3 \cdot 10^9$	48.4	-0.3	1.3
CH_4	$8.3 \cdot 10^5$	30	-0.3	1.3
C_2H_6	$1.1 \cdot 10^{12}$	30	0.1	1.65
C_3H_8	$8.6 \cdot 10^{11}$	30	0.1	1.65
C_4H_{10}	$7.4 \cdot 10^{11}$	30	0.15	1.6
C_5H_{12}	$6.4 \cdot 10^{11}$	30	0.25	1.5
C_6H_{14}	$5.7 \cdot 10^{11}$	30	0.25	1.5
C_7H_{16}	$5.1 \cdot 10^{11}$	30	0.25	1.5
C_8H_{18}	$4.6 \cdot 10^{11}$	30	0.25	1.5
C_9H_{20}	$4.2 \cdot 10^{11}$	30	0.25	1.5
$C_{10}H_{22}$	$3.8 \cdot 10^{11}$	30	0.25	1.5
CH_3OH	$3.2 \cdot 10^{11}$	30	0.25	1.5
C_2H_5OH	$1.5 \cdot 10^{12}$	30	0.15	1.6
C_6H_6	$2.0 \cdot 10^{11}$	30	-0.1	1.85
C_7H_8	$1.6 \cdot 10^{11}$	30	-0.1	1.85

[a] Units of A_0: $(mol/cm^3)^{1-a-b}/s$.
[*] Note that for methane, the constants associated with the high activation energy are only appropriate for shock tubes and turbulent flow applications

For $[O_2]$, $x_{O_2} = 2/(1 + 2 \cdot 4.76) = 0.19$

$$[O_2] = \frac{0.19 \cdot 101.325\,\text{kPa}}{8.314\ \text{kPa}\times\text{m}^3/(\text{kmol} - \text{K}) \cdot 1800\text{K}}$$
$$= 1.28 \cdot 10^{-3}\ \text{kmol/m}^3$$
$$= 1.28 \cdot 10^{-6}\text{mol/cc}$$

Similarly $x_{CH4} = 1/(1 + 2 \cdot 4.76) = 0.095$ and $[CH_4] = 6.4 \cdot 10^7$ mol/cm^3. The initial consumption rate of methane is

$$\frac{d[CH_4]}{dt} = -1.3 \cdot 10^9 \cdot \exp\left(-\frac{24,358}{1800}\right)(6.4 \cdot 10^{-7})^{-0.3}(1.28 \cdot 10^{-6})^{1.3}$$
$$= 2.72 \cdot 10^{-3}\text{mol/cc} - \text{s}$$

If the consumption is assumed constant, the amount of time to consume all the fuel is

$$\frac{[CH_4]}{-d[CH_4]/dt} = 2.35 \cdot 10^{-4}\ \text{s} = 0.24\ \text{ms}$$

Since both fuel and oxidizer decrease during combustion, the consumption rate also decreases with time. Let's estimate the consumption rate when methane is half of its original value $(0.5 \cdot 6.4 \cdot 10^{-7} = 3.2 \cdot 10^{-7}$ mol/cm$^3)$ and oxygen is $1.28 \cdot 10^{-6} - 2 \cdot (0.5 \cdot 6.4 \cdot 10^{-7}) = 6.4 \cdot 10^{-7}$ mol/cm^3 as

$$\frac{d[CH_4]}{dt} = -1.3 \cdot 10^9 \cdot \exp\left(-\frac{24,358}{1800}\right)[3.2 \cdot 10^{-7}]^{-0.3}[6.4 \cdot 10^{-7}]^{1.3}$$
$$= 1.36 \cdot 10^{-3}\text{mol/cc} - \text{s}$$

This is half of its initial value and the amount of time to consume all the fuel is

$$\frac{[CH_4]}{-d[CH_4]/dt} = 0.48\ \text{ms}$$

It is clear that the above estimates are rather crude. Luckily there is an analytical solution of this problem. For a stoichiometric methane-air mixture, the oxygen consumption rate is directly related to the methane consumption rate as

$$\frac{d[O_2]}{dt} = 2\frac{d[CH_4]}{dt}$$

$$[O_2](t) - [O_2]_0 = 2\big([CH_4](t) - [CH_4]_0\big)$$

$$[O_2](t) = [O_2]_0 - 2 \cdot ([CH_4]_0 - [CH_4](t))$$
$$= \underbrace{([O_2]_0 - 2 \cdot [CH_4]_0)}_{=0} + 2 \cdot [CH_4](t)$$
$$= 2 \cdot [CH_4](t)$$

With this expression the consumption rate of methane assuming a constant temperature of 1,800 K becomes

$$\frac{d[CH_4]}{dt} = -\dot{q}_{RxT} = -A_0 \exp\left(\frac{-E_a}{RT}\right)[CH_4]^{-0.3}[O_2]^{1.3}$$
$$= -1.3 \cdot 10^9 \exp\left(-\frac{24,358}{T(K)}\right)[CH_4]^{-0.3}(2[CH_4])^{1.3}$$
$$\frac{d[CH_4]}{dt} = -2.46 \cdot 1.3 \cdot 10^9 \cdot \exp\left(-\frac{24,358}{1800K}\right) \cdot [CH_4]$$
$$= -4245.3 \cdot [CH_4]$$

The solution of the above equation is

$$\frac{[CH_4](t)}{[CH_4]_{t=0}} = \exp(-4245.3 \cdot t)$$

The half life time, $\tau_{1/2}$, is defined as the time at which concentration of fuel is decreased to half of its initial value. The half life time of methane is about $\tau_{1/2} \sim 0.16$ ms (see Fig. 3.5). Due the exponential decrease of methane concentration, the time to 'completely' consume methane is arbitrarily set when the methane concentration decreases to 5% of its initial value

$$\tau_{0.05} \approx \frac{-\ln(0.05)}{4245.3} \text{ s} = 7.1 \cdot 10^{-4}s = 0.71 \text{ ms}$$

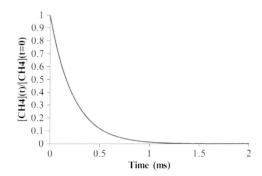

Fig. 3.5 Fuel concentration as a function of time (normalized by the initial fuel concentration)

Discussion:

If the reactor temperature drops to 300 K, the estimated time to consume all the fuel is about $1.7 \cdot 10^{22}$ million years! This is due to the strong temperature dependence of exponential term as revealed in the left table below:

T (K)	Exp $(-24,358/T)$
300	$5.47 \cdot 10^{-36}$
500	$6.96 \cdot 10^{-22}$
750	$7.85 \cdot 10^{-15}$
1,000	$2.64 \cdot 10^{-11}$
1,500	$8.86 \cdot 10^{-8}$
1,800	$1.32 \cdot 10^{-6}$
1,900	$2.71 \cdot 10^{-6}$
2,000	$5.24 \cdot 10^{-6}$

Equivalence ratio (ϕ)	Detailed chemistry (ms)	Estimates (one-step) (ms)
0.25	0.120	0.035
0.5	0.123	0.103
1.0	0.143	0.71
1.5	0.202	1.04
2.0	0.492	1.22

When temperature increases from 1,800 K to 1,900 K, the rate is doubled showing the strong temperature dependence. It is useful to gain some insights into the effect of equivalence on the consumption time at a fixed reaction temperature. For mixtures other than stoichiometric, numerical solutions are used to determine the consumption time. For rich combustion,

$$\frac{[O_2]}{-d[O_2]/dt}$$

is used to calculate consumption time because oxidizer is now the deficient species. The right table above compares the computed consumption time from numerical simulations with detailed chemistry (GRI3.0) to the estimates based on the one-step global reaction. Both results show the negative dependence of consumption time on equivalence ratio for a fixed reaction temperature. The consumption time based on 1-step chemistry depends on equivalence ratio roughly as $\propto \phi^{-2.3}$ on the lean side and $\phi^{-1.1}$ on the rich side. Remember, however, that if the reaction occurs at the adiabatic flame temperature, the rate of progress is at a maximum for stoichiometric mixtures and decreases for both lean and rich mixtures as shown in Fig. 3.4. Because of the strong temperature dependence, the trend in rate of progress with equivalence ratio follows that for the adiabatic temperature (Fig. 2.4). The consumption time would then be at a minimum for stoichiometric mixtures and would increase for either lean or rich mixtures.

3.3.2 Pressure Dependence of Rate of Progress

In addition to being strongly temperature dependent, the rate of progress is also pressure dependent through the species concentration. Starting with the general equation of the rate of progress (Eq. 3.11b) and the ideal gas relation for the concentrations the rate of progress can be expressed as

$$\dot{q}_{RxT} = A_o \exp\left(-\frac{E_a}{\hat{R}_u T}\right)[Fuel]^a[O_2]^b$$

$$= A_o \exp\left(-\frac{E_a}{\hat{R}_u T}\right)x_{fuel}^a x_{O_2}^b \left(\frac{P}{\hat{R}_u T}\right)^{a+b} \propto P^{a+b}$$

(3.17)

The rate of progress is proportional to pressure raised to the sum of the fuel and oxidizer coefficients. Based on the 1-step chemistry model in Table 3.1, the sum, $a + b$, is always positive ranging from 1.0 to 1.75. When the pressure of a combustion system is doubled, the reaction rate can increase threefold for the case $a + b = 1.75$. The corresponding consumption time decreases as

$$t_{chem} = \frac{[Fuel]}{-d[Fuel]/dt} \propto \frac{P}{P^{a+b}} \propto P^{1-(a+b)} \propto P^{-0.75}.$$

With $a + b = 1.75$, the consumption time at 1.013 MPa decreases to about 60% of its value at 101.3 kPa.

3.3.3 Heat Release Rate (HRR)

Once the consumption rate of the fuel is found, the rate of heat release, or power, of a combustion system can be calculated as:

$$HRR = -\frac{d[fuel]}{dt} \cdot M_{fuel} \cdot Q_c,$$

(3.18)

where Q_c is the heat of combustion as described in Chap. 2 ($Q_c = -Q_{rxn,p}$). The rate of heat release is a very important factor in combustion systems since it provides the heat power available for conversion into mechanical work or to be controlled if the combustion is accidental. The expression in Eq. 3.18 will be used often in the subsequent chapters.

3.3.4 Modeling of Chemical Kinetics with Detailed Description

The aforementioned 1-step overall chemistry has severe restrictions, as many intermediates exist before major products are formed. Also, multiple pathways are possible between each oxidation step making it difficult to comprehend by analytical means. Numerical modeling has become useful in providing insights into the complexities of combustion chemistry of practical fuels.

3.3.4.1   An Example of a Detailed CH₄-air Combustion Mechanism

To illustrate the complicated nature of combustion chemistry, Fig. 3.6 below is a path diagram for the combustion of methane. The reaction pathways in the bracket are those that do not involve C_2 chemistry pathways (species with two atoms, such C_2H_6, C_2H_4, C_2H_2) that are important under high pressure or rich conditions. Chemistry involving C_2 is initiated through the recombination of $CH_3 + CH_3 + M \rightarrow C_2H_6 + M$ and therefore important when pressure is high.

Table B in Appendix 4 details the important elementary steps in this mechanism for the branch of the reaction in the boxed region without C_2 chemistry. Some observations are:

- The initiation step has a large activation energy. For example, the activation temperature of $CH_4 + M \rightarrow CH_3 + H$ in step (1) is about 50,000 K. This means that it takes a significant amount of energy to abstract a hydrogen atom from methane.
- The activation energy of a 3-body recombination step is zero. For example the following reaction steps have zero activation energy

$$H + H + M \rightarrow H_2M \ (48), \quad H + OH + M \rightarrow H_2O + M \ (52),$$
$$H + O + M \rightarrow OH + M \ (53)$$

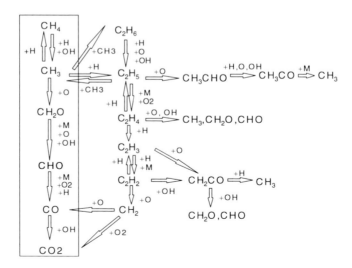

Fig. 3.6 Simplified flow diagram for methane combustion

Note that 3-body reaction rates increase with the third power of pressure and become more important at high pressures.

• To better fit the experimental data, the rate constant formula often includes an extra temperature term T^b and the general form is

$$k = A_o T^b \exp\left(-\frac{E_a}{\hat{R}_u T}\right).$$

For instance, the forward rate constant for $CO + OH \rightarrow CO_2 + H$ is $k_f = 1.51 \cdot 10^7 \cdot T^{1.3} \cdot exp(381/T)$, where the temperature dependence term is $T^{1.3}$. Note that this reaction step has a 'negative' activation temperature that is small compared to the usual activation temperature in most 2-body reaction steps. However, k_f still increases with T in the range of 300–2,000 K as sketched Fig. 3.7.

Figure 3.8 plots computed time evolution profiles of major species for stoichiometric methane-air combustion at constant $T = 1,600$ K using the GRI3.0 detailed

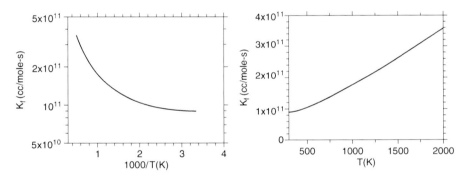

Fig. 3.7 Forward rate constant versus temperature for $CO + OH = CO_2 + H$; Left versus temperature; right versus $1,000/T(K)$ showing a weak temperature dependence

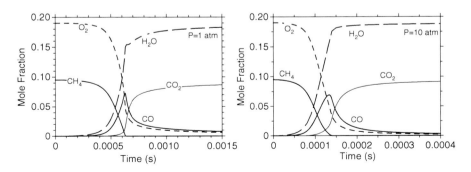

Fig. 3.8 Computed profiles of major species versus time during combustion at $P = 101.3$ kPa (*left*) and $P = 1.013$ MPa (*right*)

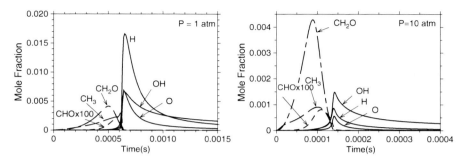

Fig. 3.9 Computed profiles of intermediate and radical species versus time during combustion at $P = 101.3$ kPa (*left*) and $P = 1.013$ MPa (*right*)

mechanism with two pressures of 101.3 kPa and 1.013 MPa. At 101.3 kPa, the major reactants, CH_4 and O_2, are consumed around 0.6 ms. Water is formed following closely the consumption of the major species. The intermediate species CO is formed and peaks around 0.65 ms when CH_4 is completely consumed. Then CO is oxidized to form CO_2 with a time scale of about 0.5 ms. The oxidation process at 1.013 MPa is similar to that at 101.3 kPa except it occurs about five times faster than at 101.3 kPa.

The corresponding time evolution profiles of intermediate species, CH_3, CH_2O, CHO, and radical species, H, OH, and O, are presented in Fig. 3.9. Consistent with the view that CH_4 is initially decomposed through step (1) of the detailed mechanism in Table B of Appendix 4, CH_3 is formed immediately and then consumed around 0.6 ms. Other intermediate species, CH_2O and CHO, also form before the major oxidation event. Radicals such as O, H, and OH, are formed in large amounts when all the fuel is consumed. When pressure increases to 1.013 MPa, the levels of intermediate and radical species decrease to about one fifth. Also noticed are shifts in the relative importance among the radicals.

Figures 3.8 and 3.9 can also be used to demonstrate the importance of chemical kinetics to pollutant formation (the subject of Chap. 9). In many practical applications, such as a car engine, there is only a finite time available for the chemical reactions to occur. This time, often referred to as the physical or residence time, is a function of the engine speed – the higher the RPM the less time the fuel and air have to complete combustion. Say for example that the engine RPM is such that the residence time of the combustion gases is 0.12 ms, meaning that the gases are exhausted from the engine and the combustion process is stopped. Assuming the pressure in the engine during combustion is 1.013 MPa, Figs. 3.8 (*right*) and 3.9 (*right*) show that it takes about 0.2 ms to completely burn the fuel. In this case, the residence time is less than the "chemistry" time and the exhaust of the engine will include CO and unburned hydrocarbons. However, if the engine were run at an RPM such that the residence time was 0.25 ms, the chemistry time would be less than the residence time allowing for more complete combustion and virtually no CO and unburned hydrocarbon emissions.

3.3.5 Partial Equilibrium

Due to the difficulty in measuring radicals in high temperatures ($\sim$>1,500 K), estimates of radical concentrations can be made by assuming that even though the combustion process is in a non-equilibrium state, a subset of the combustion reactions are in equilibrium. The combustion process is then said to be in a partial equilibrium state. The advantage is that by assuming partial equilibrium the number of intermediate reactions is reduced accordingly. For instance, if reaction step $O_2 \leftrightarrow O + O$ is assumed in an equilibrium state, one can estimate the concentration of O atom as

$$k_f[O_2] = k_b[O]^2 \quad [O] = \sqrt{\frac{k_f}{k_b}}[O_2] = \sqrt{K_c[O_2]}$$

Another reaction often assumed in equilibrium is $H_2 + OH \leftrightarrow H_2O + H$ (reaction 3 in Table B of Appendix 4) relating the concentration of [OH] to [H] as

$$[OH] = \frac{[H_2O][H]}{K_c[H_2]}.$$

In hydrogen combustion, the following set of reactions can be assumed in equilibrium at high temperatures:

$$H + O_2 \leftrightarrow OH + O \quad (R1)$$
$$O + H_2 \leftrightarrow OH + H \quad (R2)$$
$$H_2 + OH \leftrightarrow H_2O + H \quad (R3)$$

Setting the forward rates equal to backward rates, the concentrations of OH, H, and O can be expressed in terms of stable species, H_2, O_2, and H_2O, as

$$[OH] = \sqrt{K_{c,1} \cdot K_{c,2} \cdot [O_2][H_2]}$$

$$[H] = \sqrt{K_{c,1} \cdot K_{c,2} \cdot K_{c,3}^2 \cdot \frac{[H_2]^3[O_2]}{[H_2O]^2}}$$

$$[O] = K_{c,1} \cdot K_{c,3} \frac{[H_2][O_2]}{[H_2O]}$$

3.3.6 Quasi-Steady State

Intermediate combustion species are produced during the combustion process and will be consumed at the end of the combustion process. For instance, in methane combustion discussed previously, there exist many intermediate species, such as CH_3, CH_2O, and CH. The consumption rates of these intermediate species are fast in comparison to their production rates. An alternative method for estimating radicals or intermediate species is based on the assumption that the consumption rate and the production rate of a species is the same leading to the following expression:

$$\frac{d[C]}{dt} = \omega_p - \omega_c \approx 0 \quad \text{or} \quad \omega_p = \omega_c,$$

where ω_p and ω_c stand for the net production and net consumption rates respectively. As consumption rate depends on the concentration of [C], its value can be determined by solving $\omega_c([C]) - \omega_p = 0$. Iterative methods are required when $\omega_c([C])$ is a nonlinear function of [C].

Example 3.2 Consider the following two reactions (Zeldovich Mechanism) for the formation of nitric oxide as

$$N_2 + O \rightarrow NO + N \quad k_1 = 1.8 \cdot 10^{14} \exp(-38,370/T) \quad \text{(R1)}$$

$$N + O_2 \rightarrow NO + O \quad k_2 = 1.8 \cdot 10^{10} T \exp(-4,680/T) \quad \text{(R2)}$$

Assuming N atom is in a quasi-steady state, derive an expression for [N] in terms of other species.

Solution:

$$\frac{d[N]}{dt} = k_{f1}[N_2][O] - k_{f2}[N][O_2] \approx 0 \rightarrow [N] = \frac{k_{f1}[N_2][O]}{k_{f2}[O_2]}$$

With this approximation, the *NO* production rate becomes

$$\frac{d[NO]}{dt} = k_{f1}[N_2][O] + k_{f2}[N][O2] \cong 2k_{f1}[N_2][O]$$

Example 3.3 The O atom is an important species involved in the formation of thermal NO (Zeldovich Mechanism $N_2 + O \rightarrow NO + N$). Estimate the mole fraction of radical O in air when it is heated to 2,000 K.

Solution:
At 2,000 K, the reaction $O_2 \leftrightarrow 2O$ is assumed to be equilibrated. Using the equilibrium relation $k_f[O_2] = k_b[O]^2$, the concentration of O atoms is estimated as

$$[O] = \sqrt{\frac{k_f}{k_b}[O_2]} = \sqrt{K_c[O_2]} = \sqrt{K_p\left(\frac{\hat{R}_u T}{101.3 \text{ kPa}}\right)^{-1}[O_2]}$$

The value of $K_p(T)$ is computed as

$$\ln K_p(T = 2000K) = \frac{\hat{g}_{O2}^o}{\hat{R}_u T} - 2\frac{\hat{g}_O^o}{\hat{R}_u T} = 28.752 - 2(-7.059) = -14.634$$

$$K_p = (2000K) = \exp(-14{,}634) = 4.41 \cdot 10^{-7}$$

$$[O_2] = 0.21 \cdot 101 \text{ kPa}/(8.314 \cdot 10^3 \text{ kPa cm}^3/\text{mol} - K \cdot 2000K)$$

$$= 1.28 \cdot 10^{-6} \text{ mol/cc.}$$

With these values

$$[O] = \{1.28 \cdot 10^{-6} \text{ mole/cc} \cdot (82.05 \text{ cm}^3 \text{ atm/mol} - K \cdot 2000K/1 \text{ atm})^{-1} \cdot 4.41 \cdot 10^{-7}\}^{1/2}$$

$$= 1.855 \cdot 10^{-9} \text{ mol/cc.}$$

The total concentration is $6.074 \cdot 10^{-6}$ mol/cm^3 and $x_O = 3.0 \cdot 10^{-4}$.

Example 3.4 In a gas turbine burner, engineers estimate the flame temperature to be 2,200 K and wish to reduce the nitric oxide (NO) formation rate. As NO formation is very sensitive to temperature, one solution is to inject a small amount of water into the combustor so that the flame temperature is reduced. The NO production rate is modeled by the following rate equation

$$\frac{d[NO]}{dt} \approx 2k[O][N_2]$$

$$k = 1.8 \cdot 10^{14} \exp(-38{,}000/T(K)) \text{ units of of rate (mol/cc - s)}$$

In the combustor, the mole fractions of O and N_2 are $1 \cdot 10^{-3}$ and $7 \cdot 10^{-1}$ respectively. Since only a small amount of water is injected, the pressure and the concentrations of O and N_2, (i.e., [O] and [N_2]) are assumed to remain unchanged. Estimate the flame temperature with water injection at which the NO formation rate is reduced to half of that at 2,200 K.

Solution:
Formation of NO is very sensitive to temperature due to the high activation temperature. Using the scaling relation

$$\frac{d[NO]}{dt}_{water} \bigg/ \frac{d[NO]}{dt}_{dry} \approx \frac{\exp(-38{,}000/T_{water})}{\exp(-38{,}000/T_{dry})} = 0.5$$

Solving for T_{water}

$$\exp(-38{,}000/T_{water}) = 0.5 \exp(-38{,}000/T_{dry})$$

Taking the *ln* of both sides

$$-38{,}000/T_{water} = \ln(0.5) - 38{,}000/T_{dry}$$
$$1/T_{water} = 1/T_{dry} - \ln(0.5)/38{,}000$$
$$T_{water} = 2115.12 \text{ K}$$

Note that NO production rates drop by half when the temperature drops only by 85 K. The rough rule of thumb is that NO production drops by half for every 100 K drop in temperature. By combining the partial equilibrium expression for [O] and the quasi-steady state assumption for [N], the following global expression can be used to estimate the formation of thermal NO $(mol/cm^3$-s):

$$\frac{d[NO]}{dt} \cong 2k_{f1}[N_2][O] \cong 1.476 \cdot 10^{15}[N_2][O_2]^{1/2} \exp\left(-\frac{67,520}{T(K)}\right)$$

Example 3.5 When burning hydrogen, an important chain branching reaction is

$$H_2O_2 + M \rightarrow OH + OH + M$$

If hydrogen is being burned in an engine which operates at $T = 1,000$ K and $P = 4.052$ MPa (40 atm) at the end of the compression stroke, how long is the hydrogen peroxide present? Assume the pre-exponential factor of this elementary reaction to be $1.2 \cdot 10^{17}$ and the activation temperature to be 22,750 K.

Solution:
The consumption rate of hydrogen peroxide is

$$\frac{d[H_2O_2]}{dt} = -k[H_2O_2][M]$$

A general characteristic time for this reaction can be found using dimensional analysis as

$$\tau = \frac{[H_2O_2]}{|d[H_2O_2]/dt|} = \frac{[H_2O_2]}{k[H_2O_2][M]} = \frac{1}{k[M]}$$

The reaction rate constant is

$$k_f = 1.2 \cdot 10^{17} \exp(-22,750/T(K)) \quad (mol/cc)^{-1}/s$$

Plugging the expression for the rate constant into the formula for the characteristic time and rearranging:

$$\tau = 8.3 \cdot 10^{-18} \exp\left(\frac{22,750}{T(K)}\right)[M]^{-1} \text{ (s)}$$

Because M represents any molecule that collides with the hydrogen peroxide, the ideal gas law can be used to calculate its concentration:

$$\frac{n}{V} = \frac{P}{\hat{R}_u T} = \frac{4,052(\text{kPa})}{8.314(\text{kPa} \times \text{m}^3/\text{kmol - K}) \cdot 1000(\text{K})}$$

$$= 0.487 \frac{\text{kmol}}{\text{m}^3}$$

$$= 4.87 \cdot 10^{-4} \frac{\text{mol}}{\text{cc}}$$

The characteristic time is then

$$\tau = 8.3 \cdot 10^{-18} \exp\left(\frac{22,750}{1,000}\right) \frac{1}{4.87 \cdot 10^{-4}} = 1.29 \cdot 10^{-4} s = 0.129 \text{ ms}$$

Exercises

3.1 A vessel contains a stoichiometric mixture of butane and air. The vessel is at a temperature of 500 K, a pressure of 1 atm, and has a volume of 1 m^3.

(a) Given the following equation for the rate of progress: $\dot{q}_{RxT} = -A_0$ $[Fuel]^a [Oxygen]^b \exp\left(-E_a/\hat{R}_u T\right)$ and the following values: $A_0 = 8 \cdot 10^{11} \text{cc}^{2.25}/\text{mol}^{0.75} \text{ s}, E_a = 125$ kJ/mol, $a = 0.15$, and $b = 1.6$. Evaluate the rate of consumption of fuel.

(b) Evaluate the reaction rate for the same equivalence ratio, temperature and volume if the pressure were 10 atm. (Note: Remember that you can write the reaction rate in terms of pressure).

(c) Sketch a graph of ln(k) vs. 1/T. Label the slope.

3.2 Consider a constant-volume homogeneous well-mixed combustor containing a stoichiometric mixture of a hydrocarbon fuel and air. The combustor is adiabatic and there is no mass transfer in or out of the combustor. The fuel consumption rate can be described according to a single-step, global reaction:

$$\frac{d[Fuel]}{dt} = -A_0 [Fuel][O_2] \exp\left(-\frac{E_a}{\hat{R}_u T}\right)$$

where t is time [s], [Fuel] is the fuel concentration [mol/cm^3], [O_2] is the O_2 concentration [mol/cm^3], A_0 is a pre-exponential factor [cm^3/(mol-s)] of the one-step reaction, E_a is the activation energy [J/mole] of the one-step reaction, $\hat{R}_u$ is the universal gas constant [J/(mol-K)], and T is the temperature inside the combustor in K. Assume complete combustion and that the only species involved are fuel, N_2, O_2, CO_2, and H_2O. Assume that the initial pressure

(P) is 1 atm and that the initial temperature is 1,300 K. The fuel is completely consumed within 10^{-3} s. With t as the x-axis, sketch approximate plots of the following: (a) $T(t)$, (b) $P(t)$, (c) Reaction rate $(-d[Fuel](t)/dt)$,(d) $[Fuel](t)$, $[O_2](t)$, $[N_2](t)$, $[H_2O](t)$, $[CO_2](t)$.

3.3 In order to reduce the risk of handling a certain fuel, it is desired to evaluate two different additives. On the one hand, our chemistry lab has informed us that Additive **A** reduces the pre-exponential factor of the fuel by a 60%, while leaving the activation energy the same. On the other hand, the lab reports that additive **B** increases the activation energy of the fuel by a 5%, while leaving the pre-exponential factor the same. Given the above information, discuss which fuel is safer to handle at room temperature (25°C) based on the reaction rate constant (k). In addition, a graphical explanation will help. For the fuel without additives: Pre-exponential factor: $A_0 = 4.2 \cdot 10^{11}$, Activation Energy $E_a = 30$ kcal/mol.

3.4 A stoichiometric mixture of methane and air is burned in a flow reactor operating at constant temperature and pressure. The consumption rate of fuel is modeled by the following global reaction rate as

$$\frac{d[CH_4]}{dt} = \hat{r}_{CH_4} = -8.3 \cdot 10^5 \cdot \exp\left(-\frac{15,000}{T}\right)[CH_4]^{-0.3}[O_2]^{1.3}$$

units: concentration [mol/cm^3], T [K], overall rate [mol/cm^3-s].

(a) Determine the fuel consumption rate [mol/cm^3-s] when $T = 1,500$ K and $P = 1$ atm

(b) An engineer measures the mole fraction of CH_4 at the reactor exit to be 0.001. Determine the mole fraction of O_2 at the exit. Assume that combustion of methane with air forms CO_2 and H_2O only.

(c) If the reactor inlet compositions, temperature, velocity, and combustion duration remain unchanged, the mole fraction of CH_4 at the combustor exit remains the same when the reactor pressure is changed. Provide an explanation based on the above rate equation in terms of mole fractions.

3.5 In a natural gas combustor, engineers measure the flame temperature to be 2,500 K and wish to reduce the nitric oxide (NO) formation rate. As NO formation is very sensitive to temperature, one solution is to inject a small amount of water into the combustor so that the flame temperature is reduced. The NO production rate is modeled by the following rate equation

$$\frac{d[NO]}{dt} \approx 2k[O][N_2]$$

$$k = 1.8 \times 10^{14} \exp(-E_a/\hat{R}T) \quad \text{units [cc/mol - s]}$$

$$E_a = 76.24 \text{ kcal/mol}, \quad \hat{R}_u = 1.897 \text{ cal/mol - K}$$

$$\text{(units : kcal, K, mol, cm}^3\text{, and s)}$$

In the combustor, the mole fractions of O and N_2 are 1×10^{-3} and 7×10^{-1} respectively.

(a) Evaluate the NO formation rate at 2,500 K and 1 atm without water injection.
(b) Since only a small amount of water is injected, the pressure and the concentrations of O and N_2, (i.e., [O] and [N_2]) are assumed to remain unchanged. Determine the flame temperature with water injection so that the NO formation rate is reduced to half of that at 2,500 K.

3.6 Following Exercise 3.5 with a given pressure, sketch $\ln\left(\frac{d[NO]}{dt}\right)$ versus $1/T$ for the following three cases (in the range of $T = 1,000$ K to 3,000 K):

(a) With the assumption that the mole fractions of O and N_2 remain constant, derive an approximate expression for $\ln\left(\frac{d[NO]}{dt}\right)$ as function of $1/T$. Note that since T is large, $\ln\frac{1}{T} \approx \frac{1}{T}$. Sketch $\ln\left(\frac{d[N\dot{O}]}{dt}\right)$ versus $\frac{1}{T}$ for $P = 1$ atm and label the approximate slope.
(b) Repeat (a) with the same assumption but $P = 10$ atm.
(c) Repeat (a) but with the following assumptions

    i. the mole fraction of N_2 remains constant
    ii. the mole fraction of O is approximated by
        $x_O = 0.038 \exp\left(-\frac{8,000}{T}\right)$ [mol/cc]
    iii. $P = 1$ atm

3.7 A stoichiometric mixture of n-octane (C_8H_{18}) vapor and air is burned in a vessel of 1,000 cm^3. Using the following global consumption rate

$$\frac{d[C_8H_{18}]}{dt} = -5. \cdot 10^{11} \cdot \exp\left(-\frac{15,000}{T}\right) [C_8H_{18}]^{0.25} [O_2]^{1.5}$$

units: concentration [mol/cm^3], T [K], overall rate [mol/cm^3-s].

(a) Determine the initial fuel consumption rate [mol/s] when $T = 1,000$ K and $P = 1$ atm.
(b) If the reactor is kept at 1,000 K and 1 atm, estimate the time for 95% consumption of fuel based on the initial reaction rate.
(c) Repeat (b) when the pressure is doubled to 2 atm while the temperature remains unchanged at 1,000 K.

3.8 A flow reactor operates at constant pressure and temperature (isothermal at 1,000 K). A very lean mixture of n-heptane and air enters the reactor ($\phi \ll 1$). When the reactor operates at $P = 1$ atm, 50% of n-heptane remains unburned at the exit of the reactor, i.e., $[C_7H_{16}]_e/[C_7H_{16}]_i = 0.5$, where $[C_7H_{16}]_e$ is the n-heptane concentration at the exit and $[C_7H_{16}]_i$ is the concentration of n-heptane at the inlet. Using the following global consumption rate for n-heptane

$$\frac{d[C_7H_{16}]}{dt} = -3.75x10^9 \left(\frac{P}{T}\right)^2 \exp\left(\frac{2370}{T}\right)[O_2]^2[C_7H_{16}]$$

units: atm, K, mol, cc, s

estimate the percentage of n-heptane at the exit of reactor when the pressure is raised to 2 atm. The inlet mixture stoichiometry and temperature are kept the same as in the case of $P = 1$ atm. List the assumptions you make and justify them if possible.

3.9 In methane-air combustion, the global consumption rate has the following expression

$$\frac{d[CH_4]}{dt} = -8.3 \cdot 10^5 \cdot \exp\left(-\frac{15,000}{T}\right)[CH_4]^{-0.3}[O_2]^{1.3} \text{ (mol/cc - s)}$$

The negative dependence of the overall consumption rate on fuel concentration is due to the competition between the main chain branching reaction

$$H + O_2 \rightarrow OH + O \quad (R1)$$

$$k_{f1} = 5.13 \cdot 10^{16} \cdot T^{-0.816} \cdot \exp\left(-\frac{8307}{T}\right) \text{ units (mol/cc)}^{-1}/\text{s}$$

and the radical scavenge nature of the following reaction

$$CH_4 + H \rightarrow CH_3 + H_2 \quad (R2)$$

$$k_{f2} = 2.2 \cdot 10^4 \cdot T^3 \cdot \exp\left(-\frac{4403}{T}\right) \text{ units (mol/cc)}^{-1}/\text{s},$$

where temperature is in K. For a stoichiometric methane-oxygen mixture at 1,200 K and 1 atm, determine which reaction has larger rate of progress.

3.10 In hydrogen-oxygen combustion over a certain range of pressure, the explosive nature of combustion is largely controlled by the competition between the chain branching reaction

$$H + O_2 \rightarrow OH + O \quad (R1)$$

$$k_{f1} = 5.13 \cdot 10^{16} \cdot T^{-0.816} \cdot \exp\left(-\frac{8307}{T}\right) \text{ units (mol/cc)}^{-1}/\text{s}$$

and the radical recombination step

$$H + O_2 + M \rightarrow HO_2 + M \quad (R2)$$

$$k_{f2} = 3.61 \cdot 10^{17} \cdot T^{-0.72} \text{ units (mol/cc)}^{-2}/\text{s},$$

where T is in K and M represents a third body species with concentration $[M] = \frac{P}{R_u T}$. For simplicity, only forward reactions will be considered here.

(a) Derive expressions for the rate of progress for both reactions.
(b) At $T = 800$ K, determine the pressure at which the rate of progress of (R1) is equal the rate of progress of (R2).
(c) Experiments show that at a given temperature and composition, explosion occurs at low pressures but stops at high pressures. Using results from (a), provide a scientific explanation for this unexpected phenomenon.

References

1. Curran HJ, Gaffuri P, Pitz WJ, Westbrook CK (2002) A comprehensive modeling study of iso-octane oxidation. Combustion and Flame, 129:253–280.
2. Dryer FL, Glassman I (1973) High temperature oxidation of CO and CH4. Symposium (International) on Combustion 14(1):987–1003.
3. Westbrook CK, Dryer FL (1984) Chemical Kinetic Modeling of Hydrocarbon Combustion. Prog. Energy Comb. Sci. 10:1–57.

Chapter 4
Review of Transport Equations and Properties

The transport of heat and species generated by the chemical reactions is an essential aspect of most combustion processes. These transport processes can be described by the continuum mechanics approximations commonly used in fluid and heat transfer analysis of engineering problems. Additional terms in the mass, momentum, and energy conservation equations account for the effects of the chemical reactions. The following discussion briefly presents the equations governing combustion systems.[1]

4.1 Overview of Heat and Mass Transfer

In a general combustion process, heat is transferred by conduction, convection, and radiation. Conduction is the molecular transfer of energy from high to low temperature. The molecules at high temperature have a lot of energy and pass some of that energy onto the molecules at lower temperature. The rate of heat transferred (J/s or W) can be calculated by Fourier's law of heat conduction:

$$\vec{q}_{cond} = -Ak\nabla T, \qquad (4.1)$$

where k is the thermal conductivity of the material, A is the area, and ∇T is the temperature gradient.[2] Typical units of the thermal conductivity are W/m-K. Fourier's law implies that the amount of heat transferred is proportional to the temperature gradient.

Convection is the combination of two mechanisms of energy transport. The first is the transport due to molecular collisions (conduction) and the second is the transport of energy due to the bulk flow of the fluid (advection). Treating convection as a

---

[1] The equations presented in this chapter are valid under the condition where the characteristic length scale of system is larger than the mean free path of molecules, i.e., the distance between collisions of molecules.

[2] $\nabla T = \frac{\partial T}{\partial x}\vec{e}_x + \frac{\partial T}{\partial y}\vec{e}_y + \frac{\partial T}{\partial z}\vec{e}_z$ where $\vec{e}_i$ is the unit vector in i-th direction.

S. McAllister et al., *Fundamentals of Combustion Processes*,
Mechanical Engineering Series, DOI 10.1007/978-1-4419-7943-8_4,
© Springer Science+Business Media, LLC 2011

combination of conduction and bulk flow, we can apply Fourier's law of heat conduction:

$$\vec{q}_{conv} = -Ak\nabla T(u),$$ (4.2)

where the temperature gradient is a function of the fluid velocity. Because of the no-slip condition at a solid surface, the fluid forms a momentum and thermal boundary layer near the surface. If only one dimension is considered, the temperature gradient can be written as

$$-\frac{dT}{dx} \approx \frac{T_{hot} - T_{cold}}{\delta},$$ (4.3)

where δ is the thermal boundary layer thickness. If the above expression is inserted into Eq. 4.2,

$$\dot{q}_{conv} = Ak\frac{T_{hot} - T_{cold}}{\delta} = A\tilde{h}(T_{hot} - T_{cold}),$$ (4.4)

where $\tilde{h}$ is the convective heat transfer coefficient (W/m²-K) defined as the ratio of the thermal conductivity and the thermal boundary layer thickness. Equation 4.4 is called Newton's law of cooling. The convective heat transfer coefficient is either determined with similarity solutions of boundary layer equations or with experimental correlations and can be found in handbooks on heat transfer. The convective heat transfer coefficient varies with geometry and flow conditions, but many situations can be represented by a correlation of the form

$$\tilde{h} = C\frac{k}{L}\mathrm{Re}^a\mathrm{Pr}^b,$$ (4.5)

where Re is the Reynolds number, Pr is the fluid Prandtl number, L is the characteristic length, and a, b and C are empirical constants. For buoyantly dominated processes Eq. 4.5 becomes

$$\tilde{h} = C\frac{k}{L}Gr^a\mathrm{Pr}^b,$$ (4.6)

where Gr is the Grashoff number (the ratio of buoyancy to viscous force).

Radiation is energy transfer through electromagnetic waves and therefore does not require a "medium." To calculate the amount of heat transfer by radiation from a substance at temperature T to the surroundings at temperature T_∞, the following expression is used:

$$\dot{q}_{rad} = F_{12}A\varepsilon\sigma_s(T^4 - T_\infty^4),$$ (4.7)

where ε is the emissivity of the body ($0 \leq \varepsilon \leq 1$), σ_s is the Stefan-Boltzmann constant ($5.67 \cdot 10^{-8}$ W/m^2-K^4), and A the surface area (m^2) of the substance and F_{12} is a geometrical factor.

Mass is transported by advection and diffusion. Advection is the transport of species through fluid motion as described by

$$\dot{m}''_{adv} = \rho_i u = \rho y_i u \tag{4.8}$$

The double primes denote the mass flux through a unit surface area with the units of kg/m^2-s, ρ_i is the mass density (kg/m^3) of species i which is related to the overall density as $\rho_i = \rho y_i$.

Diffusion is the transport of mass due to a gradient in species concentrations. Let's consider an infinite one-dimensional domain. Initially, the left side of the domain is filled with fuel and the right side with the oxidizer as sketched in Fig. 4.1. Diffusion between fuel and oxidizer starts at the interface, creating a layer of mixture containing both fuel and oxidizer. The diffusion process is described by Fick's law[3] as

$$\dot{m}''_{D,i} = -\rho D_i \frac{\partial y_i}{\partial x}, \tag{4.9}$$

where ρ is density (kg/m^3), D_i is the diffusivity of i-th species (m^2/s), and y_i is the corresponding mass fraction. The top plot in Fig. 4.2 sketches the diffusion process from the molecular point of view where molecules from high concentration regions migrate to regions of low concentration. The concentration gradient (equivalently the mass fraction gradient) drives such movement. The time evolution of concentration is plotted on the bottom.

As time proceeds, the mixed region grows and its size (δ_D) scales with $\sqrt{D_i t}$ as seen in Fig. 4.2 where the concentration profile becomes smoother with time. Diffusion is driven primarily by species gradients and secondarily by a temperature gradient. Pressure gradients also play a role.

Fig. 4.1 Fuel and oxidizer initially separated at $x = 0$. Concentration of fuel is unity in the left domain and zero on the right

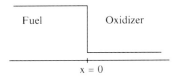

---

[3] Diffusion processes are driven dominantly by concentration gradient. Secondary mechanisms including temperature and pressure gradients also drive diffusion. For this treatment, only Fick's law is considered.

Fig. 4.2 *Top*: In a diffusion
process, molecules move
from a high concentration
region to a low concentration
region. *Bottom*: mass fraction
of fuel concentration as
function of time

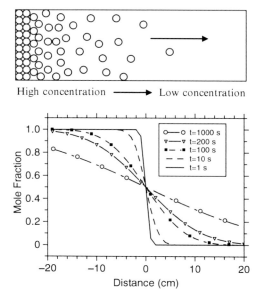

The mass of a species, i, can be created or destroyed by chemical reactions at a
rate given by

$$\dot{m}'''_{i,gen} = \hat{r}_i M_i \tag{4.10}$$

It is on a volumetric basis with units of kg/m^3-s. M_i is the molecular mass of
species i (kg/kmol), and $\hat{r}_i$ is the molar production rate with the units of kmol/m^3-s.

4.2 Conservation of Mass and Species

Because combustion does not create or destroy mass, the conservation of mass
(or continuity) equation applies[4]:

$$\frac{\partial \rho}{\partial t} + \nabla \cdot (\rho \vec{u}) = 0 \tag{4.11}$$

In one dimension with x being the coordinate, this equation reduces to

$$\frac{\partial \rho}{\partial t} + \frac{\partial (\rho \vec{u})}{\partial x} = 0 \tag{4.12}$$

---

[4] $\nabla \cdot (\rho \vec{u}) = \frac{\partial \rho u_x}{\partial x} + \frac{\partial \rho u_y}{\partial y} + \frac{\partial \rho u_z}{\partial z}$ where u_i is the velocity component in i-th direction.

Fig. 4.3 One-dimensional control volume for species conservation

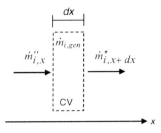

Though overall mass is conserved, combustion creates and destroys individual species. In addition to the usual set of balance laws, prediction of combustion processes requires additional relations to track each chemical species. For gaseous fuels, a simplified 1-D species conservation equation can be derived on the basis of models for advection, diffusion, and generation due to chemical reactions. Consider a one-dimensional domain with a differential width dx and unity area in Fig. 4.3.

The volume for this control volume with unity area is $V = dx \cdot 1 = dx$. Conservation of species gives

$$\frac{dm_{i,CV}}{dt} = \dot{m}''_{i,x} - \dot{m}''_{i,x+dx} + \dot{m}'''_{i,gen} \cdot dx, \qquad (4.13)$$

where the mass flux due to convection and diffusion can be expressed as

$$\dot{m}''_{i,x} = \dot{m}''_{adv} + \dot{m}''_{D,i} = \rho u y_i - \rho D_i \frac{\partial y_i}{\partial x} \qquad (4.14)$$

and

$$\dot{m}''_{i,x+dx} = \dot{m}''_{i,x} + \frac{\partial \dot{m}''_{i,x}}{\partial x} dx.$$

Therefore, Eq. 4.13 becomes

$$\frac{\partial m_{i,CV}}{\partial t} = -\frac{\partial \dot{m}''_{i,x}}{\partial x} \cdot dx + \dot{m}'''_{i,gen} \cdot dx. \qquad (4.15)$$

The mass of species i in the control volume is $m_{i,cv} = \rho_i V = \rho y_i V$. Substitution of Eq. 4.14 into Eq. 4.15 leads to

$$\frac{\partial(\rho y_i)}{\partial t} dx = -\frac{\partial}{\partial x}(\rho u y_i) dx + \frac{\partial}{\partial x}\left(\rho D_i \frac{\partial y_i}{\partial x}\right) dx + \hat{r}_i M_i dx. \qquad (4.16)$$

After eliminating dx, one obtains

$$\frac{\partial(\rho y_i)}{\partial t} + \frac{\partial(\rho u y_i)}{\partial x} = \frac{\partial}{\partial x}\left(\rho D_i \frac{\partial y_i}{\partial x}\right) + \hat{r}_i M_i \tag{4.17}$$

Using the continuity Eq. 4.12, the left hand side of Eq. 4.17 can be further simplified

$$\rho \frac{\partial y_i}{\partial t} + \rho u \frac{\partial y_i}{\partial x} = \frac{\partial}{\partial x}\left(\rho D_i \frac{\partial y_i}{\partial x}\right) + \hat{r}_i M_i \tag{4.18}$$

Assuming that ρD_i is constant,[5] Eq. 4.18 is simplified as

$$\rho \frac{\partial y_i}{\partial t} + \rho u \frac{\partial y_i}{\partial x} = \rho D_i \frac{\partial^2 y_i}{\partial x^2} + \hat{r}_i M_i \tag{4.19}$$

4.3 Conservation of Momentum

The conservation of momentum equation in a system with combustion is the same as in non-reacting systems. The x-momentum equation is given by

$$\frac{\partial(\rho u)}{\partial t} + u \frac{\partial(\rho u)}{\partial x} = -\frac{\partial P}{\partial x} + \mu \frac{\partial^2 u}{\partial x^2} + X \tag{4.20}$$

where u is the velocity and X is the body force.

4.4 Conservation of Energy

Combustion processes involve multiple physical processes including transport of reactants through fluid flows, heat and mass transfer, and chemical kinetics. For gaseous fuels, a simplified 1-D energy equation (first law of thermodynamics) can be derived on the basis of models for these processes.

4.4.1 Terms in the Conservation of Energy Equation

a. Conduction: Fourier's law of heat conduction $\dot{q}''_{cond} = -k\frac{\partial T}{\partial x}$ where k is the conductivity (W/m-K)

---

[5] For constant-pressure combustion, $\rho \propto T^{-1}$ and $D_i \propto T^{1.5}$; therefore $\rho D_i \propto T^{0.5}$. For common combustion of hydrocarbon fuels, the temperature changes by a factor of 7, the corresponding increase of ρD_i is by a factor of 2.64.

b. Advection: $\dot{q}_{conv}'' = \rho u h$, where h is specific enthalpy, u is fluid velocity, and ρ is density

c. Radiation heat loss: $\dot{q}_{rad}'' = \varepsilon \sigma_s (T^4 - T_\infty^4)$ where ε is the emissivity of the body ($\varepsilon = 1$ for blackbody), and $\sigma_s =$ Stefan-Boltzman constant $= 5.67 \cdot 10^{-8}$ (W/m²-K⁴).

d. Combustion: treated as an internal heat generation where $\dot{q}_{gen} = \hat{r}_{fuel} \hat{Q}_c V$.

e. Mass diffusion: When specific heat, c_p, and diffusivity, D, are assumed constant, energy carried by diffusion of different species is zero as shown below. First, when diffusion occurs, the molecules move on average at a velocity different from the bulk fluid velocity. The difference in velocity is called the 'diffusive' velocity, v_i, and it is related to the mean species gradient as $v_i = -\frac{D}{y_i}\frac{\partial y_i}{\partial x}$. [6]

Next the energy carried by 'diffusion' is

$$\sum_{i=1}^{K} \rho v_i y_i h_i$$

and can be expressed in terms of the species gradient as

$$\sum_{i=1}^{K} \rho v_i y_i h_i = -\sum_{i=1}^{K} \rho D \frac{\partial y_i}{\partial x} h_i$$

By using the product rule of differentiation in reverse, one has

$$\sum_{i=1}^{K} \rho v_i y_i h_i = -\sum_{i=1}^{K} \rho D \frac{\partial y_i}{\partial x} h_i = -\sum_{i=1}^{K} \rho D \frac{\partial y_i h_i}{\partial x} + \sum_{i=1}^{K} \rho D \frac{\partial h_i}{\partial x} y_i$$

$$= -\rho D \frac{\partial h}{\partial x} + \rho D \sum_{i=1}^{K} \frac{\partial h_i}{\partial x} y_i.$$

For simplicity, let's assume that c_p is constant, then we have

$$\frac{\partial h}{\partial x} = c_p \frac{\partial T}{\partial x}$$

and

$$\sum_{i=1}^{K} \rho v_i y_i h_i = -\rho D \frac{\partial h}{\partial x} + \rho D \sum_{i=1}^{K} c_p \frac{\partial T}{\partial x} y_i$$

$$= -\rho D c_p \frac{\partial T}{\partial x} + \rho D c_p \frac{\partial T}{\partial x} = 0$$

---

[6] Diffusion velocity is driven primarily by concentration gradient. Temperature gradient (thermal diffusion) and pressure gradient also contribute to diffusion velocity.

Fig. 4.4 One-dimensional
control volume for energy
conservation

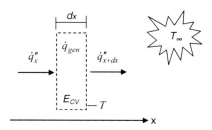

4.4.2   Derivation of a 1-D Conservation of Energy Equation

Let's consider a one-dimensional domain with a differential distance of dx and
unity area as shown in Fig. 4.4. The volume of the control volume is $V = dx$.
The first law of thermodynamics gives

$$\frac{\partial E_{cv}}{\partial t} = (\dot{q}_x'' - \dot{q}_{x+dx}'') - A\varepsilon\sigma_s(T^4 - T_\infty^4) + \hat{r}_{fuel}\hat{Q}_c V, \qquad (4.21)$$

where E_{cv} is the internal energy inside the control volume and A is area of radiating
surface, and

$$\dot{q}_x'' = \rho u h - k\frac{\partial T}{\partial x}$$

Using the thermodynamics relation

$$E_{cv} = me = m(h - Pv) = mh - PV = \rho V h - PV,$$

the above energy equation becomes

$$\frac{\partial \rho V h}{\partial t} - \frac{\partial PV}{\partial t} = (\dot{q}_x'' - \dot{q}_{x+dx}'') - A\varepsilon\sigma_s(T^4 - T_\infty^4) + \hat{r}_{fuel}\hat{Q}_c V$$

Division of the above equation by V leads to

$$\frac{\partial \rho h}{\partial t} = -\frac{\partial P}{\partial t} + \frac{(\dot{q}_x'' - \dot{q}_{x+dx}'')}{dx} - \frac{A}{V}\varepsilon\sigma_s(T^4 - T_\infty^4) + \hat{r}_{fuel}\hat{Q}_c$$

Next substituting the relation

$$\dot{q}_x'' = \rho u h - k\frac{\partial T}{\partial x},$$

taking the limit $dx \to 0$, and rearranging the results, we have

$$\frac{\partial \rho h}{\partial t} + \frac{\partial \rho u h}{\partial x} = -\frac{\partial P}{\partial t} + \frac{\partial}{\partial x}\left(k\frac{\partial T}{\partial x}\right) - \frac{A}{V}\varepsilon\sigma_s(T^4 - T_\infty^4) + \hat{r}_{fuel}\hat{Q}_c.$$

Using the continuity equation (Eq. 4.12)

$$\frac{\partial \rho}{\partial t} + \frac{\partial \rho u}{\partial x} = 0,$$

the energy equation becomes

$$\rho\frac{\partial h}{\partial t} + \rho u\frac{\partial h}{\partial x} = -\frac{\partial P}{\partial t} + \frac{\partial}{\partial x}\left(k\frac{\partial T}{\partial x}\right) - \frac{A}{V}\varepsilon\sigma_s(T^4 - T_\infty^4) + \hat{r}_{fuel}\hat{Q}_c$$

Next the total enthalpy is

$$h = \Delta h^0 + \int_{T_0}^{T} c_p(T)dT.$$

For simplicity, let's assume that c_p is constant, then we have $\frac{\partial h}{\partial t} = c_p\frac{\partial T}{\partial t}$ and $\frac{\partial h}{\partial x} = c_p\frac{\partial T}{\partial x}$.

Assuming that $\partial P/\partial t = 0$ and $k =$ constant, the simplified 1-D energy equation in terms of temperature (with constant c_p and k) is

$$\rho c_p\frac{\partial T}{\partial t} + \rho c_p u\frac{\partial T}{\partial x} = k\frac{\partial^2 T}{\partial x^2} - \frac{A}{V}\varepsilon\sigma_s(T^4 - T_\infty^4) + \hat{r}_{fuel}\hat{Q}_c \qquad (4.22)$$

The radiation heat loss term is written for a general case where A is the area of radiating body. For instance, soot particles radiate heat to surroundings. In this case, A is the total surface area of soot particles within the volume V. Remember, $k =$ thermal conductivity (W/m-K), $\varepsilon =$ emissivity of the body (~1 for black body), $\sigma_s =$ Stefan-Boltzmann constant $= 5.67 \cdot 10^{-8}$ W/m^2-K^4, $A/V =$ area to volume ratio for the radiating medium, $\hat{r}_{fuel} =$ fuel consumption rate (kmol/m^3-s), and $\hat{Q}_c =$ heat of combustion (J/kmol).

When the conservation of energy equation is applied to a control volume taken over a non-differential element, the temperature gradient through the volume may become important. There are two limiting cases to consider. One limiting case is where the temperature gradient is small throughout the entire volume. In other words, the temperature of the control volume is constant throughout. This corresponds to the lumped capacitance model of transient conduction and Eq. 4.22 can be used. The other limiting case to consider is where the temperature gradient only penetrates to a very shallow depth into the volume. In other words, the temperature of the far side of the control volume remains unchanged from the initial temperature. This corresponds to the semi-infinite model of transient conduction and this temperature gradient must be taken into consideration. Closed form solutions for the temperature profile inside a semi-infinite volume can be found in any general heat transfer text. To evaluate the significance of the temperature gradient, one can compare the ratio between the internal resistance to heat transfer and the resistance to heat transfer at the solid-gas phase boundary, i.e. the Biot number. The lumped capacitance model can be used if

$$B_i = \frac{\tilde{h} L_c}{k_s} < 0.1, \tag{4.23}$$

where L_c is the characteristic length, k_s is the conductivity of solid, and $\tilde{h}$ is the heat transfer coefficient for the interface between control volume and surroundings. However, this method requires the knowledge of the heat transfer coefficient, which may not be always known. Another method is to compare the volume thickness to the thermal diffusion length, L_d:

$$L_d = \sqrt{\alpha_s t} = \sqrt{\frac{k_s}{\rho_s c_s} t} \tag{4.24}$$

where α_s is the thermal diffusivity of the volume material. If the volume thickness is on the order of this thermal diffusion length at a given time, then the lumped capacitance approximation can be used.

4.5 Normalization of the Conservation Equations

Normalization of the governing conservation equations provides a mechanism to extract the primary parameters controlling a particular process and the relation between the different controlling mechanisms. The heuristic of the normalization process is as follows:

1. Define the characteristic quantities that are related to the dimensional variables and that define the particular process under study. Looking for these characteristic quantities in the boundary conditions is often helpful. If none are found, introduce an undefined characteristic quantity with a symbol related to the corresponding variable.
2. Define the non-dimensional variable as the ratio of the dimensional variable and its characteristic quantity.
3. Substitute the normalized variables into the conservation equations with associated boundary conditions.
4. Identify the non-dimensional groups of characteristic quantities that appear in each term of the equations. Use these groups to define any undefined characteristic quantity by equating the value of the non-dimensional group to unity.
5. Solve the resulting non-dimensional equations.

The above approach is used here to normalize the conservation equations presented above. Defining the characteristic quantities as t_c for time, l_c for the spatial variable, T_c for temperature, y_c for mass fraction, u_c for velocity and P_c for pressure, the non-dimensional variables are defined as

$$\bar{t} \equiv \frac{t}{t_c}, \bar{x} \equiv \frac{x}{l_c}, \bar{T} \equiv \frac{T}{T_c}, \bar{y}_i \equiv \frac{y_i}{y_c}, \bar{u} \equiv \frac{u}{u_c}, \bar{P} = \frac{P}{P_c}.$$

With these non-dimensional quantities, the non-dimensional forms of the conservation equations for species, momentum, and energy equations can be derived. For conservation of species, inserting the non-dimensional variables into Eq. 4.19 gives:

$$\rho \frac{\partial(y_c \bar{y}_i)}{\partial(t_c \bar{t})} + \rho(u_c \bar{u}) \frac{\partial(y_c \bar{y}_i)}{\partial(l_c \bar{x})} = \rho D_i \frac{\partial^2(y_c \bar{y}_i)}{\partial(l_c \bar{x})^2} + \hat{r}_i M_i$$

Multiplying by $l_c / \rho u_c y_c$,

$$\frac{l_c}{t_c u_c} \frac{\partial \bar{y}_i}{\partial \bar{t}} + \bar{u} \frac{\partial \bar{y}_i}{\partial \bar{x}} = D_i \frac{1}{u_c l_c} \frac{\partial^2 \bar{y}_i}{\partial \bar{x}^2} + \frac{\hat{r}_i M_i l_c}{\rho u_c y_c}$$

or

$$\frac{l_c}{t_c u_c} \frac{\partial \bar{y}_i}{\partial \bar{t}} + \bar{u} \frac{\partial \bar{y}_i}{\partial \bar{x}} = \frac{1}{LePe} \frac{\partial^2 \bar{y}_i}{\partial \bar{x}^2} + Da_i \qquad (4.25)$$

For conservation of momentum with constant density, inserting the non-dimensional variables into Eq. 4.20 with $X = \rho g$ gives:

$$\rho \frac{\partial(u_c \bar{u})}{\partial(t_c \bar{t})} + \rho(u_c \bar{u}) \frac{\partial(u_c \bar{u})}{\partial(l_c \bar{x})} = -\frac{\partial(P_c \bar{P})}{\partial(l_c \bar{x})} + \mu \frac{\partial^2(u_c \bar{u})}{\partial(l_c \bar{x})^2} + \rho g$$

Multiplying the above equation by $l_c / \rho u_c^2$ leads to

$$\frac{l_c}{t_c u_c} \frac{\partial \bar{u}}{\partial \bar{t}} + \bar{u} \frac{\partial \bar{u}}{\partial \bar{x}} = -\frac{P_c}{\rho_c u_c^2} \frac{\partial \bar{P}}{\partial \bar{x}} + \frac{\mu}{\rho u_c l_c} \frac{\partial^2 \bar{u}}{\partial \bar{x}^2} + \frac{g l_c}{u_c^2}$$

or

$$\frac{l_c}{t_c u_c} \frac{\partial \bar{u}}{\partial \bar{t}} + \bar{u} \frac{\partial \bar{u}}{\partial \bar{x}} = -\frac{P_c}{\rho_c u_c^2} \frac{\partial \bar{P}}{\partial \bar{x}} + \frac{1}{Re} \frac{\partial^2 \bar{u}}{\partial \bar{x}^2} + \frac{1}{Fr}. \qquad (4.26)$$

Similarly, for conservation of energy, inserting the non-dimensional variables into Eq. 4.22 and neglecting the radiation term gives:

$$\rho c_p \frac{\partial(T_c\bar{T})}{\partial(t_c\bar{t})} + \rho c_p(u_c\bar{u})\frac{\partial(T_c\bar{T})}{\partial(l_c\bar{x})} = k\frac{\partial^2(T_c\bar{T})}{\partial(l_c\bar{x})^2} + \hat{\dot{r}}_{fuel}\hat{Q}_c.$$

Multiplying by $l_c/\rho c_p T_c u_c$,

$$\frac{l_c}{t_c u_c}\frac{\partial\bar{T}}{\partial\bar{t}} + \bar{u}\frac{\partial\bar{T}}{\partial\bar{x}} = \frac{k}{\rho c_p}\frac{1}{l_c u_c}\frac{\partial^2\bar{T}}{\partial\bar{x}^2} + \frac{l_c\hat{\dot{r}}_{fuel}\hat{Q}_c}{\rho c_p T_c u_c}$$

or

$$\frac{l_c}{t_c u_c}\frac{\partial\bar{T}}{\partial\bar{t}} + \bar{u}\frac{\partial\bar{T}}{\partial\bar{x}} = \frac{1}{Pe}\frac{\partial^2\bar{T}}{\partial\bar{x}^2} + Da_{fuel}\frac{y_c\hat{Q}_c}{c_p T_c M_{fuel}} \qquad (4.27)$$

From the above equations it is seen that there are five primary non-dimensional groups that can be defined that determine the physics of combustion processes: the Reynolds number,

$$\mathrm{Re} = \frac{\rho u_c l_c}{\mu};$$

the Froude number,

$$Fr = \frac{u^2_c}{g l_c};$$

the Peclet number,

$$Pe = \frac{\rho c_p u_c l_c}{k} = \frac{l_c u_c}{\alpha};$$

Lewis number,

$$Le = \frac{\alpha}{D};$$

and Damköhler number,

$$Da_i = \frac{\text{flowtime}}{\text{chemistry time}} = \frac{l_c/u_c}{\left(\rho y_c \big/ \hat{\dot{r}}_i M_i\right)}.$$

By equating the group $l_c/t_c u_c$ to unity it is also deduced that the characteristic time in convective-dominated flows is $t_c = l_c/u_c$, which is normally referred to as the residence time. Similarly one can choose y_c such that

$$\frac{y_c\hat{Q}_c}{c_p T_c M_{fuel}} = 1.$$

The normalization of the boundary conditions for a particular problem may provide further non-dimensional parameters.

The Reynolds number (Re) compares inertia to viscous forces, and when it exceeds a critical value, a transition occurs from laminar to turbulent flow. As seen From Eq. 4.26, if Re is large the viscous force term becomes relatively small in the momentum equation. The flow behaves like an inviscid fluid and therefore becomes less stable. For small Re the flow is laminar and any disturbance is damped out quickly by the viscous force. The Froude number (Fr) compares forced convection versus natural convection. The Peclet number (Pe) compares convection versus conduction heat transfer. From Eq. 4.27 it is seen that when Pe is large the conduction terms becomes small and the formulation of the energy equation becomes convection dominated. Similarly, for small Pe the transfer of heat is conduction dominated. The Lewis number (Le) compares thermal diffusivity to mass diffusivity and the product of the Lewis (Le) and Pecklet (Pe) numbers compares convection versus diffusion. From Eq. 4.25 it is seen that for large $LePe$ the transport of mass is primarily by convection and that for small $LePe$ by diffusion.

The Damköhler number (Da_i) is particularly important in combustion processes. For instance, Da_{fuel} describes the relative importance of the residence time $t_c = l_c/u_c$ (the time that the fuel and oxidizer remain in the combustor) and the chemical time, $t_{chem} = \rho y_c / \dot{r}_{fuel} M_{fuel}$ (the time that it takes for the fuel and oxidizer to react). If the Damköhler number is small it is seen from Eqs. 4.25 and 4.27 that the reaction terms become small and the formulation of the process becomes that of a non-reacting system. If the Damköhler number is large the reaction terms become dominant and the only terms that remain from Eqs. 4.25 and 4.27 are the transient term and the reaction term. In these cases the reaction zone is often very thin, so this form of the equations applies to a very narrow domain with the rest described by the non-reactive equations. As it will be seen in later chapters, the Damköhler number determines a number of factors including whether complete combustion occurs and pollution formation.

4.6 Viscosity, Conductivity and Diffusivity

A simple kinetic theory of gases is often used to elucidate the dependence of transport properties on temperature and pressure [1]. The simplest theory treats the molecules as rigid spheres that interact elastically and yield the following relations:
Viscosity:

$$\mu = \frac{5}{16} \frac{\sqrt{\pi \cdot M \cdot k_B \cdot T}}{\pi \cdot d^2 \cdot} \propto \sqrt{T} \quad \text{units} : \text{kg}/(\text{m} - \text{s}), \tag{4.28}$$

Conductivity:

$$k = \frac{25}{32} \frac{c_v \sqrt{\pi \cdot M \cdot k_B \cdot T}}{\pi \cdot d^2 \cdot M} \propto \sqrt{T} \quad \text{units: J}/(\text{m} - \text{K} - \text{s}), \tag{4.29}$$

Diffusion Coefficient:

$$D = \frac{3}{8} \frac{\sqrt{\pi \cdot M \cdot k_B \cdot T}}{\pi \cdot d^2} \frac{1}{\rho} \propto \frac{T^{3/2}}{P} \quad \text{units} : \text{m}^2/\text{s}, \tag{4.30}$$

where d is the diameter of molecules, M the molecular mass, k_B the Boltzmann constant, T temperature, and ρ density. These relations provide an estimate of the dependence of transport properties on temperature and pressure. For instance, viscosity and conductivity are not dependent on pressure for ideal gases; however, diffusion has an inverse dependence on pressure. Viscosity and conductivity scale with $\sqrt{T}$ and the diffusion coefficient increases with temperature as $T^{3/2}$. For a mixture of two components, the diffusion coefficient is modified and referred to as the binary diffusion coefficients. Improved kinetic theories with real gas effects by including intermolecular attractive or repulsive forces can now be used to reasonably predict transport properties and some exemplary comparisons are shown in [2].

References

1. Bird RB, Stewart WE, Lightfoot, EN (1960) Transport Phenomena. John Wiley & Sons, New York.
2. Warnatz J, Mass U, Dibble RW (2001) Combustion, Physical and Chemical Fundamentals, Modeling and Simulation, Experiments, Pollutant Formation. Springer-Verlag Berlin, Heidelberg.

Chapter 5
Ignition Phenomena

Ignition is the mechanism leading to the onset of a vigorous combustion reaction and is characterized by a rapid increase of the species temperature. An understanding of ignition is important in a wide range of combustion processes, from designing practical combustion devices to preventing unwanted fires. Ignition of a combustible material is often classified in two ways: spontaneous ignition, also known as auto-ignition, occurs through the self heating of the reactants, whereas piloted ignition occurs with the assistance of an ignition source. From the discussion on chemical kinetics, we learned that the rate of a chemical reaction is a strong function of temperature. It follows then that the chemical reactions involved in combustion occur even at low temperatures even if only at a very slow rate. If the heat generated by the slow reactions is all lost to the surroundings, then the reactants do not ignite. However, if the heat generated by the reaction is greater than the heat losses to the surroundings, a self-heating process may occur where the temperature of the reactants increases until they spontaneously ignite. In piloted ignition, combustion is triggered by an external energy source that locally increases the temperature of the reactants until ignition. Piloted ignition can be initiated by many means such as a spark, pilot flame, friction, electrical resistance (glow plug), or a laser beam.

5.1 Autoignition (Self-ignition, Spontaneous Ignition) Based on Thermal Theory

Autoignition is of special relevance to internal combustion engine and fire safety applications. For example, in diesel engines, fuel is injected into hot air and the combustion process is initiated by autoignition. Also, in Homogeneous Charge Compression Ignition (HCCI) engines the fuel-air mixture autoignites when the pressure and temperature of the mixture reaches a certain value. In spark-ignited engines, unwanted engine knock is due to autoignition of unburned gas. The storage of combustible materials also requires attention to the possibility of autoignition. There is a limitation on the size of a haystack because the larger the haystack, the more it insulates itself. When the heat generated inside the haystack becomes larger

S. McAllister et al., *Fundamentals of Combustion Processes*,
Mechanical Engineering Series, DOI 10.1007/978-1-4419-7943-8_5,
© Springer Science+Business Media, LLC 2011

than the heat losses, spontaneous ignition may occur. Oily rags are another good example. The rags are effective insulators and will ignite unless they are well ventilated or sealed in a container to limit the oxygen supply.

To provide a basic understanding of the mechanisms leading to thermal auto-ignition, let's consider a vessel filled with a combustible mixture at a temperature T. The vessel is in contact with surroundings at T_∞. Using the energy conservation equation, we can express the temperature evolution of the combustible mixture as

$$\rho c_p \frac{\partial T}{\partial t} = \underbrace{\left(-\rho c_p u \frac{\partial T}{\partial x} + k \frac{\partial^2 T}{\partial x^2} \right)}_{\text{heatloss}} + \underbrace{\hat{r}_{fuel} \hat{Q}_c}_{\text{heatgeneation}} , \qquad (5.1)$$

Using a lumped-type formulation, the above equation can be simplified to

$$\rho c_p \frac{\partial T}{\partial t} = -\dot{q}_L''' + \dot{q}_R''', \qquad (5.2)$$

where $\dot{q}_L'''$ is the heat loss from the surfaces of the vessel and $\dot{q}_R'''$ is the heat generated per unit volume and time inside the vessel. From now on, the combustible mixture will be referred to as the system. Next we can express the heat transfer term without radiation heat loss by using an overall convective heat transfer on a per volume basis as

$$\dot{q}_L''' = \frac{\tilde{h}A}{V}(T - T_\infty), \qquad (5.3)$$

where $\tilde{h}$ is the heat transfer coefficient and A is the surface area of the vessel in contact with the surroundings. The heat generation term can be expressed in terms of a global Arrhenius reaction in the form

$$\dot{q}_R''' = A_0 [F]^a [O]^b \exp\left(-\frac{E_a}{\tilde{R}_u T_c} \right) \hat{Q}_c \qquad (5.4)$$

The combustible mixture temperature changes according to the balance between the heat generation and heat loss terms. If the heat generated is less than the heat lost, the temperature of the system will decrease. If the heat generated is more than the heat lost, the temperature of the system will increase. The limiting condition for ignition is reached when the heat losses become equal to the heat generated, $\dot{q}_L''' = \dot{q}_R'''$. Figure 5.1 sketches $\dot{q}_L'''$ and $\dot{q}_R'''$ versus temperature. We can then graphically analyze the possibility of autoignition.

Let's keep the values of $\tilde{h}, A$, and V fixed while changing the values of T_∞. Note that $\dot{q}_L'''$ depends linearly on the system temperature. Because the fuel consumption rate increases exponentially with temperature, $\dot{q}_R'''$ will be larger than $\dot{q}_L'''$ when the system temperature is above a certain value. When this occurs, the system temperature will increase rapidly due to the self-heating process. If there is no additional heat

Fig. 5.1 Volumetric heat release rate versus temperature with different surrounding temperatures

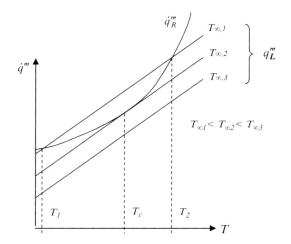

loss, the system will autoignite in time. Such a process is referred to as the thermal run-away phenomenon.

Let's start out with a sufficiently high surrounding temperature $T_{\infty,3}$. Because the heat generation would be greater than the heat loss in this case, autoignition will always occur. When the surrounding temperature decreases to a point $T_{\infty,2}$, there exists a system temperature such that both $\dot{q}_L''' = \dot{q}_R'''$ and $\frac{d\dot{q}_L'''}{dT} = \frac{d\dot{q}_R'''}{dT}$ hold at a point, T_c, as identified in Fig. 5.1. The temperature at this point, T_c, is called the critical autoignition temperature. The critical autoignition temperature is the lowest temperature at which a combustible material can ignite without the application of a flame or other means of ignition. At T_c the heat loss term just balances the heat generation term and any slight increase in temperature will trigger ignition. At temperatures below T_c the heat losses would be greater than the heat generated and ignition will not occur.

Further decreasing the ambient temperature to $T_{\infty,1}$ it is seen that the heat generation and heat losses balance each other at two different temperatures T_1 and T_2. At temperature T_1 the system is stable and ignition will not occur. A slight increase in T will lead to $\dot{q}_L''' > \dot{q}_R'''$ and the system temperature will return to T_1. Similarly, a slight decrease in T will lead to $\dot{q}_L''' < \dot{q}_R'''$ and, consequently, the system temperature will increase and return to T_1. In contrast, the system is unstable at T_2 as a slight perturbation in temperature will drive the temperature away from T_2. Thus autoignition will only occur if the mixture temperature is greater than the temperature at the onset of self-heating (T_2).

One can draw a similar diagram if we keep T_∞ fixed while changing the heat transfer process via $\tilde{h}$ or A/V. As shown in Fig. 5.2, the slope of the heat loss line changes in this case. Since both $\dot{q}_L'''$ and $\dot{q}_R'''$ depend on the system, the autoignition temperature also depends on the situation at hand. Particularly important is to notice that as A/V decreases, the autoignition temperature (T_2) decreases, explaining why it is important to limit the size of a haystack or a pile of rags to prevent their autoignition.

Fig. 5.2 Volumetric heat
release rate versus
temperature with different
heat transfer coefficients

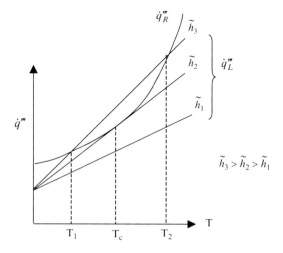

From the above graphical analysis it is seen that the critical autoignition temperature can be obtained by setting the following two equalities $\dot{q}'''_L = \dot{q}'''_R$ and $\frac{d\dot{q}'''_L}{dT} = \frac{d\dot{q}'''_R}{dT}$.

$$\dot{q}'''_L = \dot{q}'''_R \rightarrow$$

$$\frac{\tilde{h}A_s}{V}(T_c - T_\infty) = \hat{r}_{fuel}\hat{Q}_c = A_0[F]^a[O]^b \exp\left(-\frac{E_a}{\hat{R}_u T_c}\right)\hat{Q}_c \tag{5.5}$$

$$\frac{d\dot{q}'''_L}{dT} = \frac{d\dot{q}'''_R}{dT} \rightarrow$$

$$\frac{\tilde{h}A_s}{V} \cong A_0[F]^a[O]^b \exp\left(-\frac{E_a}{\hat{R}_u T_c}\right)\frac{E_a}{\hat{R}_u T_c^2}\hat{Q}_c \tag{5.6}$$

In deriving Eq. 5.5, the dependence of [F] and [O] on temperatures has been considered negligible in comparison to the exponential term under the assumption of high activation energy. This is justified when the activation energy is high as occurs in most combustion systems. Using Eqs. 5.5 and 5.6, we solve for T_c to obtain

$$T_c = T_\infty + \frac{\hat{R}_u T_c^2}{E_a} = T_\infty + \frac{T_c^2}{E_a/\hat{R}_u} = T_\infty + \frac{T_c^2}{T_a}$$

or

$$T_c = \frac{T_a - \sqrt{T_a^2 - 4T_\infty T_a}}{2} \tag{5.7}$$

Table 5.1 Autoignition temperature in air at 1 atm

Substance	Autoignition temperature (°C)
Methane	537
Ethane	472
Propane	470
n-Butane	365
n-Octane	206
Isooctane	418
Methanol	464
Ethanol	423
Acetylene	305
Carbon monoxide	609
Hydrogen	400
Gasoline	370
Diesel #2	254
Paper	232

For combustion processes with a high activation energy, $T_a >> T_c$, so Eq. 5.7 gives T_c very close to T_∞.[1] For instance, with $T_a = 10,000$ K and $T_\infty = 500$ K, Eq. 5.7 gives $T_c = 527.9$ K. Therefore a rough estimate of autoignition temperature is $T_c \approx T_\infty$ in a laboratory. In other words, when the conditions of $\dot{q}_L''' = \dot{q}_R'''$ and $\frac{d\dot{q}_L'''}{dT} = \frac{d\dot{q}_R'''}{dT}$ are satisfied, any slight perturbation in the system temperature will result in ignition. It should be noted that such a low critical temperature for ignition requires very special circumstances of heat generation and heat losses. In real life, natural variations in system temperature and heat losses cause discrepancies from the theoretical minimum ignition temperature. Therefore, Eq. 5.7 should be used only for understanding general trends.

The above thermal theory provides a qualitative understanding of the nature of the critical conditions for ignition with the major assumption that ignition is controlled by thermal energy. As discussed in Chap. 3, reactions are also induced by chain-branching reactions that release little heat. For large straight-chain molecules, such as n-heptane, combustion chemical kinetics during autoignition often exhibit two-stage ignition with a complex dependence on temperature and pressure.

Table 5.1 shows actual typical values for the autoignition temperature for a variety of fuels.

5.2 Effect of Pressure on the Autoignition Temperature

Since reaction rates change with pressure, the autoignition temperature is also a function of the system pressure. As the pressure increases, the reaction rate increases, tipping the balance between the heat generation and heat losses. If the system is at the

---

[1] Using $\sqrt{1-x} = 1 - \frac{x}{2} - \frac{x^2}{8} - \frac{x^3}{16} - \cdots$, Eq. 5.7 leads to $T_c = T_\infty \left\{ 1 + \frac{T_\infty}{T_a} + 2\left(\frac{T_\infty}{T_a}\right)^2 + \cdots \right\}$.

critical temperature for ignition, an increase in the pressure above some threshold level will result in thermal run-away and ignition. In other words, there is not only a critical temperature for ignition, but also a critical pressure for ignition. To determine how these two quantities are related, we begin by using Eq. 5.6 to solve for pressure in terms of the autoignition temperature as

$$\frac{\tilde{h}A_s}{V}\frac{\hat{R}_u T_c^2}{E_a} = \hat{Q}_c A_0 [F]^a [O]^b \exp\left(-\frac{E_a}{\hat{R}T_c}\right)$$

$$= \hat{Q}_c A_0 x_f{}^a x_{O_2}{}^b \exp\left(-\frac{E_a}{\hat{R}_u T_c}\right)\left(\frac{P_c}{\hat{R}T_c}\right)^{a+b}$$

Further expressing the critical pressure as function of critical temperature leads to

$$P_c = \hat{R}_u T_c \left(\frac{\dfrac{\tilde{h}A_s}{V}\dfrac{\hat{R}_u T_c^2}{E_a}}{\hat{Q}_c A_0 x_f{}^a x_{O_2}{}^b \exp\left(-\dfrac{E_a}{\hat{R}_u T_c}\right)}\right)^{1/(a+b)}$$

$$P_c = \hat{R}_u T_c \exp\left(\frac{E_a}{(a+b)\hat{R}_u T_c}\right)\left(\frac{\dfrac{\tilde{h}A_s}{V}\dfrac{\hat{R}_u T_c^2}{E_a}}{\hat{Q}_c A_0 x_f{}^a x_{O_2}{}^b}\right)^{1/(a+b)}$$

(5.8)

This equation was developed by Semenov and is often called the *Semenov Equation*. For most combustion reactions with high activation energy, the term $\exp\left(\frac{E_a}{(a+b)\hat{R}_u T_c}\right)$ dominates and the critical pressure decreases with increasing temperature as sketched in Fig. 5.3.

In the figure, autoignition is possible in the upper region above the line. Because the reaction rate increases with pressure (for combustion chemistry with $a+b>0$), combustion proceeds faster at high pressures.[2] It follows that the corresponding autoignition temperature decreases as pressure increases. The results of Fig. 5.3 have important implications in internal combustion engines and other combustion processes where an increase in pressure can lead to the autoignition of the fuel and a potential explosion. This will be discussed in subsequent chapters.

---

[2] For hydrogen combustion in a certain pressure region, increasing pressure leads to a decrease in the tendency of explosion. Such a behavior cannot be explained by the thermal theory presented here. Chemical kinetics plays an important role; that is, the chain branching reaction $H + O_2 \rightarrow OH + O$ competes with the chain termination step $H + O_2 + M \rightarrow HO_2 + M$ which increases with pressure at a rate faster than two-body reactions.

Fig. 5.3 Critical pressure versus temperature. Ignition is possible in the region above the curve for combustion chemistry when the global order is greater than 1

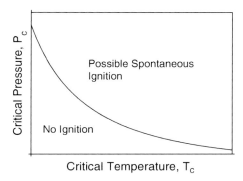

5.3 Piloted Ignition

In piloted ignition, the combustion process is initiated when an energy source locally heats the mixture to a high temperature. Burning is then sustained once the ignition source is removed. Piloted ignition can be achieved using a spark, pilot flame, electrical resistance (glow plug), friction, or any sufficiently hot source. Let's consider the case of a spark generated with a spark plug, such as in a car engine. The spark plug consists of two electrodes spaced a distance d apart. A high voltage is applied to the electrodes as shown in Fig. 5.4.

The high applied voltage creates an electric arc across the gap between the electrodes, heating the combustible mixture in between. The energy required for igniting the mixture is important for both engineering applications and explosion/fire safety. In the following, a simple analysis will be presented for estimating just how much spark energy is required to ignite the fuel mixture. This analysis assumes that the ignition energy is the energy necessary to heat the gas between the electrodes to the adiabatic flame temperature. Using the lumped form of the energy conservation equation, the following equation can be used to describe ignition with a pilot source:

$$\rho c_p V \frac{\partial T}{\partial t} = -\dot{Q}_{loss} + \hat{r}_{fuel}\hat{Q}_c V + \dot{Q}_{pilot}, \tag{5.9}$$

where $\dot{Q}_{pilot}$ is the rate of energy source from the spark (J/s) and $\dot{Q}_{loss}$ is the rate of heat loss term including heat lost to the electrodes by conduction and heat transfer to the surroundings by convection. For this analysis, we will assume that the heat generated from the combustion reaction is negligible during the ignition process, i.e., $\hat{r}_{fuel}\hat{Q}_c V = 0$. Integrating Eq. 5.9 over the period of ignition duration and assuming that the temperature after ignition reaches the adiabatic flame temperature, we have

$$E_{ignition} = \rho c_p V (T_f - T_r) + Q_{loss}, \tag{5.10}$$

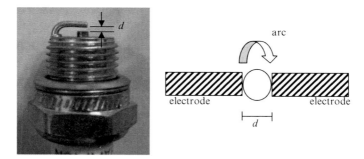

Fig. 5.4 Piloted ignition with spark plug

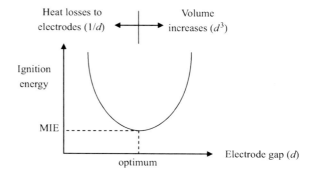

Fig. 5.5 Ignition energy as a
function of electrode gap

where $E_{ignition} = \int \dot{Q}_{pilot} dt$, $T_f =$ flame temperature, and $T_r =$ initial reactant temperature. From the above equation, we can see that the ignition energy from the spark increases with the volume of mixture and the heat losses to the surroundings. Increasing the gap between the electrodes increases the volume of mixture that must be heated, raising $E_{ignition}$. Because the heat lost by the mixture is primarily by conduction to the electrodes, decreasing the gap between the electrodes increases the heat lost. Following this line of reasoning Eq. 5.10 can be rewritten as function of the gap between the electrodes as

$$E_{ignition} \propto V + Q_{loss} \propto (c_1 \cdot d^3 + c_2/d) \qquad (5.11)$$

It follows that there is an optimal spacing of the electrodes that results in a minimum energy required for ignition (MIE – minimum ignition energy), as shown in Fig. 5.5.

This optimum electrode gap (d_{opt}) is related to the thickness of the reaction zone since it is affected by the heat losses from the incipient reaction zone to the electrode surfaces. In the next chapter it will be seen that the physics determining this gap are similar to that related to the quenching of a flame, and that consequently

Fuel	MIE (mJ)
Methane	0.30
Ethane	0.42
Propane	0.40
n-Hexane	0.29
Isooctane	0.95
Acetylene	0.03
Hydrogen	0.02
Methanol	0.21

Table 5.2 Minimum ignition energies of stoichiometric fuel/air mixtures at 1 atm and 20°C

Minimum ignition energies for a wide variety of flammable materials are listed in Appendix 6

the dimensions of the optimum electrode gap are proportional to the quenching distance. Assuming that the volume is a sphere with diameter, d, equal to the gap between the electrodes, the minimum ignition energy (MIE) (assuming no heat losses) can be estimated for a fixed mixture as

$$MIE \approx \rho c_p \frac{\pi d_{opt}^3}{6}(T_f - T_r), \qquad (5.12)$$

where

d_{opt} = optimum gap between the electrodes
T_f = flame (product) temperature
T_r = initial reactant temperature

Equation 5.12 addresses the ignition of the combustible mixture only but does not guarantee that the combustion reaction will continue to propagate through the mixture. The energy necessary for combustion propagation is generally larger than that for simple ignition and will be discussed in the premixed combustion chapter.

Typical values of the minimum ignition energy are shown in Table 5.2. One might notice that the minimum ignition energy for hydrogen is much smaller than those for other fuels. This is just one of the reasons why hydrogen is a dangerous fuel. Note also that the ignition energy is very small in comparison to the heat release from the corresponding combustion process.

Example 5.1 A spark plug has a gap of 0.1 cm (0.04 in., typical for car applications). Using air properties at $T = 300$ K and $P = 101.3$ kPa, estimate the temperature increase (ΔT) when 0.33 mJ is deposited into the gases between the spark plug gap.

Solution:
The volume occupied by the gases between the spark plug is

$$V = \frac{\pi d^3}{6} = \frac{3.1415926 \cdot 0.1^3}{6} = 5.24 \cdot 10^{-4} \, cm^3$$

and the temperature rise is

$$\Delta T \approx \frac{E_{deposited}}{\rho \cdot c_p \cdot V} = \frac{0.33 \, \text{mJ} \cdot 10^{-3} \, (\text{J/mJ})}{1.2 \cdot 10^{-3} (\text{g/cm}^3) \cdot 1.00 (\text{J/g - K}) \cdot 5.24 \cdot 10^{-4} (\text{cm}^3)} = 525 \, \text{K}$$

Note that if the input energy is increased by a factor of 10 (i.e., 3.3 mJ), the temperature can be increased by more than 5,000 K!

5.4 Condensed Fuel Ignition

An important aspect of the combustion of liquid and solid fuels is their ease of ignition. This is important not only for the utilization of the fuel in a combustor but also for safety reasons. Condensed-phase fuels burn mostly in the gas phase (flaming), although some porous materials may react on the solid surface (smoldering). For a condensed fuel to ignite and burn in the gas phase, enough fuel must vaporize so that when mixed with air, the combustible mixture falls within the flammability limits of the fuel. Ignition of the combustible mixture is then similar to the gas-phase fuel mixtures discussed above. Once the gaseous mixture above the condensed fuel ignites, a non-premixed flame is established at the surface that sustains the material burning. This process is sketched in Fig. 5.6. The gasification of liquid fuels (evaporation) is physically different than that of solid fuels (pyrolysis), and it is for this reason that they are often treated differently.

5.4.1 Fuel Vaporization

In liquid fuels, the partial pressure of fuel vapor near the liquid surface is approximately in equilibrium with the liquid phase. The saturation pressure of the liquid enables the determination of the mole fraction of fuel at the liquid surface as

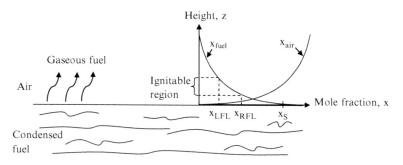

Fig. 5.6 Sketch of condensed fuel combustion

$x_s = P_{sat}(T_{sat})/P$, where P is the total pressure. Lookup tables of saturation pressures as function of temperature can be found for many combustion fuels. However, if such a table is unavailable, it is reasonable to use the Clausius-Clapeyron equation or the Antoine equation (see Exercise 5.5). For solid fuels, determining the mole fraction of gaseous fuel above the surface is more complex. As mentioned, the vaporization of solid fuels isn't merely a change of phase, but a chemical decomposition reaction called pyrolysis. The rate of pyrolysis per volume of solid fuel is estimated by:

$$\dot{m}_F''' = A_0 \exp\left(-\frac{E_a}{RT}\right) \tag{5.13}$$

where A_0 is a pre-exponential factor and E_a is the activation energy for pyrolysis, both of which are properties of the material. Note the similarity to the Arrhenius reaction rate developed in Chap. 3. Because of its Arrhenius nature and typically high activation energy, the rate of pyrolysis is highly temperature dependent and is very slow at low temperatures. At a sufficiently high temperature, the pyrolysis rate dramatically increases and the corresponding temperature is referred to as the pyrolysis temperature. If the temperature profile in the solid is known, the mass flux of fuel leaving the fuel surface can be calculated as

$$\dot{m}_F'' = \int_0^{\delta_{py}} A_0 \exp\left(-\frac{E_a}{RT}\right) dx \tag{5.14}$$

where δ_{py} is the depth of the solid heated layer.

Over time, concentration gradients of fuel and air form over the condensed fuel surface as shown in Fig. 5.6. The gaseous fuel can both diffuse and buoyantly convect up into the surrounding air, so that the fuel mole fraction decreases with height. Conversely, the air diffuses back toward the condensed surface, so that the mole fraction of air increases with height. Logically, there is a region above the surface where both gaseous fuel and air coexist within the flammability limits. Below this region, the mixture is too rich to ignite; above this region, the mixture is too lean to ignite. A combustion reaction can then be ignited if a spark or pilot were to exist in the flammable region above the surface.

5.4.2 Important Physiochemical Properties

The lower the evaporation temperature of a liquid fuel, the easier it will ignite. Two commonly used terms for describing the ignition properties of a liquid fuel are the flash point and fire point. *Flash point* is defined as the minimum liquid temperature at which a combustion reaction (flame) is seen (flashing) with the assistance of a spark or a pilot flame. The flash point is then the liquid temperature that is sufficiently high to form a mixture above the pool that is just at the lean flammability limit. The flame merely "flashes" because the heat release rate of the establishing

Fig. 5.7 Flash point and fire
point

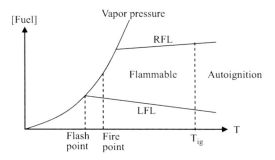

flame is insufficient to overcome the rate of heat losses to the surroundings. Some
flash point data is found in Appendix 8. *Fire point* refers to the minimum liquid
temperature for sustained burning of the liquid fuel. At the fire point, the heat release
rate of the establishing flame balances the rate of heat losses to the surroundings.
It should be noted that the concept is similar to that referred above for gaseous fuels
separating the mechanism of ignition from that of ignition leading to propagation of
the incipient combustion reaction. As discussed in Chap. 3, the heat release rate of a
combustion reaction increases with equivalence ratio. The fire point often occurs at a
higher temperature than the flash point because more fuel is in the gas phase,
increasing the equivalence ratio above the liquid pool. Figure 5.7 shows the flash
point and fire point in relation to the saturation temperature at various vapor
pressures and flammability limits of the mixture above the surface. Note that if the
fuel temperature is sufficiently high, autoignition may occur. Solid fuels are
typically less volatile than liquid fuels, so solid fuels usually are more difficult to
ignite than liquid fuels. As in liquid fuels, the terms "fire point" and "flash point" can
also be used to describe the ignition of solid fuels. For solid fuels, the fire point is
frequently referred to as simply the ignition temperature.

5.4.3 *Characteristic Times in Condensed Fuel Ignition*

As it was explained above, the ignition of a condensed fuel requires the gasification of
the fuel, mixing of the fuel vapor and oxidizer, and ignition of the mixture. Each one
of these processes requires some amount of time. Their combined times determine the
time of ignition. The time of ignition, often referred to as the ignition delay time, is
important in a number of combustion processes, particularly fuel fire safety.

   If the temperature of gasification of the fuel is higher than room temperature,
the fuel must be heated to its gasification temperature before it can ignite.
An expression for the fuel heating time t_g can be found by performing an energy
balance on the material.

   As discussed in Chap. 4, there are two simplifying assumptions about the temperature gradient inside the material that can be made in a transient conduction analysis.

Fig. 5.8 Energy balance for
semi-infinite fuel

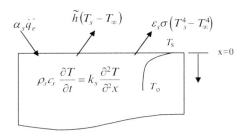

One assumption is that the temperature of the material is uniform throughout, corresponding to the lumped capacitance model. Alternatively, the far side of the material can remain constant at the initial temperature, corresponding to the semi-infinite model. Proceeding with the semi-infinite model and assuming a constant surface heat flux and material properties, the following energy balance, boundary conditions, and initial condition apply as sketched in Fig. 5.8.

$$\text{1D Energy Equation: } \rho_s c_s \frac{\partial T}{\partial t} = k_s \frac{\partial^2 T}{\partial x^2}$$

$$\text{Boundary Conditions: } -k_s \frac{\partial T}{\partial x}\bigg|_{x=0} = \dot{q}''_s$$

$$T(\infty, t) = T_0$$

$$\text{Initial Condition: } T(x, t \leqslant 0) = T_0$$

Here, x is the distance from the surface and $\dot{q}''_s$ is the total surface heat flux. In general, the surface heat flux can include the radiation from an external source ($\dot{q}''_e$), convective heat losses to the cold ambient air ($\dot{q}''_{conv.cool} = \tilde{h}(T_s - T_\infty)$), and surface re-radiation heat losses ($q''_{sr} = \varepsilon_s \sigma(T_s^4 - T_\infty^4)$) so the total surface heat flux is given by $\dot{q}''_s = -\dot{q}''_{conv,cool} - \dot{q}''_{sr} + \alpha_s \dot{q}''_e$ where α_s is the fraction of the external source reaching the surface.

The general solution for this problem is [2]:

$$T(x,t) - T_0 = \frac{2\dot{q}''_s (\alpha_s t/\pi)^{1/2}}{k_s} \exp\left(\frac{-x^2}{4\alpha_s t}\right) - \frac{\dot{q}''_s x}{k_s} erfc\left(\frac{x}{2\sqrt{\alpha_s t}}\right), \tag{5.15}$$

where $erfc(\xi) = 1 - erf(\xi)$ and $erf(\xi)$ is the error function which is zero when $\xi = 0$ and 1 when $\xi = \infty$. Since the fuel is heated from above, the ignition temperature would be satisfied first at the surface. The time it takes for the surface ($x = 0$) to reach the ignition temperature ($T = T_{ig}$) is found to be:

$$T_{ig} - T_0 = \frac{2\dot{q}''_s (\alpha_s t_g/\pi)^{1/2}}{k_s}$$

Rearranging:

$$t_g = \frac{\pi}{4} k_s \rho_s c_s \frac{(T_{ig} - T_0)^2}{\dot{q}''^2_s} \tag{5.16}$$

To get a feel for just how long this time takes for a thick solid such as wood, let's plug in some typical values for wood exposed to heat flux from an adjacent fire:

$$k_s \rho_s c_s = 0.67 \left(\left[kW /m^2 \text{ - } K \right]^2 \times s \right)$$

$$T_{ig} = 354°C$$

$$\dot{q}''_s = 20 \text{ kW/m}^2$$

$$t_g = \frac{\pi}{4} 0.67 \left(\left[kW/m^2 \text{ - } K \right]^2 \times s \right) \frac{(354 - 25)^2 (K^2)}{20^2 (kW/m^2)^2} = 142.4 \text{ s}$$

Once these pyrolysis gases are formed, they must mix to form a flammable mixture. A conservative estimate of this mixing time is obtained by assuming that the vapors mix purely by diffusion. The diffusion time can be estimated from

$$t_{mix} = \frac{L^2}{D}$$

where L is the diffusion distance and D is the diffusivity. Again, to get a feel for just how long this step takes, again we will plug in some typical values. In this case, let's assume a boundary layer forms over the heated surface due to natural convection:

$$\delta \approx 3 \text{ mm}$$

$$D \approx 1 \times 10^{-5} \frac{m^2}{s}$$

$$t_{mix} \approx 0.9 \text{ s}$$

The last step in the process is the chemistry. The chemical time can be estimated using the same method described in Chap. 3:

$$t_{chem} = \frac{[Fuel]_i}{-d[Fuel]/dt}$$

Once again, let's plug in some typical values to get a feel for the time that this step takes. If we assume that the gases consist primarily of methane, ignition occurs at the lean flammability limit, and that the reaction occurs at an average temperature of 1,600 K:

Stoichiometric methane-air combustion is

$$CH_4 + 2(O_2 + 3.76N_2) \rightarrow CO_2 + 2H_2O + 7.52N_2$$

The global rate of reaction progress is

$$\dot{q}_{RxT} = A_0 \exp\left(-\frac{E_a}{\hat{R}T}\right) [Fuel]^a [O_2]^b.$$

Using values from Table 3.1 in Chap. 3, $A_0 = 8.3 \cdot 10^5$, $E_a = 30$ kcal/mol, $a = -0.3$, $b = 1.3$, and $E_a/\hat{R}_u = 15,101$ K, the global consumption rate for methane is

$$\frac{d[CH_4]}{dt} = -\dot{q}_{RxT} = -8.3 \cdot 10^5 \cdot \exp\left(-\frac{15,101}{T\,(\text{K})}\right) [CH_4]^{-0.3} [O_2]^{1.3}$$

Next the concentrations of methane and oxygen are evaluated at $T = 473$ K (a typical pyrolysis temperature) using the ideal gas law

$$P_i V = N_i \hat{R} T$$

$$[C_i] = \frac{N_i}{V} = \frac{P_i}{\hat{R}T} = \frac{P x_i}{\hat{R}T}$$

At the lean flammability limit, the equivalence ratio is approximately 0.5, so for $[O_2]$, $x_{O2} = (2/0.5)/(1 + (2/0.5) \cdot 4.76) = 0.2$ and

$$[O_2] = \frac{101.3 \cdot 10^3\,(\text{Pa}) \cdot 0.2}{8.314\,(\text{Pa} \times \text{m}^3/\text{mol} - \text{K}) \cdot 1600\,\text{K}} = 5.15 \cdot 10^{-6}\,\text{mol/cc}$$

Similarly $x_{CH4} = 1/(1 + (2/0.5) \cdot 4.76) = 0.05$ and $[CH_4] = 1.28 \cdot 10^{-6}\,\text{mol/cc}$. The consumption rate of methane is

$$\frac{d[CH_4]}{dt} = -8.3 \cdot 10^5 \cdot \exp\left(-\frac{15,101}{1,600}\right) (1.28 \cdot 10^{-6})^{-0.3} (5.15 \cdot 10^{-6})^{1.3} \cdot$$

$$= 5.18 \cdot 10^{-4}\,\text{mol/cc} - \text{s}$$

Assuming that the reaction is irreversible, the amount of time to consume all the fuel is

$$t_{chem} = \frac{[CH_4]}{-d[CH_4]/dt} = 0.0025\,\text{s} = 2.5\,\text{ms}$$

By comparing the above times for the gasification, mixing, and chemistry process, it is clear that the gasification time for a solid fuel such as wood is much greater than the mixing and chemistry times. It is for this reason that the solid fuel ignition time is generally estimated by the gasification (pyrolysis) time, or

$$t_{ig} \approx t_g \tag{5.17}$$

Example 5.2 A cigarette just lit a fire in a trash can which is now providing an external radiant heat flux of 35 kW/m^2 on some nearby curtains. How long will it take before the curtains also catch on fire? Assume the curtains are cooled by natural convection ($\tilde{h} = 10$ W/m^2–K) and the rest of the room remains at 25°C. The curtains are 0.5 mm thick with $\varepsilon = \alpha = 0.9$.

Solution:

We will treat the curtains as cotton fabric so that the material properties can be found in Table 5.3. First let us calculate the Biot number to determine which assumption of transient conduction is appropriate.

$$Bi = \frac{\tilde{h}L}{k} = \frac{10\left(\text{W}/\text{m}^2\text{K}\right) \cdot 0.0005(\text{m})}{0.06(\text{W}/\text{mK})} = 0.083$$

The Biot number is smaller than the threshold value of 0.1, so the lumped capacitance model can be used. As an estimate of the ignition time of the curtains, the lumped capacitance energy balance is

$$\text{Energy in} = \text{Energy stored}$$

$$\dot{q}''_s = \rho_s c_s d \frac{dT}{dt}$$

$$\text{Initial Condition: } T(t = 0) = T_o$$

Table 5.3 Solid material properties (From Quintiere [5] unless noted)

Material	k (W/mK)	ρ (kg/m^3)	c (J/kgK)	Effective $k\rho c$ ((kW/m^{2}K)^{2}s)	T_{ig} (piloted) (°C)	L_v (kJ/g)
Carpet	0.074[a]	350[a]		0.25	435	
Cotton	0.06[b]	80[b]	1,300[b]		254[c]	
Douglas fir	0.11[e]	502	2,720[e]	0.25	384	1.81[e]
Maple	0.17[e]	741	2,400[e]	0.67	354	
Paper	0.18[b]	930[b]	1,340[b]		229[c]	3.6
Plywood	0.12	540	2,500	0.16	390	
Fire retardant plywood				0.76	620	0.95[d]
Rigid polyurethane	0.02	32	1,300	0.03	378[c]	1.52[d]
Redwood		354		0.22	375	3.14[d]
Red oak	0.17[e]	753	2,400[e]	1	305	
Polypropylene	0.23	1,060	2,080	0.51	374	2.03[d]

All wood properties measured across the grain

Use effective $k\rho c$ for semi-infinitely thick solids

L_v: heat of vaporization

[a] From National Institute of Standards and Technology [4]

[b] From Incropera et al. [3]

[c] From Babrauskas [1]

[d] From Drysdale [2]

[e] From National Fire Protection Association [6]

where d is the thickness of the material and $\dot{q}''_s$ is the total surface heat flux. Assuming the surface heat flux and material properties are constant and that curtains ignite when heated to the ignition temperature (T_{ig}), we use separation of variables and integrate:

$$\int_0^{t_{ig}} \dot{q}''_s dt = \rho_s c_s d \int_{T_0}^{T_{ig}} dT$$

$$\dot{q}''_s (t_{ig} - 0) = \rho_s c_s d (T_{ig} - T_0)$$

$$t_{ig} = \frac{\rho_s c_s d (T_{ig} - T_0)}{\dot{q}''_s}$$

The surface heat flux includes the external radiant flux from the burning trash can, the cooling due to natural convection, and the cooling due to re-radiation. The surface heat flux is then:

$$\dot{q}''_s = \alpha \dot{q}''_{ext} - \dot{q}''_{conv} - \dot{q}''_{reradiation}$$

$$\dot{q}''_s = \alpha \dot{q}''_{ext} - \tilde{h}(T_s - T_\infty) - \varepsilon\sigma (T_s^4 - T_\infty^4)$$

As the solid heats up, the amount of heat losses by convection and radiation will change. In deriving the expressions for the ignition time, however, we assumed that the surface heat flux was constant. The heat losses will only range from 0 kW/m^2 at the initiation of the fire to

$$\dot{q}''_{conv,max} = 10 \frac{\text{kW}}{\text{m}^2\text{K}} (254 - 25)\text{K} = 2.29 \frac{\text{kW}}{\text{m}^2}$$

$$\dot{q}''_{rad,max} = 0.9 \left(5.67 \times 10^{-8} \frac{\text{W}}{\text{m}^2\text{K}^4} \right) \left[(254 + 273)^4 \text{K}^4 - (25 + 273)^4 \text{K}^4 \right]$$

$$= 3.53 \frac{\text{kW}}{\text{m}^2}$$

when the surface temperature reaches the ignition temperature $T_s = T_{ig} = 254°\text{C}$. The total heat losses are at most 5.82 kW/m^2, only 18% of the heat flux due to the trash can fire. Because the external radiant heat flux is so large compared to the heat losses, we will disregard the heat loss terms and assume that the total surface heat flux is due solely to the external radiant flux. Note that this assumption can only be made when this heat flux is large relative to the heat losses, which may not always be the case as we will see in the next section. The ignition time is then:

$$t_{ig} = \frac{\rho_s c_s d (T_{ig} - T_\infty)}{\dot{q}''_s} = \frac{\rho_s c_s d (T_{ig} - T_\infty)}{\alpha \dot{q}''_{ext}}$$

$$t_{ig} = \frac{\left(80 \, \frac{\text{kg}}{\text{m}^3}\right)\left(1300 \, \frac{\text{J}}{\text{kg K}}\right)(0.002 \, \text{m})(254 - 25)\text{K}}{0.9\left(35000 \, \frac{\text{W}}{\text{m}^2}\right)} = 1.51 \, \text{s}$$

5.4.4 Critical Heat Flux for Ignition

From Eq. 5.16, the ignition time is a function of the net heat flux on the surface. For a high level of heat flux, the ignition time will be relatively short. Conversely, for a low level of heat flux, the ignition time will be relatively long. However, it is possible that the net heat flux on the solid is not sufficient to heat the material to its ignition point. It follows that there is a critical level of external heat flux that must be applied to the solid to offset the heat losses enough to eventually reach the ignition temperature. This level of heat flux is called the "critical heat flux" (CHF) for ignition. Figure 5.9 below shows some typical ignition time trends as a function of the external heat flux level for different convective cooling velocities. As the external heat flux decreases, the ignition time increases. As the velocity of the convective flow increases, more heat is lost from the material and ignition takes longer. The asymptotes on the curve represent the critical heat flux (CHF) for ignition. As shown, any external heat fluxes less than this value will not result in an ignition. The CHF for ignition is a function of the convective cooling velocity because of the surface energy balance on the solid. More convective cooling requires a higher external heat flux to heat the solid to its ignition point.

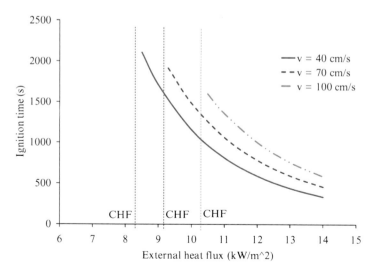

Fig. 5.9 Ignition time as a function of external heat flux for three flow velocities

Example 5.3 For a material with an ignition temperature of $T_{ig} = 315°C$, what is the critical heat flux for ignition if the material is cooled by natural convection (assume $\tilde{h} = 10$ W/m^2–K) in an environment at $T_\infty = 30°C$? Assume the emissivity = adsorptivity = 0.9.

Solution:

By definition, the critical heat flux for ignition is the minimum heat flux capable of heating a material to its ignition point. At the extreme limit of an infinite ignition time, the material temperature reaches a steady value equal to the ignition temperature. The problem can then be treated as a steady state heat conduction problem. Additionally, for such a long heating time, even the thickest of materials will behave as a thermally thin solid. Performing an energy balance for a thermally thin solid, at ignition

$$\text{heat loss} = \text{heat gain}$$

$$\text{heat loss} = \dot{q}''_{conv} + \dot{q}''_{reradiation}$$

$$\text{heat gain} = \dot{q}''_{ext} = \dot{q}''_{critical}$$

$$\dot{q}''_{conv} = \tilde{h}(T_{ig} - T_\infty) = \left(10\frac{\text{W}}{\text{m}^2\text{K}}\right)(315°C - 30°C) = 2850\frac{\text{W}}{\text{m}^2} = 2.85\frac{\text{kW}}{\text{m}^2}$$

$$\dot{q}''_{reradiation} = \varepsilon\sigma\left(T_{ig}^4 - T_\infty^4\right)$$

$$= 0.9\left(5.67 \times 10^{-8}\frac{\text{W}}{\text{m}^2\text{K}^4}\right)\left[(315 + 273)^4\text{K}^4 - (30 + 273)^4\text{K}^4\right]$$

$$= 5669.9\frac{\text{W}}{\text{m}^2} = 5.7\frac{\text{kW}}{\text{m}^2}$$

$$\dot{q}''_{crit} = \dot{q}''_{conv} + \dot{q}''_{reradiation} = (2.85 + 5.67)\frac{\text{kW}}{\text{m}^2} = 8.52\frac{\text{kW}}{\text{m}^2}$$

The critical heat flux for ignition is extremely dependent on the ambient conditions and varies with the convective cooling conditions and the amount of heat lost via reradiation to the environment. At temperatures near ignition, the losses due to reradiation can be greater than those due to convection and cannot be ignored. Notice in this analysis that the material's properties (such as thermal conductivity and density) were not used directly. The critical heat flux was calculated strictly by using an energy balance and would hold for any material in this situation with the same ignition temperature. Why do some materials ignite and some don't in the same conditions? The ignition temperature varies quite widely between materials (see Table 5.3) and can even be a function of the environmental conditions.

Exercises

5.1 For spontaneous ignition (autoignition), how is the critical temperature defined? How is the critical pressure defined? Show the conditions and equations to solve for these two variables. Sketch a qualitative plot of critical temperature and pressure for spontaneous ignition.

5.2 Consider a spherical vessel (constant volume) having a radius of 10 cm. It contains a stoichiometric mixture of methane and air at 1 atm. The system is initially at temperature T_i. The heat losses to the surroundings per unit volume of the vessel are $\dot{q}_L''' = \frac{A_S}{V} \tilde{h}(T - T_\infty)$, where T is the temperature, V is the volume of the vessel, A_S is its surface area, $\tilde{h}$ is the heat transfer coefficient (15 W/m²-K), and T_∞ is the ambient temperature (300 K). The rate of heat generation per unit volume is $\dot{q}_G''' = Q_c \hat{r}$ where Q_c is the heat of combustion (MJ/mol) and $\hat{r}$ is the fuel consumption rate [mol/(m³-s)].

   a. Calculate the heat of combustion of the mixture Q_c.

   b. For $\tilde{h} = 15$ W/m² - K, plot $\dot{q}_L'''$ and $\dot{q}_G'''$ as a function of the system's initial temperature T_i for $T_i \geqslant 300$ K. You do not have to calculate how the system evolves in time, focus only on its initial state.

   c. For $\tilde{h} = 15$ W/m² - K, what is the lowest initial temperature at which the rate of heat production by combustion offsets the heat losses?

   d. Calculate the autoignition temperature of the system (T_c).

5.3 Plot the autoignition temperature versus the number of carbon atoms for those straight chain hydrocarbon fuels listed in Table 5.1. Discuss any trends.

5.4 Determine the ratio between the minimum ignition energy and the heat release for a 400 cc spark-ignition piston engine running with a stoichiometric isooctane-air mixture at ambient conditions.

5.5 In the chemical industry, a fitted equation called the Antoine equation with three parameters is often used as $\log P = A - \frac{B}{T+C}$ or $\ln P = A - \frac{B}{T+C}$, where A, B, and C are parameters fitted from data. Write a program to find the vapor pressure of a given chemical species at a specified temperature based on the following Antoine equation.

$$log(P) = A - B/(T + C),$$

where log is the common (base 10) logarithm, the coefficients A, B, and C for a few select species are tabulated in Table 5.4 (values for other species are found

Table 5.4 Exercise 5.5: Antoine equation coefficients

Fuel	Formula	A	B	C	T_{min} (°C)	T_{max} (°C)
Methane	CH_4	6.69561	405.420	267.777	−181	−152
Ethane	C_2H_6	6.83452	663.700	256.470	−143	−75
Propane	C_3H_8	6.80398	803.810	246.990	−108	−25
Butane	C_4H_{10}	6.80896	935.860	238.730	−78	19
Pentane	C_5H_{12}	6.87632	1075.780	233.205	−50	58

in Appendix 7). P is expressed in *mmHg*, T is expressed in Celsius, and the valid temperature range ($T_{min} < T < T_{max}$) is also given.

Note that it is inappropriate to use the Antoine equation when the temperature is outside the range given for the coefficients (A, B, and C), for pressures in excess of 1 MPa, or when the components differ in nature (for example a mixture of propanol/water).

5.6 A 2 cm thick plywood is subject to a uniform heat flux of 50 kW/m^2. Estimate the time it takes for the plywood to catch fire.

References

1. Babrauskas V (2003) Ignition Handbook: Principles and applications to fire safety engineering, fire investigation, risk management and forensic science. Fire Science Publishers, Issaquah
2. Drysdale D (1998) An Introduction to Fire Dynamics, 2nd edition. John Wiley & Sons, New York
3. Incropera FP, DeWitt DP, Bergman TL, Lavine AS (2006) Fundamentals of Heat and Mass Transfer, 6th edition. John Wiley & Sons, New York
4. National Institute of Standards and Technology http://srdata.nist.gov/insulation/
5. Quintiere JG (2006) Fundamentals of Fire Phenomena. John Wiley & Sons, San Francisco
6. (2008) SFPE Handbook of Fire Protection Engineering, 4th edition. National Fire Protection Association, Quincy

Chapter 6
Premixed Flames

As their name implies, premixed flames refer to the combustion mode that takes place when a fuel and oxidizer have been mixed prior to their burning. Premixed flames are present in many practical combustion devices. Two such applications are shown in Fig. 6.1: a home heating furnace and a lean premixed "can combustor" in a power-generating gas turbine. In premixed flame combustors, the fuel and oxidizer are mixed thoroughly before being introduced into the combustor. Combustion is initiated either by ignition from a spark or by a pilot flame, creating a 'flame' that propagates into the unburned mixture. It is important to understand the characteristics of such a propagating flame in order to design a proper burner. Some relevant engineering questions arise, such as: How fast will the flame consume the unburned mixture? How will flame propagation change with operating conditions such as equivalence ratio, temperature, and pressure? From a fire protection viewpoint, how can flame propagation be stopped?

6.1 Physical Processes in a Premixed Flame

In a duct containing a premixed mixture of fuel and oxidizer, it can be observed that after ignition, a flame propagates into the unburned mixture as sketched in Fig. 6.2. The lower part of the sketch is a close up view of the structure of the flame. The combustion reaction zone, or "flame" is quite thin, usually a few millimeters for hydrocarbon fuels in ambient conditions. In the preheat zone, the temperature of the reactants increases gradually from the unburned mixture temperature to an elevated temperature near the reaction zone. As the reactant temperature approaches the ignition temperature of the fuel, the chemical reactions become rapid, marking the front of the combustion reaction zone (flame). Inside the flame, the reaction rate increases rapidly and then decreases as fuel and oxidizer are consumed and products produced. Because of the species concentration gradient, the reactants diffuse toward the reaction zone and their concentrations in the preheat zone decrease as they approach the reaction zone. The temperature of the products is close to the adiabatic flame temperature. Various species in the reaction zone become molecularly excited at high temperature and emit radiation at different

S. McAllister et al., *Fundamentals of Combustion Processes*,
Mechanical Engineering Series, DOI 10.1007/978-1-4419-7943-8_6,
© Springer Science+Business Media, LLC 2011

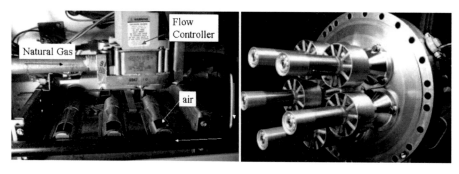

Fig. 6.1 Premixed flame applications. *Left* – home furnace; *Right* – GE Dry Low NO_x combustor for power generation (Reprinted with permission from GE Energy)

Fig. 6.2 Sketch of a premixed flame propagating in a duct from *right* to *left*

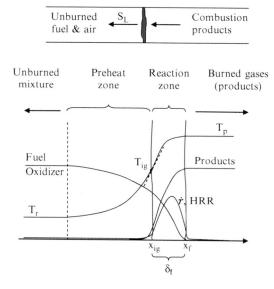

wavelengths that give flames different colors. For lean mixtures of hydrocarbon fuels and air, the bluish color is due to radiation from excited CH radicals, while radiation from CO_2, water vapor, and soot particles produces a reddish orange color. For rich mixtures, a greenish color from excited C_2 molecules is also observed.

Flame propagation through the unburned mixture depends on two consecutive processes. First, the heat produced in the reaction zone is transferred upstream, heating the incoming unburned mixture up to the ignition temperature. Second, the preheated reactants chemically react in the reaction zone. Both processes are equally important and therefore one expects that the flame speed will depend on both transport and chemical reaction properties.

6.1.1 Derivation of Flame Speed and Thickness

A simple 'thermal' theory (similar to Mallard and Le Chatelier's [10]) is useful for estimating the flame speed, flame thickness, and their dependence on operating conditions. Let's consider the preheat zone first. Since the temperature is lower than the autoignition temperature, chemical reactions are negligible. Consider a control volume around the preheat zone up to the location where temperature reaches the ignition temperature (right side of preheat zone in Fig. 6.2). The steady-state energy equation

$$\rho c_p u \frac{\partial T}{\partial x} = k \frac{\partial^2 T}{\partial x^2}$$

is integrated from the beginning of preheat zone to the location where temperature reaches T_{ig}.

$$\int_0^{x_{ig}} \rho c_p u \frac{\partial T}{\partial x} dx = \int_0^{x_{ig}} k \frac{\partial^2 T}{\partial x^2} dx$$

$$\rho_r c_p S_L (T_{ig} - T_r) = k \frac{\partial T}{\partial x}\Big|_{x_{ig}} \approx k \frac{T_p - T_{ig}}{\delta}, \tag{6.1}$$

S_L is the flame propagation speed into the unburned mixture ($u = S_L$), T_{ig} is the ignition temperature, T_r and ρ_r are respectively the temperature and density of the reactant mixture, c_p is the specific heat (assumed constant), k is thermal conductivity (assumed constant), and T_p is the temperature of the combustion products in the burned zone. The temperature gradient has been approximated by $(T_p - T_{ig})/\delta$ where δ is the thickness of reaction zone, normally referred to as the "flame thickness." By considering the overall energy balance for a control volume including both the preheat and reaction zones, integration of the energy equation leads to

$$\int_0^{x_f} \rho c_p u \frac{\partial T}{\partial x} dx = \int_0^{x_f} k \frac{\partial^2 T}{\partial x^2} dx + \int_0^{x_f} \hat{r}_{fuel} \hat{Q}_c dx$$

$$\rho_r c_p S_L (T_p - T_r) = 0 + \delta \cdot \hat{r}_{fuel, ave} \hat{Q}_c$$

$$\underbrace{\rho_r S_L c_p (T_p - T_r)}_{\text{convective energy}} = \underbrace{\delta \cdot \hat{r}_{fuel, ave} \hat{Q}_c}_{\text{energy from combustion}}, \tag{6.2}$$

where $\hat{r}_{fuel, ave}$ is the average magnitude of fuel consumption rate over the entire flame, and $\hat{Q}_c$ is the heat release per unit mole of fuel burned. With Eqs. 6.1 and 6.2, one can solve for the two unknowns S_L and δ, leading to

$$\rho_r S_L c_p (T_{ig} - T_r) \cdot \rho_r S_L c_p (T_p - T_r) = k(T_p - T_{ig})\hat{r}_{fuel,ave}\hat{Q}_c$$

$$\rho_r S_L c_p = \left\{ \frac{k(T_p - T_{ig})\hat{r}_{fuel,ave}\hat{Q}_c}{(T_{ig} - T_r)(T_p - T_r)} \right\}^{1/2}$$

$$S_L = \left\{ \frac{k(T_p - T_{ig})\hat{r}_{fuel,ave}\hat{Q}_c}{\rho_r c_p (T_{ig} - T_r)\rho_r c_p (T_p - T_r)} \right\}^{1/2}$$

The heat of combustion is approximately related to the flame temperature by $\hat{Q}_c \cdot [Fuel]_r = \rho_r c_p (T_p - T_r)$, where $[Fuel]_r$ is the fuel concentration (mol/cc) in the fresh mixture ($\hat{Q}_c$ has the unit of kJ/mol). The flame speed then becomes

$$S_L = \left\{ \frac{k(T_p - T_{ig})\hat{r}_{fuel,ave}/[Fuel]_r}{\rho_r c_p (T_{ig} - T_r)} \right\}^{1/2} = \left\{ \frac{\alpha}{\tau_{chem}} \frac{(T_p - T_{ig})}{(T_{ig} - T_r)} \right\}^{1/2}, \qquad (6.3)$$

Where $\alpha = k/\rho c_p$ is the thermal diffusivity (cm^2/s) and $\tau_{chem} \equiv [Fuel]_r/\hat{r}_{fuel,ave}$ is the time scale of chemical kinetics. Using $\hat{Q}_c \cdot [Fuel]_r = \rho_r c_p (T_p - T_r)$, Eq. 6.2 becomes

$$S_L = \frac{\delta}{\tau_{chem}} \quad \text{or} \quad \delta = S_L \cdot \tau_{chem} \qquad (6.4)$$

Equation 6.4 suggests that for a given flame speed, the flame thickness is proportional to the time scale of chemical kinetics. If chemistry is fast, the flame thickness is expected to be small. Substituting Eq. 6.3 into Eq. 6.4 one has

$$\delta = S_L \cdot \tau_{chem} = \left\{ \frac{\alpha}{\tau_{chem}} \frac{(T_p - T_{ig})}{(T_{ig} - T_r)} \right\}^{1/2} \tau_{chem}$$

$$\delta = \left\{ \alpha \cdot \tau_{chem} \frac{(T_p - T_{ig})}{(T_{ig} - T_r)} \right\}^{1/2} \qquad (6.5)$$

The flame thickness is often correlated to flame speed through the thermal diffusivity. This correlation is obtained by multiplying Eqs. 6.3 and 6.5 leading to

$$\delta \cdot S_L = \alpha \frac{(T_p - T_{ig})}{(T_{ig} - T_r)} \qquad (6.6)$$

The right hand side of Eq. 6.6 depends on the thermodynamics of the combustion system. For a given fuel, one can estimate the right hand side. For methane-air combustion at ambient conditions, $T_r = 300$ K, $T_p = 2{,}250$ K, and $T_{ig} \sim 810$ K, so $\delta \cdot S_L \approx 3.5\alpha$. Since the average fuel consumption rate, $\hat{r}_{fuel,ave}$, has a strong temperature dependence, the choice of temperature for evaluating the average fuel consumption rate has a strong impact on the outcome; hence Eqs. 6.3–6.5

provide only a rough estimate of S_L and δ. However, Eq. 6.3 is valuable in providing insight into the dependence of flame speed on various parameters, including transport properties, temperature, pressure, and reaction rate order. For an order of magnitude estimate, we will use the reaction thickness, δ, to represent flame thickness. One must recognize that these equations were derived from a simple analysis to provide an order of magnitude assessment. More accurate solutions are now routinely solved using detailed chemistry and transport equations for one-dimensional flames. For most hydrocarbon fuels, the flame speed of a stoichiometric mixture at the reference state is about 40 cm/s. However, the flame speed of hydrogen flame is 220 cm/s, about five times faster.

Example 6.1 Using the one-step reaction (Table 3.1) and the simple thermal theory of Eq. 6.3, estimate the laminar burning velocity of a stoichiometric propane-air mixture initially at 300 K and 1 atm. The adiabatic flame temperature is 2,240 K and the ignition temperature is 743 K.

Solution:
Equation 6.3 reads

$$S_L = \left\{ \frac{k(T_p - T_{ig})\hat{r}_{fuel,ave}/[Fuel]_r}{\rho_r c_p (T_{ig} - T_r)} \right\}^{1/2} = \left\{ \frac{\alpha}{\tau_{chem}} \frac{(T_p - T_{ig})}{(T_{ig} - T_r)} \right\}^{1/2}$$

The overall one-step description of propane-air combustion is

$$C_3H_8 + 5(O_2 + 3.76N_2) = 3CO_2 + 4H_2O + 18.8N_2$$

The total concentration of reactants including N_2 is evaluated at $T_r = 300$ K as

$$[\text{reactants}] = \frac{P}{\hat{R}_u T} = \frac{1(\text{atm})}{82.0574 \ (\text{atm} \times \text{cm}^3/\text{mol} - \text{K}) \cdot 300(\text{K})}$$
$$= 4.06 \cdot 10^{-5} (\text{mol/cc})$$

$[C_3H_8] = x_{C3H8} [\text{reactants}];$

$$x_{C_3H_8} = \frac{1}{1 + 5 \cdot (1 + 3.76)} = 0.0403$$

$[C_3H_8] = 0.0403 \cdot 4.06 \cdot 10^{-5} = 1.64 \cdot 10^{-6}$ (mol/cc); $[O_2] = 5[C_3H_8] = 8.18 \cdot 10^{-6}$ (mol/cc)

Table 3.1 gives

$$\dot{q}_{RxT} = 8.6 \cdot 10^{11} \exp\left(-\frac{30,000}{1.987 \cdot T(K)}\right) [C_3H_8]^{0.1} [O_2]^{1.65} \ (\text{mol/cc} - \text{s})$$

With $T_p = 2,240$ K, $T_{ig} = 743$ K, $T_r = 300$ K, we need to estimate α and τ_{chem}. Since both α and τ_{chem} depend on temperature (especially the reaction rate), one needs to determine the approximate temperatures to evaluate these two quantities. For α, we can use the average temperature between the reactants and products as $T_{1,ave} = (T_p + T_r)/2 = 1,270$ K. Since most of the mixture is air, we will use air properties (listed in Appendix 2) to estimate α. From the Appendix 2: $k = 7.85 \cdot 10^{-5}$ kW/m-K, $\rho = 0.2824$ kg/m^3, $c_p = 1.182$ kJ/kg-K, $\alpha = k/(\rho \cdot c_p) = 7.85 \cdot 10^{-5}$ kW/m-K/(0.2824 kg/m^3 $\cdot$ 1.182 kJ/kg-K) $= 23.52 \cdot 10^{-5}$ m^2/s $= 2.35$ cm^2/s.

Next, the chemical time scale is estimated on the basis of the average reaction rate. Since chemical reactions are very sensitive to temperature, we will try using $T_{ave} = 1,270$ K. Also, because the reactant concentrations decrease with time, we will assume that the average reactant concentrations are half of their initial value.

$$\dot{q}_{RxT} = 8.6 \cdot 10^{11} \exp\left(-\frac{30,000}{1.987 \cdot T_{2,ave}(K)}\right)\left(\frac{[C_3H_8]}{2}\right)^{0.1}\left(\frac{[O_2]}{2}\right)^{1.65}$$

$$\hat{r}_{C_3H_8,ave} = \dot{q}_{RxT} = 8.6 \cdot 10^{11} \exp\left(-\frac{30,000}{1.987 \cdot 1270}\right)\left(\frac{[C_3H_8]}{2}\right)^{0.1}\left(\frac{[O_2]}{2}\right)^{1.65}$$

$$= 8.6 \cdot 10^{11} \cdot 6.87 \cdot 10^{-6} \cdot \left(\frac{1.64 \cdot 10^{-6}}{2}\frac{300}{1270}\right)^{0.1}\left(\frac{8.18 \cdot 10^{-6}}{2}\frac{300K}{1270K}\right)^{1.65}$$

$$= 1.5 \cdot 10^{-4} \text{ (mol/cc} - \text{s)}$$

Note that the ratio 300 K/1,270 K accounts for the decrease in concentration due to temperature change under constant pressure by the ideal gas law.

$$\tau_{chem} \equiv [Fuel]_r/\hat{r}_{fuel,ave}$$
$$= 1.64 \cdot 10^{-6}(\text{mol/cc})/1.5 \cdot 10^{-4} \text{ (mol/cc} - \text{s)}$$
$$= 1.1 \cdot 10^{-2}\text{s}$$

$$S_L = \left\{\frac{\alpha}{\tau_{chem}}\frac{(T_p - T_{ig})}{(T_{ig} - T_r)}\right\}^{1/2}$$
$$= \left\{\frac{2.35(\text{cm}^2/\text{s})}{1.1 \cdot 10^{-2}\text{s}}\frac{(2240 - 743)(K)}{(743 - 300)(K)}\right\}^{1/2}$$
$$= 26.9 \text{ cm/s}$$

Alternatively, we can use $T_{2,ave} = (T_{ig} + T_p)/2 = 1,490$ K and repeat the above process leading to

$$\hat{r}_{C_3H_8,ave} = 6.56 \cdot 10^{-4} \text{ (mol/cc} - \text{s)},$$

$$\tau_{chem} = 1.65 \cdot 10^{-6}(\text{mol/cc})/6.56 \cdot 10^{-4}(\text{mol/cc} - \text{s}) = 2.5 \cdot 10^{-3}\text{s}$$

$$S_L = \left\{ \frac{\alpha}{\tau_{chem}} \frac{(T_p - T_{ig})}{(T_{ig} - T_r)} \right\}^{1/2}$$

$$= \left\{ \frac{2.35(\mathrm{cm}^2/\mathrm{s})}{2.5 \cdot 10^{-3}\mathrm{s}} \frac{(2240 - 743)(\mathrm{K})}{(743 - 300)(\mathrm{K})} \right\}^{1/2}$$

$$= 56.4 \text{ cm} / \text{s}$$

Note that the measured value is about 38.9 cm/s. Simplified thermal theory thus provides only a rough estimate.

6.1.2 Measurements of the Flame Speed

Bunsen burners are frequently used for the determination of laminar flame speed. As presented in the left of Fig. 6.3, the Bunsen burner has a vertical metal tube through which gaseous fuel-air mixture is introduced. Air is drawn in through air holes near the base of the tube and mixes with the gaseous fuel. The combustible mixture is ignited and burns at the tube's upper opening. The flow rate of air is controlled by an adjustable collar on the side of the metal tube. If the mixture at the exit of the burner tube falls within the flammability limits, a premixed flame can be established. If the equivalence ratio of this mixture is greater than one but still below the rich flammability limit (RFL), the mixture is combustible and a rich premixed flame can be established with a cone shape as depicted in the middle figure. Since the unburned mixture does not contain enough oxidizer to react all of the fuel, the products downstream of the rich premixed flame contain reactive species from incomplete combustion. Consequently, the reactive species from the inner rich premixed flame form an outer diffusion flame as they mix with the surrounding air. This is seen as an outer cone in the picture.

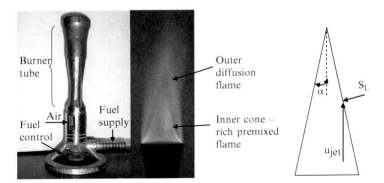

Fig. 6.3 *Left*: Bunsen burner; *Middle*: Rich premixed cone with outer diffusion flame; *Right*: Sketch of inner rich premixed flame allowing determination of flame speed

The Bunsen flame is stationary relative to a laboratory observer. Therefore, the cone angle is determined by the balance of the local fluid speed with the flame propagation speed as sketched in the right of Fig. 6.3. Using geometric relations, one can determine the flame speed as $S_L = u_{jet} \, sin(\alpha)$, where α is the angle between the premixed flame (slanted) and the vertical centerline. Several factors can influence the accuracy of this technique: (1) the flame shape along the edge may not be straight due to heat loss to the burner, (2) effects of stretch[1] on the flame that may not be uniform, (3) a boundary layer is formed in the inner surface of the metal tube that contributes to the distortion of a perfect cone shape, and (4) buoyancy effects may be important.

Because the laminar flame speed is a fundamentally important feature of many combustion systems, measurements have been gradually improved leading to a consistent determination of flame speeds. These improvements include laser techniques for measuring flow ahead of the flame and an opposed flame burner for setting the stretch rate. Since the important effect of stretch on flame speed has been recognized, systematic methods to measure flame speeds of weakly stretched flames have been used to extrapolate flame speeds at zero stretch. Figure 6.4 shows a converging trend in experimentally-determined flame speeds as techniques and science in combustion engineering have improved over the years.

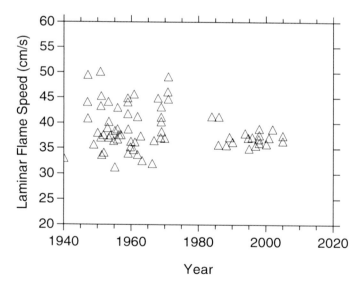

Fig. 6.4 Measured highest flame speeds of methane-air mixtures at ambient condition versus year showing a convergent trend (Reprinted with permission from Law [9])

---

[1] Imagining the flame being a material surface, the effect of aerodynamics from flow field on a flame can increase the flame surface. Such a stretch effect can cause flame speed to deviate from a planar flame.

6.1.3   Structure of Premixed Flames

Due to the small thickness of premixed flames (a few millimeters at 1 atm), it is difficult to measure the species concentrations accurately. Computations of premixed flames with detailed chemistry and transport have been useful in illustrating the structure of a typical premixed flame. Figure 6.5 presents the predicted structure of a laminar stoichiometric methane-air premixed flame initially at

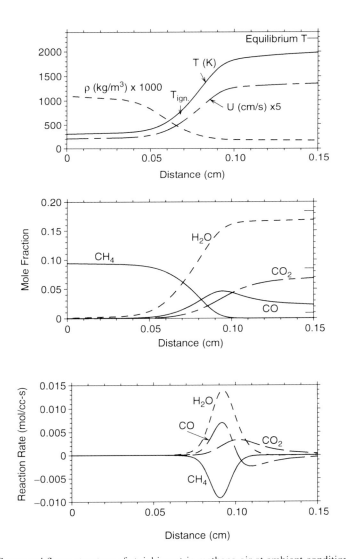

Fig. 6.5 Computed flame structure of stoichiometric methane-air at ambient condition

ambient conditions using a detailed methane mechanism.[2] The unburned mixture flows from the left to the right in the figure. The top plot shows the profiles of temperature, density, and fluid speed relative to the unburned mixture versus distance. The fluid density decreases from about 1.13 kg/m^3 in the unburned mixture to about 0.17 kg/m^3 in the burned zone. The unburned fluid speed relative to the flame is about 39 cm/s and the corresponding fluid speed in the burned zone is about 270 cm/s. The arrow in the plot marks the location where the temperature reaches the autoignition temperature (537°C ≈ 810 K). The flame thickness can be determined on the basis of the temperature rise. For instance, one can define two reference points when temperature reaches 10% and 90% of the total temperature rise. For the current example, these two points are $T_{10\%} = (T_p - T_r)*0.1 + T_r = 495$ K and $T_{90\%} = (T_p - T_r)*0.9 + T_r = 2,055$ K. Based on these two points, the computed preheat and reaction zone thickness is 1.4 mm. The chemical time scale is $\tau_{chem} = \delta/S_L = 3.6$ ms. Also indicated on the right vertical axis is the equilibrium flame temperature (~2,250 K). The peak flame temperature shown in this limited region is about 2,000 K, but computer results show that the temperature reaches 2,250 K about 5 cm further downstream. The time to reach the equilibrium state can be estimated as ~5 (cm)/270 (cm/s) ~ 0.019 s = 19 ms.

The middle plot presents the predicted profiles of the major species (CH_4, H_2O, CO_2) and the main intermediate specie (CO). Their equilibrium values are marked on the right vertical axis. The bottom plot, which presents the predicted net reaction rates for the major species and CO, shows that the reaction zone thickness is about 0.025 cm. As expected, methane has negative net reaction rates throughout the flame since it is consumed. The two major species, H_2O and CO_2, have positive net reaction rates throughout the flame. CO, as an intermediate species, has positive net rates over the region between 0.075 and 0.1 cm; then it has negative rates beyond 0.1 cm in the hot product zone.

The corresponding profiles of selected radical concentrations and their net production rates are plotted in Fig. 6.6. The methyl radical, CH_3, is the first intermediate specie that is produced from the decomposition of CH_4 in the region from 0.07 to 0.09 cm. Consumption of CH_3 starts when the radical species, OH, H, and O, rise at 0.09 cm. Since the majority of CO is oxidized through CO + OH = CO_2 + H, the location where radicals start to increase correlates well with the location where CO begins to decrease. Among the three radicals, OH, H, and O, the H radical diffuses fastest into the unburned zone due to its high diffusivity (i.e., low molecular mass). NO is a pollutant specie that has low concentration but a strong influence on the environment. It will be a topic of discussion in a later chapter. NO is produced via a thermal route with a rate strongly correlated to production of radical species O and OH.

---

[2] Chemkin II software "PREMIX" was used in the computation with GRI30 detailed methane mechanism.

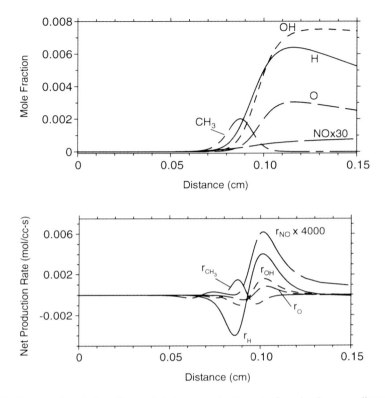

Fig. 6.6 Computed radical profiles and their net production rate for a laminar one-dimensional stoichiometric methane-air premixed flame initially at ambient conditions

6.1.4 Dependence of Flame Speed on Equivalence Ratio, Temperature and Pressure

Since the flame speed depends on the chemical reaction rate, one expects a strong dependence of S_L on temperature and consequently on equivalence ratio. On the left of Fig. 6.7 is a plot of flame temperatures of several fuels versus equivalence ratio showing that the peak flame temperatures occur at a slightly rich mixture. The main reason for the flame temperature's peak at a slightly rich condition is the relation between the heat of combustion and heat capacity of the products. Both of these decline when the equivalence ratio exceeds unity, but the heat capacity decreases slightly faster than heat of combustion between $\phi = 1$ and the peak rich mixture. One expects that the flame speed dependence on ϕ will be similar to the temperature dependence on ϕ. The right plot of Fig. 6.7 presents measured flame speeds of a methane-air flame at ambient conditions. Indeed, the peak value is slightly on the rich side.

The influence of the fresh gas temperature, T_r, on the flame speed is through several effects. Increasing temperature leads to faster chemical reactions, thus the

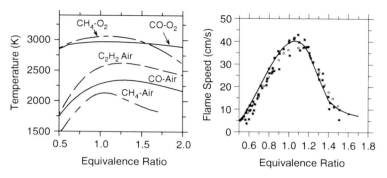

Fig. 6.7 *Left*: peak flame temperatures versus equivalence ratio. *Right*: measured flame speed of methane-air mixture versus equivalence ratio (Reprinted with permission from Bosschaart and de Goey [4]; line computed results with GRI 30 mechanism)

Fig. 6.8 Flame speed of propane-air versus equivalence at 1 atm with various initial temperatures

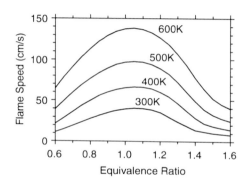

chemistry time is shorter and the flame speed is higher. For ideal gases, the thermal diffusivity has the following dependence on temperature and pressure[3]

$$\alpha = \frac{k(T)}{\rho c_p} = \frac{RTk(T)}{Pc_p} \propto T^{1.5}P^{-1} \tag{6.7}$$

An increase in temperature will increase the thermal diffusivity; hence a higher flame speed will result. Figure 6.8 shows the experimental data of laminar propane-air premixed flames with different unburned gas temperatures. As theory predicts, higher initial temperatures yield faster flame speeds.

Next we consider the effect of pressure on flame speed. For most hydrocarbon fuels, increasing pressure actually leads to a *decrease* in flame speed. Again, guided by Eq. 6.3, we examine the pressure dependence of the individual parameters.

---

[3] Conductivity, k, scales roughly as $\propto \sqrt{T}$; diffusivity, D, scales with $\propto \sqrt{T^3}/P$; viscosity $\mu \propto \sqrt{T}$.

Thermal diffusivity is inversely proportional to pressure as $\alpha \propto P^{-1}$. The flame temperature usually increases slightly with pressure as less dissociation occurs at high pressure; this effect is not significant and will not be included. The effect of pressure on the chemistry time can be analyzed by considering the definition of the chemistry time scale

$$\tau_{chem} \equiv [Fuel]_r/\hat{r}_{fuel,ave} \propto P/P^{(a+b)} \propto P^{1-a-b},$$

where a and b are the exponents of fuel and oxidizer used in the one-step global reaction step. With the above information, the flame speed has the following pressure dependence

$$S_L = \left\{ \frac{\alpha}{\tau_{chem}} \frac{(T_p - T_{ig})}{(T_{ig} - T_r)} \right\}^{1/2} \propto \sqrt{P^{-1}/P^{1-a-b}} \propto P^{((a+b)/2)-1} \propto P^{(n/2)-1}, \quad (6.8)$$

where $n = a + b$ is the total order of the chemical reaction. If the overall reaction order equals 2, then the flame is insensitive to pressure. For hydrocarbon flames, the overall order is less than 2, causing negative pressure dependence as shown in Fig. 6.9 for methane-air combustion. This may cause difficulties for combustion applications at high pressures.

Fortunately, for most hydrocarbon fuels, flame speed is more sensitive to temperature than pressure, so increasing the unburned gas temperature can offset the flame speed reduction due to pressure. In both gas turbine engines and internal combustion engines, the air/fuel mixture is compressed to an elevated temperature before ignition. For many engineering applications, an empirical formula is used to correlate the flame speed based on the flame speed at a reference state (often at ambient conditions). For instance, automobile engineers may use a correlation such as

$$S_L(\phi, T, P) = S_{L,ref}(\phi) \left(\frac{T_r}{T_{ref}} \right)^{\alpha} \left(\frac{P}{P_{ref}} \right)^{\beta} (1 - 2.5\psi), \quad (6.9)$$

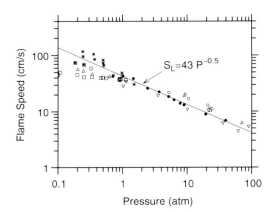

Fig. 6.9 Flame speed of stoichiometric methane-air mixture as function of pressure showing a decreasing trend (Reprinted with permission from Andrews and Bradley [1])

where

$$T_{ref} = 300K, P_{ref} = 1atm$$
$$S_{L,ref}(\phi) = Z \cdot W \cdot \phi^{\eta} \cdot \exp[-\xi(\phi - 1.075)^2].$$

In the above relation, ψ is the mass fraction of residual burned gases, ϕ is the equivalence ratio, and the other coefficients are listed in Table 6.1 for isooctane and ethanol.

The effect of inert dilution on flame speed can be demonstrated by keeping the reactants the same but using different diluent species as illustrated in Fig. 6.10. For air, the ratio of N_2 to O_2 is 3.76. By replacing N_2 by either argon or helium, the flame speeds are found to increase. Flames diluted with helium show the highest flame speeds. With different diluent species, the peak flame temperatures as well as thermal diffusivities are different. Table 6.2 lists computed values of adiabatic flame temperature and thermal diffusivity for stoichiometric mixtures. When N_2 is replaced by Ar, the flame temperature increases because argon has lower heat capacity, c_p. However, the change in thermal diffusivity is negligible; therefore the flame speed increases. When He is used as the dilution species, the flame temperature is the same as when the mixture is diluted with Ar since these

Table 6.1 Empirical coefficients for laminar flame speed [2]

Fuel	Z	W (cm/s)	η	ξ	α	β
C_8H_{18}	1	46.58	-0.326	4.48	1.56	-0.22
C_2H_5OH	1	46.50	0.250	6.34	1.75	$-0.17/\sqrt{\varphi}$
C_8H_{18}+ C_2H_5OH	$1 + 0.07X_E^{0.35}$	46.58	-0.326	4.48	$1.56 + 0.23X_E^{0.35}$	$X_G\beta_G + X_E\beta_E$[a]

[a] X_E = volume percentage of ethanol in fuel mixture, %; X_G = volume percentage of isooctane in fuel mixture, %; $\beta_E = \beta$ value for ethanol; $\beta_G = \beta$ value for isooctane

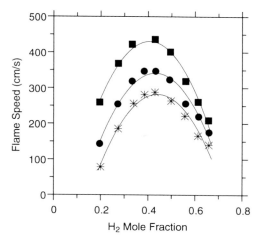

Fig. 6.10 Laminar flame speeds for atmospheric H_2/O_2 flames diluted with (*):N_2, ($\bullet$) AR, or ($\blacksquare$) He; Ratio of N_2: O_2 = AR:O_2 = He: O_2 = 3.76:1 (Reprinted with permission from Kwon and Faeth [8])

Table 6.2 Computed adiabatic flame temperatures and values of thermal diffusivity at 1 atm

Mixture	Adiabatic flame temperature (K)	Thermal diffusivity, α, at 1,300 K (cm²/s)
$H_2/O_2/N_2$	2,384	2.65
$H_2/O_2/Ar$	2,641	2.59
$H_2/O_2/He$	2,641	12.59

noble gases have the same heat capacity. However, due to the low molecular mass, the thermal diffusivity of helium is larger than that of argon and the flame speed increases further.

6.1.5 Dependence of Flame Thickness on Equivalence Ratio, Temperature and Pressure

Typically, flame thickness is about a few mm at ambient conditions. Since flame thickness scales as

$$\delta \approx \frac{\alpha}{S_L}, \tag{6.10}$$

its dependence on ϕ, T, and P can be deduced from the corresponding S_L dependence. Because the flame speed peaks near stoichiometric conditions and decreases in rich and lean mixtures, the flame thickness will have a U-shape dependence on ϕ. When the unburned gas temperature increases, one expects a smaller flame thickness. The pressure dependence is found using Eqs. 6.7 and 6.8 as

$$\delta \propto P^{-1} P^{-(a+b)/2+1} \propto P^{-(a+b)/2} \propto P^{-n/2} \tag{6.11}$$

For most fuels, the overall reaction order is positive ($n \sim 1$–1.5); therefore, flame thickness decreases with pressure. This has an important safety implication in preventing unwanted explosions as explained below.

6.2 Flammability Limits

As the combustible mixture gets too rich or too lean, the flame temperature decreases and consequently, flame speed drops significantly as sketched in Fig. 6.11. Eventually, the flame cannot propagate when the equivalence ratio is larger than an upper limit or smaller than a lower limit. These two limits are referred to as the rich and the lean flammability limits (RFL and LFL respectively), and they are often expressed as fuel percentage by volume in the mixture. These limits are also referred to as explosion limits in some engineering applications. For hydrocarbon fuels,

Fig. 6.11 Sketch of lean flammability limit (*LFL*) and rich flammability limit (*RFL*) (Reprinted with permission from Bosschaart and de Goey [4])

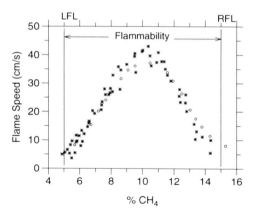

Table 6.3 Flammability at standard conditions (% of fuel by volume in mixture)

Fuel vapor	Lean limit	Rich limit	Fuel vapor	Lean limit	Rich limit
Hydrogen (H_2)	4	75	Isopropyl	2	12
Methane (CH_4)	5	15	Ethanol (C_2H_5OH)	3.3	19
Gasoline	1.4	7.6	n-Heptane (C_7H_{16})	1.2	6.7
Diesel	0.3	10	Iso-octane (C_8H_{18})	1	6.0
Ethane (C_2H_6)	3.0	12.4	Propane (C_3H_8)	2.1	9.5
n-Butane (C_4H_{10})	1.8	8.4	n-Pentane (C_5H_{12})	1.4	7.8
n-Hexane (C_6H_{14})	1.2	7.4	Dimethylether (C_2H_6O)	3.4	27

the mixture at the RFL contains about twice the amount of fuel compared to stoichiometric conditions. At the LFL, the mixture contains about half of the fuel as at stoichiometric. The flammability limits are often measured at ambient pressure using a tube with a spark plug at one end. When the temperature and pressure change, the flammability limits will also change because they affect the rate of the reaction. Adding inert or dilution gases to a combustible mixture will reduce the flammable region. Table 6.3 lists the flammability limits of some common fuels, and Appendix 5 contains a list of flammability limits of combustible gas mixtures in air or oxygen.

The information on flammability limits is quite useful in fire safety. For instance, flammability limits help in determining if storing a fuel in a tank is safe or not. Gasoline, for example, is quite volatile and therefore the vapor fills the gaseous space in storage tanks. The vapor pressure of gasoline varies with the season; the normal range is 48.2–103 kPa (7.0–15 psi) at ambient temperatures around 25°C. At the lower limit, the percentage of gasoline in the ullage[4] is about 48.2–101 kPa $\approx$ 48%, which is too rich to combust (the flammability limits of gasoline are 1.4%

---

[4] *Ullage* is widely used in industrial or marine settings to describe the empty space in large tanks or holds used to store or carry liquids.

and 7.5% by volume). However, when the tank is opened, the rich gasoline vapor starts to mix with surrounding air creating flammable gas mixtures. One must therefore exercise caution when opening a storage tank containing gasoline. Since the vapor pressure depends on temperature, the gasoline mixture in the storage tank may become flammable when the weather is really cold.

In contrast, diesel fuel and kerosene have low vapor pressure - about 0.05 kPa, or about 0.05% of air by volume in ambient conditions. This is below the lower flammability limit of No. 2 diesel (about 0.3% by volume). The upper limit is 10% by volume. Therefore, it is safe to store diesel fuels in a container at room temperatures around 25°C. Again, if the temperature increases, the vapor pressure can increase, leading to a flammable mixture of diesel fuel and air.

6.2.1 *Effects of Temperature and Pressure on Flammability Limits*

When either temperature or pressure increases, the range of flammable equivalence ratios widens. The effects of temperature and pressure on flammability limits are presented in Fig. 6.12. The left plot shows that RFL increases with temperature while LFL decreases with temperature; therefore the flammable region bounded by the RFL and LFL increases with temperature. Similar trends are observed for the effect of pressure on flammability limits as shown on the right plot of Fig. 6.12. For methane, the pressure is seen to have a more profound effect on the RFL than on the LFL.

6.3 Flame Quenching

A flame approaching a conducting material loses heat to the material, reducing the temperature of the reaction and consequently its reaction rate. If the heat losses are significant, the reaction may not be able to continue and the flame is quenched.

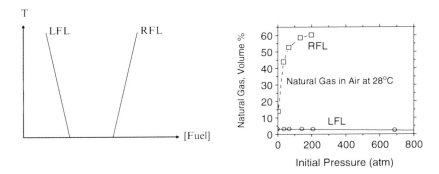

Fig. 6.12 Effect of temperature and pressure [13] on flammability limits

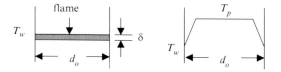

Fig. 6.13 *Left*: Sketch of a premixed flame propagating in a channel separated by two walls with distance, d_0. *Right*: temperature profile

The main physical effect lies in the balance between the heat generated by the combustion reaction and the heat lost to the adjacent material. Firemen pouring water on a fire is one of many examples of flame quenching encountered in life.

Flame quenching has many implications in combustion processes, from fire safety to pollutant emissions. An important parameter in the flame quenching process is the minimum distance at which a flame can approach a material surface before quenching. This distance is called the "quenching distance" and determines such parameters as the spacing in flame arrestors or the amount of unburned fuel left in the walls of an engine cylinder. Here, a simple analysis is used to determine the quenching distance. Let's consider a flame propagating into a channel with two walls separated by a distance d_0 in a two-dimensional region with unity depth as illustrated in Fig. 6.13.

The energy balance includes

Energy generated by the flame:

$$\dot{Q}_{generation} = V \cdot \dot{Q}''' = \delta \cdot d_0 \cdot 1 \cdot \hat{\dot{r}}_{fuel} \cdot \hat{Q}_c, \tag{6.12}$$

and *Energy loss via walls*:

$$\dot{Q}_{loss} = 2\delta \cdot 1 \cdot k \frac{T_p - T_{wall}}{d_0} \tag{6.13}$$

The criterion for flame quenching is $\dot{Q}_{loss} \geq \dot{Q}_{generation}$. By setting $\dot{Q}_{loss} = \dot{Q}_{generation}$, we have

$$\delta \cdot d_0 \cdot 1 \cdot \hat{\dot{r}}_{fuel} \cdot \hat{Q}_c = 2\delta \cdot 1 \cdot k \frac{T_p - T_{wall}}{d_0}$$

Solving for d_0:

$$d_0 = \sqrt{\frac{2k(T_p - T_{wall})}{\hat{\dot{r}}_{fuel} \cdot \hat{Q}_c}} \tag{6.14}$$

Equation 6.14 provides general guidance on the factors that influence d_0. The Flame arrestor shown in Fig. 6.14 is designed to stop unwanted flame propagation through a gas delivery system. Flammable gases pass through a metal grid, or mesh, which is generally designed with spacing smaller than the quenching distance for the conditions under consideration.

Fig. 6.14 Pictures of flame arrestors. *Left*: outside view, *Right*: inside of flame arrestor with screen in center, surrounded by small holes

It is useful to re-express the quenching distance in terms of the chemistry time so that we can identify any correlation between d_0 and the flame thickness δ. Again, using the relation $\hat{Q}_c \cdot [Fuel] = \rho_r c_p (T_p - T_r)$, we have $\hat{Q}_c = \rho_r c_p (T_p - T_r)/[Fuel]$. Substitution of this into Eq. 6.14 leads to

$$d_0 = \sqrt{\frac{2k[Fuel](T_p - T_{wall})}{\hat{r}_{fuel} \cdot \rho_r c_p (T_p - T_r)}} = \sqrt{2\alpha\tau_{chem}\frac{(T_p - T_{wall})}{(T_p - T_r)}} \tag{6.15}$$

$$\cong \sqrt{2\alpha\tau_{chem}} \text{ when } T_{wall} \approx T_r$$

Comparing Eq. 6.15 to that for flame thickness in Eq. 6.5, one obtains

$$d_0 \cong \sqrt{2\frac{(T_{ig} - T_r)}{(T_p - T_{ig})}}\delta = O(\delta) \tag{6.16}$$

This implies that quenching distance, d_0, is of the same order of magnitude as the flame thickness, i.e., several mm at ambient conditions. More importantly, d_0 has the same dependence on mixture, temperature, and pressure as δ. As shown in Fig. 6.15, the U-shape dependence of d_0 on equivalence ratio is similar to that for δ. Using the relation $\delta \propto P^{-n/2}$, one expects $d_0 \propto P^{-n/2}$ and such dependence is sketched in Fig. 6.16.

Experimental data of premixed flames against walls suggest the following relation

$$d_0 \cong 8\frac{\alpha}{S_L} \tag{6.17}$$

Using Eq. 6.6, we get

$$d_0 \cong 8\frac{T_{ig} - T_r}{T_p - T_{ig}}\delta \tag{6.18}$$

For methane-air combustion, $d_0 \sim 2.66\ \delta$ as shown in Fig. 6.15.

Fig. 6.15 Flame thickness
and quenching distance of
methane air versus
equivalence ratio (Reprinted
with permission from
Andrews and Bradley [1])

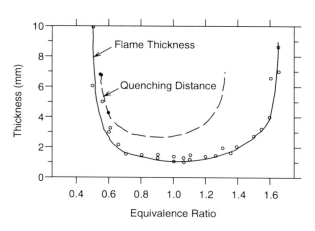

Fig. 6.16 Dependence of
quenching distance on
pressure (Reprinted with
permission from Green and
Agnew [7])

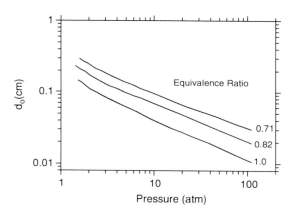

6.4 Minimum Energy for Sustained Ignition and Flame Propagation

In addition to the dependence of ignition on flame temperature as stated in Eq. 5.12, the success of an ignition process depends strongly on the mixture's ability to support flame propagation. Equation 5.12 can be extended to incorporate such effects leading to the following empirical approximation (for $u' < 2S_L$)

$$
\begin{aligned}
MIE &\approx \rho c_p \frac{\pi \delta^3}{6} (T_f - T_r) \left(\frac{10\alpha}{\delta \cdot (S_L - 0.16u')} \right)^3 \\
&= \frac{\rho c_p \pi}{6} (T_f - T_r) \left(\frac{10\alpha}{(S_L - 0.16u')} \right)^3,
\end{aligned}
\tag{6.19}
$$

Fig. 6.17 Minimum ignition energy variation with mixture composition with different turbulence velocities as computed by Eq. 6.19

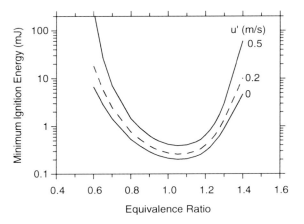

where α is the thermal diffusivity ($k/\rho c_p$), δ is the flame thickness, S_L is the laminar flame speed, and u' is the characteristic turbulence velocity. Note that Eq. 6.19 does not depend on the gap between electrodes and that α, ρ, and c_p are evaluated using properties of the reactants. Both the flame temperature and flame speed are functions of equivalence ratio, ϕ, with a bell shape. Due the cubic power of $1/(S_L-0.16u')$, the minimum ignition energy has a U-shape dependence on equivalence ratio. Figure 6.17 presents results obtained from Eq. 6.19 for methane-air combustion at ambient conditions with three turbulence velocities. The minimum ignition energy for methane shows a minimum of approximately 0.2 mJ at near stoichiometric conditions without turbulence; this estimate is reasonable in comparison with the experimental value of 0.3 mJ. Turbulence increases both flame propagation speed and heat transfer; however, the increase in heat transfer dominates the required energy for ignition. Hence, the net effect of turbulence increases the minimum ignition energy. For too lean or too rich mixtures, the mixture cannot be ignited and these two limits are called the lean and rich flammability limits.

Using Eq. 6.19, the variation of the MIE with combustion conditions can also be seen. Since $\alpha \propto P^{-1}$ and $S_L \propto P^{(n/2)-1}$, the pressure dependence of the MIE is $MIE \propto \left(P^{-1}/P^{n/2-1}\right)^3 \approx P^{-3n/2}$. For most hydrocarbon fuels, the minimum ignition energy decreases with pressure as exemplified in Fig. 6.18.

As temperature increases, density decreases while both the flame speed and the fuel vapor pressure increase. Hence, the fuel temperature can have a profound effect on MIE. For jet fuel, Fig. 6.19 indicates that an increase of 25°C results in almost a five order of magnitude reduction in MIE. Note that the LFL for jet fuel is about 3%, and near the LFL a large amount of energy is required to ignite the jet fuel-air mixture.

This drastic reduction in MIE is due primarily to the increase in vapor pressure of the jet fuel and the resulting equivalence ratio increase. Table 6.4 lists the effect of temperature on spark-ignition energy normalized by the value at 298 K for

Fig. 6.18 Minimum spark-ignition energy versus pressure showing a decreasing trend (Reprinted with permission from Blanc et al. [3])

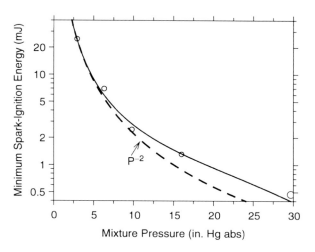

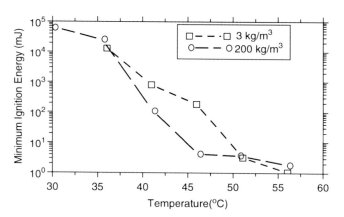

Fig. 6.19 Minimum ignition energy of jet fuel A versus temperature at 0.585 atm showing a drastic reduction with temperature [14]

Table 6.4 Effect of temperature on spark-ignition energy at 1 atm [6]

Fuel	T (K)	MIE(T)/MIE(298)	Fuel	T (K)	MIE(T)/MIE(298)
n-Heptane	298	1	Iso-octane	298	1
	373	0.46		373	0.41
	444	30.22		444	0.18
n-Pentane	243	5.76	Propane	233	1.58
	253	1.86		243	1.31
	298	1.0		253	1.14
	373	0.53		331	0.57
	444	0.30		356	0.49
				373	0.47
				477	0.19

Fig. 6.20 Minimum ignition energy versus 298 K/T (K) showing a linear correlation on a semi-log plot

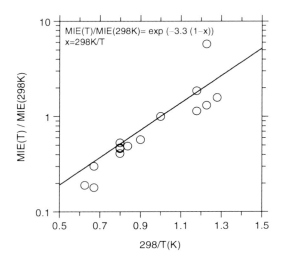

several fuels [6]. Data from the table are plotted in Fig. 6.20, showing a clear correlation between normalized MIE and $1/T$ on a semi-log scale as

$$MIE(T) = MIE(T_{298}) \exp\left(-3.3\left[1 - \frac{298}{T(K)}\right]\right) \qquad (6.20)$$

6.5 Turbulent Premixed Flames

Experimental observations reveal that premixed flames in turbulent flows propagate faster than their counterparts in laminar flows. The enhancement in flame propagation speed can be significant; turbulent flames can propagate two orders of magnitude faster than laminar flames.

6.5.1 Eddy Diffusivity

In turbulent flows, the transport processes of momentum, heat, and mass are enhanced by the motion of turbulent eddies. In analogy to laminar flows, the concept of 'eddy' diffusivity is introduced to represent the enhanced transport by turbulent eddies. For instance, in turbulent boundary layers, the following equations can be used to 'model' the effect of turbulence on transport as

Momentum Transfer

$$\tau_{total} = \rho(v + \varepsilon_M)\frac{\partial \bar{u}}{\partial y} \qquad (6.21)$$

Heat Transfer

$$q''_{total} = -\rho c_p (\alpha + \varepsilon_H) \frac{\partial \bar{T}}{\partial y} \tag{6.22}$$

Mass Transfer

$$m''_{total} = -\rho (D + \varepsilon_m) \frac{\partial \bar{Y}}{\partial y} \tag{6.23}$$

where the over bar signifies averaged values, and ε_M, ε_H, and ε_m denote the eddy diffusivities for momentum, heat, and mass transfer respectively. The transport coefficients are increased by the amount of turbulent diffusivity. In contrast to transport properties (ν, α, D) in laminar flows, eddy diffusivities are not properties of fluids. Eddy diffusivities depend on the turbulent flow itself. However, the simple eddy diffusivity concept permits us to have a rough estimate of the effect of turbulence on flame propagation.

6.5.2   Turbulent Flame Speed

The effect of turbulence on flame propagation may be classified based upon the type of interaction between turbulence and the flame. Several regimes can be classified on the basis of length, velocity, and chemical time scales. For instance, two interaction regimes have been proposed for the enhancement of flame speeds in turbulent flows:

1. Increased transport processes of heat and mass by small-scale turbulence.
2. Increased surface area due to wrinkling of the flame by large turbulent eddies.

Under the first regime, the scale of turbulence is small (less than the flame thickness), yet powerful enough to penetrate the preheat zone of a premixed flame. From Eq. 6.3, the laminar flame speed depends on transport properties as $S_L \sim \sqrt{\alpha} \sim \sqrt{D}$. Accordingly for turbulent flames, we have $S_T \sim \sqrt{D + \varepsilon}$. With a crude model for the eddy diffusivity as $\varepsilon \sim 0.01 \cdot D \cdot Re$, the ratio of turbulent flame speed to laminar flame speed at high Reynolds number is

$$\frac{S_T}{S_L} \sim \sqrt{\frac{D + \varepsilon}{D}} = \sqrt{\frac{D + 0.01 \cdot D \cdot Re}{D}} \tag{6.24}$$
$$= \sqrt{1 + 0.01 \cdot Re} \sim 0.1 \cdot Re^{1/2} \sim \sqrt{u'}$$

where $Re = u'l/\nu$ with u' being the characteristic turbulence velocity and l the associated length scale.

Next we consider the second regime: flame wrinkling by turbulence. Under this regime, turbulence is weak and its length scale is larger than the flame thickness. Turbulence affects the flame by 'wrinkling' the flame surface while the interior flame structure is the same as that of a laminar flame. Hence this regime is conventionally referred to as the wrinkled flamelet regime. The ratio of turbulent flame speed to laminar flame speed is proportional to the ratio of flame areas as

$$\frac{S_T}{S_L} \sim \frac{A_{turbulent}}{A_{la\,min\,ar}}. \quad laminar$$

One simple model to account for the wrinkled flame surface is

$$A_{turblent} \sim (1 + c_{emp}\frac{u'}{S_L})A_{la\,min\,ar} \qquad empirical\ constr \qquad (6.25)$$

where c_{emp} is an empirical constant. With this crude model, we have

$$\frac{S_T}{S_L} \sim \frac{A_{turbulent}}{A_{la\,min\,ar}} = 1 + c_{emp}\frac{u'}{S_L} \qquad (6.26)$$

Note that the dependence of $S_T/S_L \sim u'$ on the turbulence velocity is linear in the second regime while $S_T/S_L \sim \sqrt{u'}$ in the first regime. When turbulence is too powerful, such as when u' is much larger than S_L, flame extinction can occur; that is, the effect of aerodynamic strain rate causes the premixed flames to extinguish. For recent advancements in turbulent combustion, several books are available on this topic [5, 11, and 12].

6.6   Summary

rate of fuel burning

Flame speed:

$$S_L = \left\{\alpha\frac{|\hat{r}_{fuel}|}{[Fuel]}\frac{T_p - T_{ig}}{T_{ig} - T_r}\right\}^{1/2} = \left\{\frac{\alpha}{\tau_{chem}}\frac{T_p - T_{ig}}{T_{ig} - T_r}\right\}^{1/2} \text{where } \tau_{chem} = \frac{[Fuel]}{\hat{r}_{fuel,ave}}$$

Flame thickness:

$$\delta = S_L\tau_{chem} = \left\{\alpha\tau_{chem}\frac{T_p - T_{ig}}{T_{ig} - T_r}\right\}^{1/2} = \frac{\alpha}{S_L}\frac{T_p - T_{ig}}{T_{ig} - T_r}$$

T_p = flame (product) temperature
T_{ig} = ignition temperature (~ autoignition temperature)
T_r = reactant temperature

Flame quenching distance between parallel plates:

$$d_0 = 8 \frac{T_{ig} - T_r}{T_p - T_{ig}} \delta \approx 2\delta$$

Pressure effects:
With the global rate of progress expressed as

$$\dot{q}_{RxT} = A_o \exp\left(-\frac{E_a}{\bar{R}T}\right)[Fuel]^a[O_2]^b,$$

the following expressions can be derived, where $a + b = n$ is the total order of the reaction.

Effect of pressure on flame speed: $S_L \propto P^{(a+b)/2-1}$
Effect of pressure on flame thickness: $\delta \propto P^{-(a+b)/2}$
(Note that since $(a + b)$ is normally larger than zero, flame thickness is found to decrease with pressure for most hydrocarbon fuels).
Effect of pressure on Minimum Ignition Energy: $E_{ign} \propto P^{-3(a+b)/2+1}$

Exercises

6.1 For a propane/air adiabatic laminar premixed flame with single-step global kinetics, calculate the laminar flame speed S_L and flame thickness δ for an equivalence ratio $\phi = 0.7$. Assume a pressure of 1 atm, an unburned gas temperature of 300 K, a mean molecular weight of 29 g/mol, an average specific heat of 1.2 kJ/kg-K, an average thermal conductivity of 0.09 W/m-K, and a heat of combustion of 46 MJ/kg. The kinetics parameters you will need for propane (C_3H_8) are $a = 0.1$, $b = 1.65$, $T_{ig} = 743$ K, $E = 125.6$ kJ/mol, and $A = 8.6 \times 10^{11}$ cm$^{2.25}$/(s-mole$^{0.75}$). When calculating the reaction rate, be sure to evaluate the molar concentrations in units of moles/cm^3.

6.2 For a stoichiometric adiabatic laminar premixed propane flame with single-step global kinetics propagating through a gaseous mixture of fuel, oxygen, and nitrogen, how does the reaction rate $\mathbf{R}$ vary with the ratio $X_{N_2}^\infty/X_{O_2}^\infty$ where $X_{N_2}^\infty$ is the ambient nitrogen concentration and $X_{O_2}^\infty$ is the ambient oxygen concentration? In other words, indicate the proportionality $R \propto f(\psi)$ where $\psi = X_{N_2}^\infty/X_{O_2}^\infty$. Does the reaction rate increase or decrease with increasing ψ and why?

6.3 A flame arrestor (a plate with small circular holes) is to be installed in the outlet of a vessel containing a stoichiometric mixture of propane and air, initially at 20°C and 1 atm, to prevent the potential of flame propagation (flashback) to the interior of the vessel. (a) Calculate the diameter of the flame arrestor holes. (b) Based on your previous calculations, estimate the hole diameter if the pressure is 5 atm. (c) From a safety point of view, would you change the hole diameter of the flame arrestor if the mixture is made richer or leaner? (explain).

6.4 The pilot light has blown out on your gas heater at home. Your heater is defective so natural gas continues to enter your home. The natural gas (assume 100% methane) enters at a rate of 30 L/s. If your house has a volume of 350 m^3, how long will it be before your house is in danger of blowing up (lean limit)? How much longer until it is no longer in danger of blowing up (rich limit)? Assume the gases are always perfectly mixed and that methane is flammable in air for methane concentrations between 5% and 15% by volume.

References

1. Andrews GE, Bradley D (1972) The burning velocity of methane-air mixtures. Combustion and Flame 19(2):275-288.
2. Bayraktar H (2005) Experimental and theoretical investigation of using gasoline-ethanol blends in spark-ignition engines. Renewable Energy 30:1733-1747.
3. Blanc MV, Guest PG, von Elbe G, Lewis B (1947) Ignition of explosive gas mixtures by electric sparks. I. Minimum ignition energies and quenching distances of mixtures of methane, oxygen, and inert gases. Journal of Chemical Physics 15(11): 798-802 (1947).
4. Bosschaart KJ, de Goey LPH (2003) Detailed analysis of the heat flux method for measuring burning velocity. Combustion and Flame 132:170–180.
5. Cant RS, Mastorakos E (2008) An Introduction to Turbulent Reacting Flows. London Imperial College Press, London.
6. Fenn JB (1951) Lean Flammability limit and minimum spark ignition energy. Industrial & Engineering Chemistry 43(12):2865-2868.
7. Green KA, Agnew JT (1970) Quenching distances of propane-air flames in a constant-volume bomb. Combustion and Flame 15:189-191.
8. Kwon OC, Faeth GM (2001) Flame/stretch interactions of premixed hydrogen-fueled flames: measurements and predictions. Combustion and Flame 124: 590-610.
9. Law CK (2007) Combustion at a Crossroads: status and prospects. Proceedings of the Combustion Institute 31:1-29.
10. Mallard E, Le Chatelier H (1883) Combustion des melanges gaseux explosives. Annals of Mines 4:379-568.
11. Peters N (2000) Turbulent Combustion. Cambridge University Press, Cambridge.
12. Poinsot T, Veynante D (2005) Theoretical and Numerical Combustion. R.T. Edwards, Inc, Philadelphia.
13. Zabetakis MG (1965) Flammability characteristics of combustible gases and vapors. Bulletin 627, Bureau of Mines, Pittsburgh.
14. (1998) A review of the flammability hazard of Jet A fuel vapor in civil aircraft fuel tanks. DOT/FAA/AR-98/26.

Chapter 7
Non-premixed Flames (Diffusion Flames)

In many combustion processes, the fuel and oxidizer are separated before entering the reaction zone where they mix and burn. The combustion reactions in such cases are called "non-premixed flames," or traditionally, "diffusion flames" because the transport of fuel and oxidizer into the reaction zone occurs primarily by diffusion. A candle flame is perhaps the most familiar example of a non-premixed (diffusion) flame. Many combustors operate in the non-premixed burning mode, often for safety reasons. Since the fuel and oxidizer are not premixed, the risk of sudden combustion (explosion) is eliminated. Chemical reactions between fuel and oxidizer occur only at the molecular level, so "mixing" between fuel and oxidizer must take place before combustion. In non-premixed combustion the fuel and oxidizer are transported independently to the reaction zone, primarily by diffusion, where mixing of the fuel and oxidizer occurs prior to their reaction. Often the chemical reactions are fast, hence the burning rate is limited by the transport and mixing process rather than by the chemical kinetics. Consequently, greater flame stability can be maintained. This stable characteristic makes diffusion flames attractive for many applications, notably aircraft gas-turbine engines.

7.1   Description of a Candle Flame

A candle, as shown in Fig. 7.1, illustrates the complicated physical and chemical processes involved in non-premixed combustion. The flame surface is where vaporized fuel and oxygen mix, forming a stoichiometric mixture. At the flame surface, combustion leads to high temperatures that sustain the flame. The elements of the process are:

- Heat from the flame melts wax at the base of the candle flame.
- Liquid wax moves upward by capillary action, through the wick towards the flame.
- Heat from the flame vaporizes the liquid wax.
- Wax vapors migrate toward the flame surface, breaking down into smaller hydrocarbons.
- Ambient oxygen migrates toward the flame surface by diffusion and convection.

S. McAllister et al., *Fundamentals of Combustion Processes*,
Mechanical Engineering Series, DOI 10.1007/978-1-4419-7943-8_7,
© Springer Science+Business Media, LLC 2011

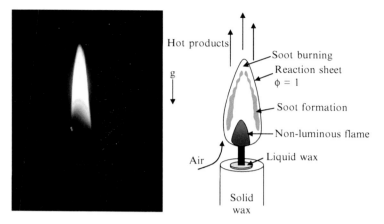

Fig. 7.1 *Left*: The simple appearance of a candle flame masks complicated processes. *Right*: Associated physical processes and the effect of buoyancy on a typical candle flame

Buoyant convection develops when the hot, less dense air around the flame rises as sketched in right plot of Fig. 7.1. This buoyant convective flow simultaneously transports oxygen to the flame and combustion products away from the flame. The resulting flame is shaped like a teardrop; elongated in the direction opposite to the gravitational force that is pointed downward. The flame's yellow section is the result of the solid particles of soot—formed between the flame and the wick—burning as they move through the flame.

7.2 Structure of Non-premixed Laminar Free Jet Flames

Non-premixed jet flames are well characterized and are very helpful in understanding the important characteristics of a typical non-premixed flame including its structure, flame location, flame temperature, and overall flame length (flame height). The right of Fig. 7.2 shows non-premixed jet flames using ethylene, JP-8, and methane. The fuel is issued from a nozzle into surrounding air. Combustion is initiated by a pilot and once the flame is stabilized, the ignition source is removed. The characteristics of a jet flame are similar to that of a candle flame except in the case of a jet flame, the fuel is already gasified and is injected into the air at a predetermined speed. The left of Fig. 7.2 presents a typical temperature distribution for a non-premixed free jet flame obtained from computer simulation. Only half of the jet is shown here as the jet is assumed to be axisymmetric. The fuel is issued from a pipe of 1 cm diameter and the overall flame height is about 2.5 cm. The measured species mole fractions and temperature along a horizontal line are shown in Fig. 7.3.

As Fig. 7.3 shows, the mass fraction of fuel decreases from unity at the centerline to zero at the flame location. Beyond $r > r_{Flame}$, the fuel mass fraction is zero

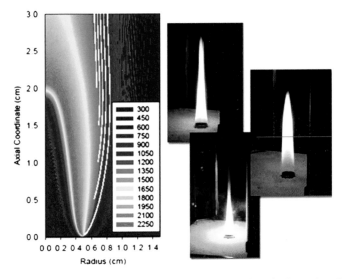

Fig. 7.2 *Left*: Computed temperature distribution of a non-premixed jet flame (graphic courtesy of Dr. Linda Blevins). *Right*: Laminar diffusion flames in air of ethylene (*left*), JP-8 surrogate (*center*), and methane (*right*) (Reprinted with permission from Sandia National Laboratories)

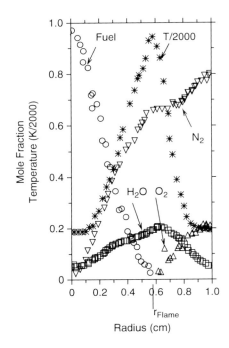

Fig. 7.3 Experimental data of species and temperature profiles in a laminar flame. The flame sheet is located approximately at 0.6 cm from the centerline (Reproduced with permission from Smyth et al. [4])

because chemistry is so fast that all of the fuel is consumed at the flame surface. The mass fraction of oxidizer decreases from its value in the surrounding fluids to zero at r_{Flame}. There is no oxidizer in the region where $r < r_{Flame}$. The product species have nonzero values at the centerline due to accumulation of products from upstream. The mass fraction of products has its maximum located at r_{Flame}. Since product species and heat have similar transport and production processes, temperature has a profile similar to that of product mass fraction. As will be estimated later, chemical kinetics are usually much faster than diffusion processes, so the reaction zone is concentrated near $r \approx r_{Flame}$. Only in this area do fuel and oxidizer co-exist prior to reaction. Temperature is highest here, leading to fast chemical reactions. As in premixed flames, the different species become molecularly excited and emit visible radiation, giving the color of the flame. The outer zone of the reaction is of a bluish color due to the radiation of CH radicals. The inner zone of the reaction is reddish due to C_2 and soot radiation. Generally, the later dominates, giving most diffusion flame reactions the same color as is commonly observed in candle flames.

The mass fraction gradients resulting from the consumption of fuel and oxidizer at the reaction zone drive the diffusion transport of fuel and oxidizer toward the flame where they mix and react. The mass flux of the fuel or oxidizer toward the reaction zone is determined by Fick's law of mass diffusion. If one of the mass gradients, let's say oxygen, increases for any reason, then the mass flux of oxygen into the reaction zone will increase. Since there is added oxygen in the reaction, more fuel will be consumed and the reaction will move toward the fuel side, increasing the gradient of fuel mass fraction. A similar event will occur if the fuel concentration is increased. As a consequence, the flame will always position itself such that the mass fluxes of fuel and oxidizer entering the reaction zone are at stoichiometric conditions. This is an important aspect of diffusion flames since it determines their shape and, as will be seen later, their emission characteristics.

7.3 Laminar Jet Flame Height (L_f)

The length, or height, of a non-premixed flame is an important property indicating the size of a flame. Current computer simulations can accurately predict diffusion flame structure and behavior; however, some of the parameters controlling the behavior of non-premixed jet flames can be determined simply by using non-dimensional analysis. Considering a simple free jet flame sketched in Fig. 7.4, a non-dimensional analysis of the species and energy equations using various scales characteristic of the flame is developed below.

Energy equation:

$$\rho u c_p \frac{\partial T}{\partial x} = \frac{k}{r} \frac{\partial}{\partial r} \left(r \frac{\partial T}{\partial r} \right) + \hat{r}_{fuel} \hat{Q}_c \tag{7.1}$$

Fig. 7.4 Sketch of a simple free jet flame

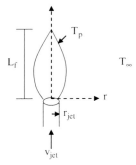

Species (fuel) mass fraction:

$$\rho u \frac{\partial y_f}{\partial x} = \frac{\rho D}{r} \frac{\partial}{\partial r} \left(r \frac{\partial y_f}{\partial r} \right) + \hat{r}_{fuel} M_f, \tag{7.2}$$

where $\hat{r}_{fuel}$ is the fuel consumption rate ($mol/cm^3 - s$), $\hat{Q}_c$ is the heat of combustion per mole of fuel burned, and M_f is the molecular mass of fuel. Defining non-dimensional quantities as

$$\bar{x} \equiv \frac{x}{L_f}, \bar{r} \equiv \frac{r}{r_{jet}}, \bar{T} \equiv \frac{T - T_\infty}{T_p - T_\infty}, \bar{y}_f \equiv \frac{y_f}{y_{f,s}}, \bar{u} \equiv \frac{u}{V_{jet}},$$

where $y_{f,s}$ denotes the fuel mass fraction of a stoichiometric mixture, r_{jet} is the fuel jet radius, and V_{jet} the jet velocity. With these non-dimensional quantities and the two relations: $\hat{Q}_c[fuel]_s = \rho c_p(T_p - T_\infty)$ and $[fuel]_s = \rho y_{fs}/M_f$, Eqs. 7.1 and 7.2 reduce to

$$\frac{V_{jet}}{L_f} \bar{u} \frac{\partial \bar{T}}{\partial \bar{x}} = \frac{\alpha}{r_{jet}^2} \frac{1}{\bar{r}} \frac{\partial}{\partial \bar{r}} \left(\bar{r} \frac{\partial \bar{T}}{\partial \bar{r}} \right) + \frac{\hat{r}_{fuel}}{[fuel]_s},$$

$$\text{or} \quad \frac{1}{\tau_{conv}} \bar{u} \frac{\partial \bar{T}}{\partial \bar{x}} = \frac{1}{\tau_{diff}} \frac{1}{\bar{r}} \frac{\partial}{\partial \bar{r}} \left(\bar{r} \frac{\partial \bar{T}}{\partial \bar{r}} \right) + \frac{1}{\tau_{chem}} \tag{7.3}$$

$$\bar{u} \frac{\partial \bar{T}}{\partial \bar{x}} = \underbrace{\frac{\tau_{conv}}{\tau_{diff}}}_{group1} \frac{1}{\bar{r}} \frac{\partial}{\partial \bar{r}} \left(\bar{r} \frac{\partial \bar{T}}{\partial \bar{r}} \right) + \underbrace{\frac{\tau_{conv}}{\tau_{chem}}}_{group2}$$

and

$$\bar{u} \frac{\partial \bar{y}_f}{\partial \bar{x}} = \underbrace{\frac{\tau_{conv}}{\tau_{diff}}}_{group1} \frac{1}{\bar{r}} \frac{\partial}{\partial \bar{r}} \left(\bar{r} \frac{\partial \bar{y}_f}{\partial \bar{r}} \right) + \underbrace{\frac{\tau_{conv}}{\tau_{chem}}}_{group2} \tag{7.4}$$

There are two distinct groups appearing in Eqs. 7.3 and 7.4. Let's examine the time scales associated with each group. $\tau_{conv} = L_f/V_{jet}$ represents the convective time scale for the jet flame; $\tau_{diff} = r_{jet}^2/D$ is the diffusive time scale for the oxidizer to diffuse to the jet centerline; $\tau_{chem} = [\text{fuel}]_S/\hat{r}_{fuel}$ is the chemistry time. Group 2 contains the ratio between the convective time and chemistry time. This ratio is referred to as the Damköhler number. It becomes infinity for infinitely fast chemistry, indicating that transport processes control the characteristics of these flames. Group 1 is the ratio between the convective time and the diffusive time. At the flame tip, these two times are approximately equal such that

$$L_f \propto \frac{V_{jet}r_{jet}^2}{D} \propto \frac{\dot{V}_{fuel}}{D}. \tag{7.5}$$

For a given fuel and oxidizer (i.e., fixed mass diffusivity D), Eq. 7.5 implies that the flame height increases linearly with the volumetric flow rate ($\dot{V}_{fuel}$). Such a linear dependence is indeed observed in experiments. When the surrounding oxidizer stream contains inert gases, the simple estimate of diffusion time as $\tau_{diff} = r_{jet}^2/D$ is insufficient to account for the dilution effects. For instance, the photos in Fig. 7.5 show methane jet flames with three different surrounding fluids: air, 50% oxygen/50% nitrogen, and pure oxygen. It is clear that the jet flame heights with pure oxygen are much shorter than those with air as a surrounding fluid. Therefore, the flame height also depends on the fuel/oxidizer type through the overall stoichiometry as will be discussed in Section 7.4.

Example 7.1 Estimate the different time scales for a methane non-premixed jet flame with the following information: $L_f = 50$ mm, $\dot{V}_{fuel} = 5.0$ cc/s, $r_{jet} = 0.50$ cm, $P = 1$ atm, $T_\infty = 300$ K.

Solution:
Using $V_{jet} = \dot{V}_{fuel}/\pi r_{jet}^2 = 6.4$ cm/s and diffusivity of air evaluated at 1,000 K, $D = 0.2$ cm^2/s

(a) diffusion time $\tau_{diffusion} \approx r_{jet}^2/D = 1.25$ s
(b) convective time $\tau_{convective} \approx L_f/V_{jet} = 0.79$ s
(c) chemistry time $\tau_{chemistry} \approx [Fuel]/\hat{r}_{fuel}$

$$CH_4 + 2(O_2 + 3.76N_2) \rightarrow CO_2 + 2H_2O + 7.52N_2$$

Fig. 7.5 Natural gas diffusion jet flames surrounded by different gas mixtures: *Left*: air; *Middle*: 50% oxygen/50% nitrogen; *Right*: 100% oxygen (Reproduced with permission from Lee et al. [3])

$x_{CH4} = 0.095$ and $x_{O2} = 0.19$ and we estimate the rate at the peak temperature $T = 2,300$ K as

$$[CH_4] = x_{CH_4} \frac{P}{R_u T} = 0.095 \frac{1}{82.05 \cdot 2300} = 5.48 \cdot 10^{-7} \text{ mol/cc}$$

$$[O_2] = 2[CH_4] = 1.1 \cdot 10^{-6} \text{mol/cc}$$

$$\frac{d[CH_4]}{dt} = -1.3 \cdot 10^9 \cdot \exp\left(-\frac{48,400}{1.987 \cdot 2300}\right) \cdot (5.5 \cdot 10^{-7})^{-0.3} \cdot (1.1 \cdot 10^{-6})^{1.3}$$

$$= 0.0443 \text{ mol/cc} - \text{s}$$

$$\tau_{chemistry} \approx [Fuel]/\hat{r}_{fuel} = \frac{5.5 \cdot 10^{-7} \text{mol/cc}}{0.0443 \text{mol/cc} - \text{s}} = 1.24 \cdot 10^{-5} \text{ s}$$

Damköhler number $\approx 10^5$, confirming that the combustion process is limited by diffusion.

7.4 Empirical Correlations for Laminar Flame Height

The flame height also depends on the fuel type through its stoichiometry. This is not accounted for in Eq. 7.5 above. For practical estimation of flame height, a semi-empirical correlation can be used:

$$L_f = \frac{\dot{V}_{fuel}(T_\infty/T_f)}{4\pi D_\infty \ln(1 + 1/S)} \left(\frac{T_\infty}{T_p}\right)^{0.67}$$

$$\approx \frac{\dot{V}_{fuel}(T_\infty/T_f)}{4\pi D_\infty} \left(\frac{T_\infty}{T_p}\right)^{0.67} \cdot S \text{ when } S \text{ is large}$$

(7.6)

where T_∞ = oxidizer temperature (K)

T_p = mean flame temperature (K)

T_f = fuel temperature (K)

S = molar stoichiometric air/fuel ratio

D_∞ = mean diffusion coefficient evaluated at T_∞ (m^2/s)

$\dot{V}_{fuel}$ = volumetric flow rate of fuel (m^3/s)

L_f = flame height (m)

The molar stoichiometric air/fuel ratio S is evaluated as

$$S = \begin{cases} 4.76 \cdot \left(\alpha + \dfrac{\beta}{4} - \dfrac{\gamma}{2}\right) & \text{for fuel } C_\alpha H_\beta O_\gamma \text{ burning with air} \\ (1 + x_{N_2}/x_{O_2}) \cdot \left(\alpha + \dfrac{\beta}{4} - \dfrac{\gamma}{2}\right) & \text{buring with variable } O_2 \text{ content} \end{cases}$$

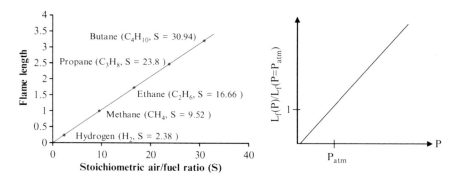

Fig. 7.6 Flame height increases with fuel complexity and with ambient pressure

In addition to the embedded physics in Eq. 7.5, Eq. 7.6 also includes the dependence of L_f on fuel type. When S is large, L_f scales linearly with S. Since $D_\infty \sim 1/P$, L_f increases with pressure linearly. These dependences are sketched in Fig. 7.6.

Example 7.2 Estimate the flame height of a laminar propane jet flame at $P = 1$ atm and $T_f = T_{air} = 300$ K. The mass flow rate of fuel is $2.7 \cdot 10^{-6}$ kg/s and the density of propane is 1.8 kg/m^3. The flame temperature is assumed to be 2,400 K and the mean diffusivity is $2.84 \cdot 10^{-5}$ m^2/s.

Solution:

Using $L_f = \frac{\dot{V}_{fuel}(T_\infty/T_f)}{4\pi D_\infty \ln(1+1/S)} \left(\frac{T_\infty}{T_p}\right)^{0.67}$ where $S = 4.76(3 + 8/4 - 0/2) = 23.8$, $T_f = T_\infty = 300$ K,

$$\dot{V}_{fuel} = 2.7 \cdot 10^{-6}(\text{kg/s})/2.8(\text{kg/m}^3) = 1.5 \cdot 10^{-6}(\text{m}^3/\text{s})$$

$$L_f = \frac{1.5 \cdot 10^{-6}(300/300)}{4\pi 2. \cdot 10^{-5} \ln(1 + 1/23.8)} \left(\frac{300}{2400}\right)^{0.67} = 0.036\text{m} = 3.6\text{cm}$$

Example 7.3 A methane non-premixed free jet is used as a pilot flame in a furnace. Estimate the fuel volumetric flow rate and heat release rate with the following information: $L_f = 5$ cm, $P = 1$ atm, $T_\infty = T_f = 300$ K, and $T_p = 2,400$ K.

Solution:
Using the diffusivity at $T = 300$ K (0.2 cm^2/s) and $S = 2 \cdot 4.76 = 9.52$, the volumetric flow rate is

$$\dot{V}_{fuel} = \frac{L_f 4\pi D_\infty \ln(1 + 1/S)}{(T_\infty/T_f)\left(\frac{T_\infty}{T_p}\right)^{0.67}}$$

$$= \frac{5 \cdot 4 \cdot 3.1415926 \cdot 0.2 \cdot \ln(1 + 1/9.52)}{1(1/8)^{0.67}}$$

$$= 5.06 \text{ cc/s} = 5.06 \cdot 10^{-3}\text{Liter/s}$$

The heat release rate is determined as follows. Using the ideal gas law $V/N = R_u T/P = 24.65$ L/mol. The mass flow rate of the jet flame is

$$\dot{m}_{fuel} = \dot{V}_{fuel}/24.65 \cdot M_{CH4} = 2.03 \cdot 10^{-4} \text{mol/s} \cdot 16 = 3.25 \cdot 10^{-3} \text{g/s}$$

With LHV $= 50.058$ J/g, the heat release rate is

$$LHV \cdot \dot{m}_{fuel} = 50.058 \text{ J/g} \cdot 3.25 \cdot 10^{-3} \text{g/s} = 162.6 \text{ J/s} = 162.6 \text{ W}$$

7.5   Burke-Schumann Jet Diffusion Flame

When a jet of fuel is issued into a tube, the amount of oxidizer available for combustion is controlled by the volumetric flow rate of the surrounding fluids. Unlike a jet issued into an infinite surrounding fluid, the entrainment of oxidizer into the jet is limited. Such a flame is sketched in Fig. 7.7 where r_{fuel} and r_{tube} are the radii of the inner fuel jet and the outer tube respectively.

In this particular confined flame, the volumetric flow rates of the fuel and surrounding fluid are fixed, while the oxygen content (y_{O2}) of the surrounding fluid is varied to create different flame shapes. Let's consider different situations for a general hydrocarbon/oxygen system such as

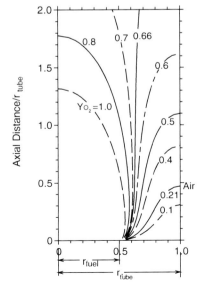

Fig. 7.7 Burke-Schumann diffusion flame: the shape of the jet flame depends on the oxidizer content in the coflowing fluids between r_{fuel} and r_{tube}

$$C_\alpha H_\beta O_\gamma + \left(\alpha + \frac{\beta}{4} - \frac{\gamma}{2}\right)\left(O_2 + \frac{x_{N_2}}{x_{O_2}}N_2\right) \rightarrow$$
$$\alpha CO_2 + \frac{\beta}{2}H_2O + \left(\alpha + \frac{\beta}{4} - \frac{\gamma}{2}\right)\frac{x_{N_2}}{x_{O_2}}N_2 \tag{7.7}$$

where the content of oxygen in the surrounding fluids is varied by adjusting the ratio x_{N2}/x_{O2} ($= 3.76$ for air), where x_i denotes the mole fraction of the i-th species. The surrounding fluids will be referred to as the oxidizer stream. The mass fraction of oxygen in the oxidizer stream is

$$y_{O_2} = \frac{M_{O_2}}{M_{O_2} + \frac{x_{N_2}}{x_{O_2}}M_{N_2}} = \frac{1}{1 + \frac{x_{N_2}}{x_{O_2}}\frac{M_{N_2}}{M_{O_2}}} \tag{7.8}$$

The ratio x_{N_2}/x_{O_2} can be expressed in terms of y_{O_2} as

$$\frac{x_{N_2}}{x_{O_2}} = \left(\frac{1}{y_{O_2}} - 1\right)\frac{M_{O_2}}{M_{N_2}} \tag{7.9}$$

The stoichiometric oxygen/fuel ratio (OFR_{st}) based on moles (volume) is

$$OFR_{st} = \left(\frac{\dot{n}_{O_2}}{\dot{n}_{fuel}}\right)_{sto} = \left(\frac{\dot{V}_{O_2}}{\dot{V}_{fuel}}\right)_{st} = \alpha + \frac{\beta}{4} - \frac{\gamma}{2} \tag{7.10}$$

The volumetric flow rate of oxygen is

$$\dot{V}_{O_2} = x_{O_2} \cdot \dot{V}_{oxidizer} = \frac{x_{O_2}}{x_{O_2} + x_{N_2}}\dot{V}_{oxidizer} = \frac{\dot{V}_{oxidizer}}{1 + x_{N_2}/x_{O_2}}, \tag{7.11}$$

where $\dot{V}_{oxidizer}$ is the volumetric flow rate of the oxidizer stream with the units of (cc/s). Since the jet contains 100% fuel, the oxygen/fuel ratio (OFR) based on molar (volumetric) flow rate is

$$OFR = \left(\frac{\dot{n}_{O_2}}{\dot{n}_{fuel}}\right) = \frac{\dot{V}_{oxidizer}/\dot{V}_{fuel}}{1 + x_{N_2}/x_{O_2}}$$
$$= \frac{\dot{V}_{oxidizer}}{\dot{V}_{fuel}}\frac{y_{O_2}}{y_{O_2} + (1 - y_{O_2})M_{O_2}/M_{N_2}} \tag{7.12}$$

Different flame shapes are developed depending on the ratio OFR/OFR$_{st}$ as follows:

(1) When OFR/OFR$_{st} > 1$, the oxidizer stream supplies more oxygen than needed for stoichiometric combustion. The flame is called "over ventilated" and it has a shape similar to a free jet flame as all the fuel will be consumed. In Fig. 7.7, the

ratio $\dot{V}_{oxidizer}/\dot{V}_{fuel}$ is fixed, and over-ventilated flames are developed when $y_{O2} > 0.66$.

(2) When $OFR/OFR_{st} = 1$, the oxidizer stream supplies just the right amount of oxygen for stoichiometric combustion. The flame surface becomes parallel to the axial direction as seen in Fig. 7.7 with $y_{O2} = 0.66$.

(3) When $OFR/OFR_{st} < 1$, the oxidizer stream supplies less oxygen than needed for stoichiometric combustion. The flame is called "under ventilated" and it has a shape similar to the mouth of a trumpet, as not all the fuel is consumed. In Fig. 7.7, the ratio $\dot{V}_{oxidizer}/\dot{V}_{fuel}$ is fixed and under-ventilated flames are developed when $y_{O2} < 0.66$.

Note that in most combustion systems, air is used as the oxidizer stream. According to Eq. 7.12, OFR/OFR_{st} can be changed by changing the ratio $\dot{V}_{oxidizer}/\dot{V}_{fuel}$ for a given fuel.

Example 7.4 Determine the flame shape of a methane Burke-Schumann diffusion flame with air as the oxidizer stream. The volumetric flow rates are: $\dot{V}_{oxidizer} = 23$ cc/s and $\dot{V}_{fuel} = 5$ cc/s. The fuel and oxidizer streams have the same temperature and pressure.

Solution:

$$OFR = \left(\frac{\dot{n}_{O_2}}{\dot{n}_{fuel}}\right) = \frac{\dot{V}_{oxidizer}/\dot{V}_{fuel}}{1 + x_{N_2}/x_{O_2}} = \frac{23/5}{1 + 3.76} = 0.966$$

$$OFR_{st} = \left(\frac{\dot{n}_{O_2}}{\dot{n}_{fuel}}\right)_{sto} = \alpha + \frac{\beta}{4} - \frac{\gamma}{2} = 1 + \frac{4}{4} - 0 = 2$$

Since $OFR/OFR_{st} < 1$, the flame is under ventilated.

7.6 Turbulent Jet Flames

As the Reynolds number of the jet flame, $Re = 2 \cdot V_{jet} \cdot r_{jet}/v$, increases to a critical value, the laminar jet flame becomes unstable, eventually transitioning into a turbulent flame. The jet starts the transition process to full turbulence when the Reynolds number is large ($\sim 10^3$ [2]). Figure 7.8 sketches experimental observations of the evolution of the flame height versus jet velocity. Before the jet becomes unstable, the flame height increases linearly with jet velocity. When the jet becomes unstable, the flame height stops growing and starts to decrease. As the jet becomes fully turbulent, the flame height is independent of jet velocity.

The following rationale is used to explain such an observation. Since turbulence enhances mixing between the fuel and oxidizer, a similar expression as Eq. 7.5 is used to scale the flame height as

Fig. 7.8 Flame height versus
jet nozzle velocity. Height has
a linear dependence when the
jet velocity is below a certain
value. The flame height
becomes independent of jet
velocity when the velocity
is sufficiently high and
reaches the fully turbulent
regime (Reproduced with
permission from Hottel
and Hawthorne [2])

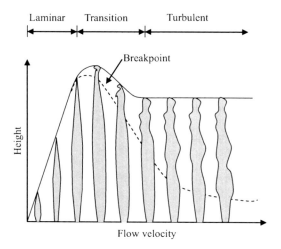

$$L_f \propto \frac{r_{jet}^2 V_{jet}}{D_t}, \tag{7.13}$$

where D_t is turbulent diffusivity, which is the only difference between Eq. 7.13 and Eq. 7.5. It is theorized that D_t has the following scaling relation

$$D_t \propto r_{jet} V_{jet}. \tag{7.14}$$

This relation is based on dimensional analysis, as D_t should have the dimension of Length^2/time. The jet is characterized by two important physical parameters, namely its size and velocity. With Eqs. 7.14 and 7.13 becomes

$$L_f \propto \frac{r_{jet}^2 V_{jet}}{r_{jet} V_{jet}} \propto r_{jet}. \tag{7.15}$$

The following empirical formula has been developed for the estimation of turbulent jet flames with hydrocarbon fuels burning with air:

$$\frac{L_f}{d_{jet}} = 6 \left(\frac{1}{f_s} + 1 \right) \sqrt{\frac{\rho_{fuel}}{\rho_{flame}}}, \tag{7.16}$$

where ρ_{fuel} and ρ_{flame} are the densities of fuel and flame, and f_s is the stoichiometric fuel-air mass ratio.

Example 7.5 Estimate the flame length of a fully developed turbulent methane-air jet flame. The adiabatic flame temperature is 2,400 K and the temperatures of fuel and air are 300 K. The diameter of the fuel jet is 7 mm.

Solution:

Use Eq. 7.16: $\frac{L_f}{d_{jet}} = 6\left(\frac{1}{f_s} + 1\right)\sqrt{\frac{\rho_{fuel}}{\rho_{flame}}}$.

The stoichiometry of methane-air combustion is

$$CH_4 + 2(O_2 + 3.76N_2) \rightarrow CO_2 + 2H_2O + 3.76N_2$$

$$f_s = \frac{16}{2(32 + 3.76 \cdot 28)} = 0.058$$

At 300 K,

$$\rho_{CH_4} = \frac{P}{\hat{R}_{CH_4}T} = \frac{1atm}{82.0574(atm \cdot cm^3/mol - K)/16(g/mol) \cdot 300(K)}$$
$$= 6.5 \cdot 10^{-4}g/cm^3 = 0.65kg/m^3$$

At the flame, the mixture consists of mostly air; therefore we estimate the density simply by scaling the density of air at 300 K to 2,400 K as

$$\rho_{flame} = \rho_{air@300K}\frac{300K}{2400K} = 0.1475kg/m^3$$

$$\frac{L_f}{d_{jet}} = 6\left(\frac{1}{0.058} + 1\right)\sqrt{\frac{0.65kg/m^3}{0.1475kg/m^3}} = 229.8$$

$$\rightarrow L_f = 229.8 \; d_{jet} = 1608.3 \; mm = 1.608 \; m$$

7.6.1 Lift-Off Height (h) and Blowout Limit

Experimentally, it is observed that when the velocity of a jet increases to a point, the flame lifts off of the nozzle. Further increase in jet velocity leads to total flame blow out. This effect is related to the fact that when the jet velocity is increased, the lower portion of the flame that anchors the flame to the jet nozzle cannot propagate against the flow. Because there is a gap between the reaction and the nozzle tip, the fuel and air mix together and the flame in this area is similar to a premixed one. Thus it is expected that the conditions for lift off should be determined by the relative magnitude of the jet velocity and the premixed flame speed. It is found experimentally that the lift-off height, h, and the blow out jet velocity are correlated by the following semi-empirical expression proposed by Gautam [1]:

Lift-off height:

$$h = 50 \cdot v_{jet}\frac{V_{jet}}{S_{L,max}^2}\left(\frac{\rho_{jet}}{\rho_\infty}\right)^{1.5} \tag{7.17}$$

Blowout jet velocity:

$$V_{jet,blowout} = S_{L,max} \left(\frac{\rho_\infty}{\rho_{jet}}\right)^{1.5} 0.17 Re_H (1 - 3.5 \cdot 10^{-6} Re_H), \qquad (7.18)$$

where

$$Re_H = \frac{\rho_{jet} S_{L,max} H}{\mu_{jet}},$$

and

$$H = 4 d_{jet} \left[\frac{y_{f,jet}}{y_{f,sto}} \left(\frac{\rho_{jet}}{\rho_\infty}\right)^{0.5} - 5.8\right],$$

$y_{f,jet}$ is the mass fraction of fuel from the jet and $y_{f,sto}$ is the mass fraction of fuel in the stoichiometric mixture. v_{jet} is the kinematic viscosity of the jet fluid and $S_{L,max}$ is the maximum laminar flame speed.

7.7 Condensed Fuel Fires

Another important type of non-premixed flames is encountered in fires, both of liquid and solid fuels. The fuel is initially in a condensed phase, and prior to burning with air it must be gasified by heat from an external source or heat from the fire itself after ignited. The gasified fuel is convected/diffused outward where it reacts with air in the same fashion as a jet flame. Examples of these types of flames are the fires that may occur after an oil spill or a wildland fire. When a liquid fuel is spilled from a storage tank, it forms a pool on the ground as shown in Fig. 7.9. In the presence of an ignition source, this pool ignites and forms a pool fire characterized by non-premixed flames. Heat from the flames is transferred back to the fuel primarily by radiation, causing the fuel to vaporize. The vaporized fuel is transported upward primarily by buoyancy where it reacts with the air, forming a diffusion flame. A similar process occurs with solid fuels, although the gasification

Fig. 7.9 A liquid pool fire (Sandia National Laboratories) and forest fire (USDA Forest Service) serve as examples of condensed-fuel non-premixed flames

of a solid fuel, such the wood in a forest fire in Fig. 7.9, is more complex and requires more energy than that of a liquid fuel. Typically, the formation of fuel vapors from a liquid pool is characterized by a change of phase, whereas the fuel vapors from a solid fuel are formed by a chemical decomposition reaction due to high temperatures.

The rate of heat release from a fire involving condensed fuels is not calculated as simply as it is for gaseous fuels. With gaseous fuels, simply knowing information about the chemical kinetics and heat of combustion is sufficient. The rate of heat release from the combustion of a condensed fuel is also highly dependent on how quickly the fuel vapor is produced. The amount of heat released per unit area of fuel is then

$$\dot{Q}'' = \dot{m}'' Q_c, \qquad (7.19)$$

where Q_c is the heat of combustion of the fuel vapors and $\dot{m}''$ is the rate of fuel generation per unit surface area. Fuel vapors are produced when the condensed phase reaches a high enough temperature. In other words, it is necessary to know the rate of heat transferred to the solid, which is no longer simply fuel dependent, but also situation dependent. For a particular fire, an energy balance can be performed on the condensed fuel to estimate the mass of fuel generated per unit area:

$$\dot{m}'' = \frac{\dot{q}''_s}{L_v} \qquad (7.20)$$

where $\dot{q}''_s$ is the total heat flux to the surface condensed fuel and L_v is the heat required to gasify the fuel. Note that the total surface heat flux can include convection $\dot{q}''_{conv}$, surface reradiation $\dot{q}''_{sr}$, flame radiation $\dot{q}''_{fr}$, and any other source of external radiant heating $\alpha \dot{q}''_e$. The total surface heat flux can then be expressed as $\dot{q}''_s = \dot{q}''_{conv} - \dot{q}''_{sr} + \dot{q}''_{fr} + \alpha \dot{q}''_e$.

Exercises

7.1 Consider a laminar methane diffusion flame stabilized on a circular burner. The pressure is 1 atm and the ambient temperature is 25°C.

(a) For a fixed fuel mass flow rate, how does the flame height vary with ambient pressure? Hint: the diffusivity is inversely proportional to pressure.

(b) If the height of the diffusion flame is L_f, qualitatively sketch the axial (centerline) profiles of the following quantities from the base of the diffusion flame to a height of $2L_f$: temperature, methane, and carbon dioxide concentrations.

7.2 Following exercise 7.1, if the height of the diffusion flame is L_f, qualitatively sketch the radial profiles of the following quantities at heights of $L_f/4$ and $L_f/2$: temperature, carbon dioxide concentration, and methane concentration. Assume that in both cases the flame sheet is located at radius r_f (radius is the distance from the centerline). If a quantity would be higher at one height make sure this is clearly indicated.

7.3 Consider the classic Burke-Schumann laminar jet flame with C_3H_8 as the fuel and standard air as the oxidizer. Both propane and air enter the burner at the standard temperature and pressure. Sketch the flame shape for the following conditions: $\dot{Q}_{fuel} = 1$ cm^3/s and $\dot{Q}_{air} = 20$ cm^3/s, where $\dot{Q}_{fuel}$ and $\dot{Q}_{air}$ are the volumetric flow rates for the fuel and air.

7.4 (a) Consider a laminar diffusion flame stabilized on a circular burner. The burner Reynolds number is $Re_d = V_{jet}d_{jet}/\nu$ where V_{jet} is the exit velocity of the fuel from the burner, d_{jet} is the burner diameter, and ν is the kinematic viscosity that is assumed to be equal to D, the effective diffusivity. For a fixed burner exit velocity and kinematic viscosity, sketch the flame height as a function of the burner Reynolds number.

(b) Now consider a turbulent diffusion flame stabilized on a circular burner. Assume that the following empirical relation holds for the turbulent diffusivity: $D_t \propto V_{jet}d_{jet}$. For a fixed burner exit velocity and kinematic viscosity, sketch the flame height as function of Reynolds number.

7.5 A burner operates with a nonpremixed (diffusion) propane jet flame enclosed in a box. The box is designed for safe operation at $P = 1$ atm. The operator wishes to increase the pressure to $P = 2$ atm with the same burner. The fuel and air temperatures are kept the same. In order to avoid flame impingement (flame hitting the box), suggest what the operator should do for the following two cases assuming that the peak flame temperature remains the same:

(a) the flame is laminar.

(b) the flame is turbulent.

References

1. Gautam T (1984) Lift-off heights and visible lengths of vertical turbulent jet diffusion flames in still air. Comb. Sci. Tech. 41:17–29.
2. Hottel HC, Hawthorne WR (1949) Diffusion in laminar jet flames. Symposium on Combustion and Flame, and Explosion Phenomena 3(1):254–266.
3. Lee KO, Megaridis CM, Zelepouga S, Saveliev AV, Kennedy LA, Charon O, Ammouri F (2000) Soot formation effects of oxygen concentration in the oxidizer stream of laminar coannular nonpremixed methane/air flames. Combustion and Flame 121:322–333.
4. Smyth KC, Miller JH, Dorfman RC, Mallard WG, Santoro RJ (1985) Soot inception in a methane/air diffusion flame as characterized by detailed species profiles. Combustion and Flame 62(2):157–181.

Chapter 8
Droplet Evaporation and Combustion

Liquid fuels are widely used in various combustion systems for their ease of transport and storage. Due to their high energy content, liquid fuels are the most common fuels in transportation applications. Before combustion can take place, liquid fuel must be vaporized and mixed with the oxidizer. To achieve this goal, liquid fuel is often injected into the oxidizer (normally air) forming a liquid spray. Figure 8.1 sketches the main physical processes occurring in a liquid fuel spray. Once the liquid fuel is injected into the combustor through the injector, the liquid spray begins to undergo various physical processes and interacts dynamically with the turbulent fluid inside the combustor. Soon after injection, the liquid fuel breaks up into droplets, forming a spray. Droplets then collide and coalesce, producing droplets of different sizes. Due to the high density of liquid fuel, the momentum of the liquid spray has a profound impact on local flow fields, creating turbulence and gas entrainment. In the case of engines, droplet spray may impinge on the wall surfaces due to the tight confinement inside the intake manifold or cylinders. Liquid films can form on the wall surfaces and then may evaporate. In piston engines, droplet combustion may occur through multiple transient events including preheating, gasification, ignition, flame propagation, formation of diffusion flames, and, ultimately, burn-out. As such, droplets can be considered the building block for providing fuel vapor in combustion systems. Understanding of single-droplet evaporation and combustion processes therefore provides important guidance in design of practical burners.

8.1 Droplet Vaporization in Quiescent Air

The simplest theoretical case of single-droplet evaporation consists of a liquid droplet surrounded by gas with no motion relative to the droplet. For this analysis, consider a droplet of initial diameter D_0 suddenly exposed to higher temperature (T_a) quiescent air. The following assumptions are made:

1. Buoyancy is unimportant, i.e., the thermal layer around the droplet is spherical.

S. McAllister et al., *Fundamentals of Combustion Processes*,
Mechanical Engineering Series, DOI 10.1007/978-1-4419-7943-8_8,
© Springer Science+Business Media, LLC 2011

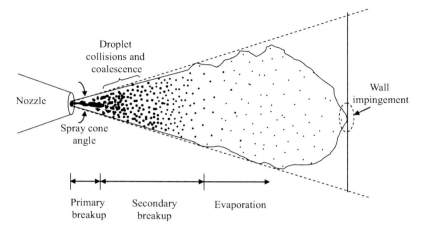

Fig. 8.1 Sketch of a diesel spray into engine with the main physical processes

Fig. 8.2 Sketch of processes involved in evaporation of a spherical droplet

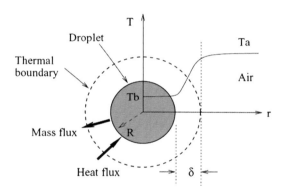

2. By the lumped capacitance formulation, the temperature in the droplet is uniformly equal to the liquid saturation temperature (boiling point) T_b. If the droplet temperature is initially at a lower temperature T_0, the droplet needs to be heated from T_0 to T_b. Once the droplet reaches T_b, its temperature remains unchanged. This heating period is discussed in Sect. 8.4.
3. Surrounding air is at constant pressure so that the liquid vapor density and the heat of vaporization remain constant during the entire evaporation process.

As presented in Fig. 8.2, an energy analysis of the spherical droplet leads to

$$-\frac{d}{dt}\left\{\underbrace{\rho_l \frac{4}{3}\pi\left(\frac{D}{2}\right)^3}_{\text{mass of droplet}} h_{fg}\right\} = \underbrace{\pi D^2}_{\substack{\text{surface area}}} \cdot \underbrace{q''_s}_{\substack{\text{heat flux per unit area}}}, \tag{8.1}$$

where ρ_l is the droplet density (liquid), D is the diameter of droplet, h_{fg} is the heat of vaporization at T_b, and q_s'' is the heat flux to the droplet surface. The negative sign is needed due to the decrease of D with time.

The heat flux towards the surface is determined by heat conduction as

$$q_s'' = k\frac{dT}{dr}\Big|_s \approx k\frac{T_a - T_b}{\delta}, \tag{8.2}$$

where k is thermal conductivity and δ is the thickness of thermal layer surrounding the droplet. The value of δ depends on the physical properties of the problem, but it is proportional to the characteristic length of the process, the droplet diameter. As an approximation, we set $\delta = C_l D$ and substitute this into Eq. 8.2 leading to

$$-\frac{d}{dt}\left\{\underbrace{\rho_l \frac{4}{3}\pi\left(\frac{D}{2}\right)^3}_{\text{mass of droplet}} h_{fg}\right\} = \underbrace{\pi D^2}_{\text{surface area}} k\frac{T_a - T_b}{C_l D}$$

$$\rho_l \frac{1}{6}\pi h_{fg}\frac{dD^3}{dt} = -\pi D k\frac{T_a - T_b}{C_1}$$

$$\rho_l \frac{1}{6}\pi h_{fg}3D^2\frac{dD}{dt} = -\pi D k\frac{T_a - T_b}{C_1} \tag{8.3}$$

$$2D\frac{dD}{dt} = -\frac{4k(T_a - T_b)}{\rho_l h_{fg} C_1}$$

$$\frac{dD^2}{dt} = -\beta_0 \text{ where } \beta_0 \equiv \frac{4k(T_a - T_b)}{\rho_l h_{fg} C_1}$$

The term β_0, on the right hand side of Eq. 8.3 is called the "vaporization constant" since it is fixed at a given air temperature. The constant C_l is here assumed for simplicity purposes to be 1/2, i.e., the thermal layer is equal to the radius of the droplet. Equation 8.4 gives the time evolution of droplet diameter as

$$D^2 = D_0^2 - \beta_0 t. \tag{8.4}$$

Equation 8.4 is traditionally referred to as the "D squared" law (D^2-law). The lifetime of a droplet with initial diameter D_0 is then obtained from Eq. 8.5 as

$$t_{life} = \frac{D_0^2}{\beta_0} \tag{8.5}$$

Figure 8.3 sketches experimental measurement of D^2 of a droplet initially at T_0 ($<T_b$) versus time showing an initial flat period that corresponds to the initial

Fig. 8.3 Evolution of droplet
size (square of diameter) vs.
time showing the '*D squared
law*'

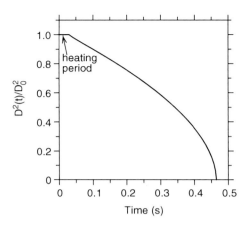

heating of the droplet before it starts to evaporate at (T_b). The instantaneous
evaporation rate of the droplet can be determined by

$$\dot{m}_l = \frac{d}{dt}\left\{\rho_l \frac{\pi}{6} D^3\right\} = \rho_l \frac{\pi}{6} 3D^2 \frac{dD}{dt} = \rho_l \frac{\pi}{4} D \frac{dD^2}{dt} \qquad (8.6)$$

Using Eq. 8.3, we get an evaporation rate that decreases with time.

$$\dot{m}_l = -\frac{\pi}{4}\rho_l D\beta_0 = -\frac{\pi}{4}\rho_l \beta_0 \sqrt{D_0^2 - \beta_0 t} \qquad (8.7)$$

Example 8.1 An ethanol droplet of initial size of 100 μm (1 μm = 10^{-6} m) is
exposed to quiescent hot air at $T_{air} = 500$ K and $P = 1$ atm. Estimate the droplet
lifetime.

Solution:
The lifetime is given by $t_{life} = \frac{D_0^2}{\beta_0}$. We need to estimate

$$\beta_0 = \frac{4k(T_a - T_b)}{\rho_l h_{fg} C_1}$$

with the following approximations.

1. The conductivity is a function of the mixture between the fuel and air, and the
   following empirical formula is found to give good results:

$$k(\bar{T}) = 0.4 \cdot k_{fuel}(\bar{T}) + 0.6 \cdot k_{air}(\bar{T}),$$

$$\bar{T} = (T_a + T_b)/2 = (500K + 351K)/2 = 425K$$

From Appendix 9, the thermal conductivity of ethanol at 425 K is
~0.0283 W/m–K. With the conductivity of air at k_{air} ($T = 425$ K) ~0.033

(W/m–K), we have $k \sim 0.0311$ W/m$-$K. Note that if fuel conductivity is not given, we may approximate k by air conductivity.

2. $h_{fg} = 797.34$ k J/kg
3. $\rho_l = 757$kg/m^3
4. $C_1 = 0.5$

$$\beta_0 = \frac{4k(T_a - T_b)}{\rho_l h_{fg} C_1} = \frac{4 \cdot 0.0311 \text{W/m} - \text{K} \cdot (500\text{K} - 351\text{K})}{757 \text{ kg/m}^3 \cdot 797.3 \text{ kJ/kg} \cdot 1000 \text{ J/kJ} \cdot 0.5}$$

$$= 6.142 \cdot 10^{-8} \text{m}^2/\text{s}$$

$$= 6.142 \cdot 10^4 (\mu\text{m})^2/\text{s}$$

$$t_{life} = \frac{D_0^2}{\beta_0} = \frac{100^2}{6.142 \cdot 10^4} = 0.163s = 163\text{ms}$$

Note: Numerical simulation of a single droplet gives 171 ms. This is in good agreement with the above estimate.

8.1.1 Droplet Vaporization in Convective Flow

In most applications, droplets are injected into a combustor with a relative velocity, u_d, with respect to the air. As sketched in Fig. 8.4, a convective boundary layer is formed around the droplet. The convective heat transfer to the droplet is given by

$$q_s'' = \tilde{h}(T_a - T_b), \tag{8.8}$$

where $\tilde{h}$ is the convective heat transfer coefficient. For a sphere, $\tilde{h}$ is obtained from the Nusselt number correlation as

$$Nu = \frac{\tilde{h}D}{k} = 2 + 0.4 \cdot \text{Re}_D^{1/2}\text{Pr}^{1/3}, \tag{8.9}$$

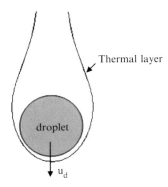

Fig. 8.4 Droplet evaporation in a convective flow showing that a convective boundary layer is formed around the droplet

where Re_D is the Reynolds number based on u_d and droplet diameter, Pr is the Prandtl number, i.e.,

$$Pr = \frac{v}{\alpha} = \frac{\text{viscous diffusion rate}}{\text{thermal diffusion rate}} = \frac{c_p\mu}{k}.$$

Following similar procedures in deriving Eq. 8.3, we have

$$-\frac{d}{dt}\left\{\underbrace{\rho_l \frac{4}{3}\pi\left(\frac{D}{2}\right)^3}_{\text{mass of droplet}} h_{fg}\right\} = \underbrace{\pi D^2}_{\text{surface area}} \tilde{h}(T_a - T_b)$$

$$\frac{dD^2}{dt} = -2C_1\beta_0 - \beta \tag{8.10}$$

where

$$\beta \equiv \frac{1.6 k Re_D^{1/2} Pr^{1/3}(T_a - T_b)}{\rho_l h_{fg}}$$

Assuming an average Reynolds number and treating it as a constant, integration of Eq. 8.10 gives

$$D^2 = D_0^2 - 2C_1\beta_0 t - \beta t \tag{8.11}$$

In reality, the Reynolds number will decrease with diameter, so in this analysis, the average Reynolds number can be approximated using half the initial diameter as $\bar{Re}_D \approx \rho D_0 u_d/(2\mu)$. Figure 8.5 plots the predicted evolution of ethanol droplet sizes versus time, showing that an increase in relative velocity leads to faster droplet evaporation.

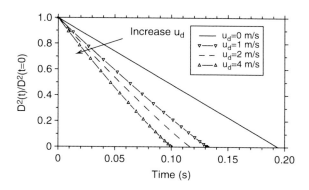

Fig. 8.5 Effect of relative velocity (slip velocity), u_d, on evaporation rate of an ethanol droplet. Rate is faster when u_d increases ($P = 1$ atm, $T_a = 600$ K, initial droplet diameter $= 150$ μm)

Example 8.2 Estimate the droplet lifetime for dodecane in air at $P = 1$ atm, $T_a = 700$ K, and a relative velocity of 2.8 m/s. The initial droplet size is 101.6 μm.

Solution:
Using the mean temperature $\bar{T} = (700K + 489K)/2 \sim 600K$.

1. Using air conductivity at 600 K, $k \sim 0.0456\,W/m - K$
2. viscosity $\mu = 3.030 \cdot 10^{-5}\,kg/m - s$
3. air density $\rho_{air} = 0.588\,kg/m^3$
4. $Pr = 0.751$
5. liquid density $\rho_l = 749\,kg/m^3$
6. $h_{fg} = 256\,kJ/kg$

$$2C_1\beta_0 = \frac{8k(T_a - T_b)}{\rho_l h_{fg}} = \frac{8 \cdot 0.0456 \cdot W/m - K \cdot (700K - 489K)}{749\,kg/m^3 \cdot 256\,kJ/kg \cdot 1000\,J/kJ}$$
$$= 4.018 \cdot 10^5 (\mu m)^2/s$$

$$\bar{Re}_D = \frac{\rho_{air}D_0/2 \cdot u_d}{\mu_{air}}$$
$$= \frac{0.588kg/m^3 \cdot 101.6/2 \cdot 10^{-6}m \cdot 2.8m/s}{3.03 \cdot 10^{-5}\,kg/m - s} = 2.75$$

$$\beta \equiv \frac{1.6k\bar{Re}_D^{1/2}Pr^{1/3}(T_a - T_b)}{\rho_l h_{fg}}$$
$$= \frac{1.6 \cdot (2.75)^{1/2}(0.751)^{1/3} \cdot (700K - 489K) \cdot 0.0456\,W/m - K}{749\,kg/m^3 \cdot 256\,kJ/kg \cdot 1000\,J/kJ}$$
$$= 12.10 \cdot 10^4 (\mu m)^2/s$$

$$t_{life} = \frac{D_0^2}{2C_1\beta_0 + \beta} = \frac{(101.6)^2(\mu m)^2}{(4.018 \cdot 10^5 + 12.10 \cdot 10^4)(\mu m)^2/s}$$
$$= 0.020\ s = 20\ ms$$

Note: Numerical results give 27 ms.

Example 8.3 Repeat Example 8.1 with $u_d = 1$ m/s and 10 m/s.

Solution:
Using air properties at 400 K, $Pr = 0.788$
With $u_d = 1$ m/s, $\bar{Re}_D = 0.99$

$$2C_1\beta_0 = \frac{8k(T_a - T_b)}{\rho_l h_{fg}} = 6.142 \cdot 10^4 (\mu m)^2/s$$

$$\beta \equiv \frac{1.6k\bar{Re}_D^{1/2}Pr^{1/3}(T_a - T_b)}{\rho_l h_{fg}}$$

$$= \frac{1.6 \cdot 0.0311\,\text{W/m} - \text{K}(0.99)^{1/2}(0.788)^{1/3} \cdot (500\text{K} - 351\text{K})}{757\ \text{kg/m}^3 \cdot 797.34\ \text{kJ/kg} \cdot 1000\ \text{J/kJ}}$$

$$= 1.13 \cdot 10^4 (\mu\text{m})^2/\text{s}$$

$$t_{life} = \frac{D_0^2}{2C_1\beta_0 + \beta} = \frac{(100)^2(\mu\text{m})^2}{(6,142 \cdot 10^4 + 1.13 \cdot 10^4)(\mu\text{m})^2/\text{s}} = 0.138\text{s}$$

Note: Numerical results give 0.165 s.
With $u_d = 10$ m/s, $\bar{Re}_D = 9.9$

$$\beta = \frac{1.6 \cdot 0.0311\,\text{W/m} - \text{K}(9.9)^{1/2}(0.788)^{1/3} \cdot (500\text{K} - 351\text{K})}{757\ \text{kg/m}^3 \cdot 797\ \text{kJ/kg} \cdot 1000\ \text{J/kJ}}$$

$$= 3.56 \cdot 10^4 (\mu\text{m})^2/\text{s}$$

$$t_{life} = \frac{D_0^2}{2C_1\beta_0 + \beta} = \frac{(100)^2(\mu\text{m})^2}{(6.142 \cdot 10^4 + 3.56 \cdot 10^4)(\mu\text{m})^2/\text{s}} = 0.103\ \text{s}$$

Note: Numerical results give 0.152 s.

8.2   Droplet Combustion

If the air temperature is high enough or a spark is present while a droplet is evaporating, the vapor/air mixture around the droplet may ignite. Once ignited, a non-premixed (diffusion) flame will establish around the droplet. Heat transfer from the flame to the droplet surface will accelerate the evaporation of the liquid. The fuel vapor is diffused radially outward toward the flame where it reacts with the air that has diffused radially inward as sketched in Fig. 8.6.

The droplet burning process is similar to that of droplet evaporation, but the ambient temperature is replaced by the flame temperature. Denoting the thermal boundary thickness by δ_f, Eq. 8.2 becomes

$$q_s'' = k\frac{dT}{dr}\Big|_s \approx k\frac{T_f - T_b}{\delta_f} \approx k\frac{T_f - T_b}{C_2 D} \tag{8.12}$$

where T_f is the flame temperature and C_2 is a parameter similar to C_1. Substituting Eq. 8.12 into Eq. 8.1, we obtain

$$\frac{dD^2}{dt} = -\beta_0' \quad \text{where} \quad \beta_0' \equiv \frac{4k(T_f - T_b)}{\rho_l h_{fg} C_2}, \tag{8.13}$$

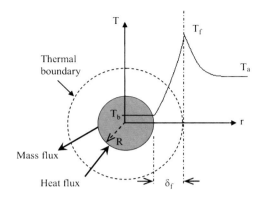

Fig. 8.6 Droplet combustion with a diffusion flame established at δ_f off the liquid surface

where β'_0 is called the droplet "burning constant." Similar to the evaporation case, integration of Eq. 8.13 leads to

$$D^2 = D_0^2 - \beta'_0 t. \tag{8.14}$$

This also has the form of the "D-squared" law except β_0 has been replaced by β'_0.

Example 8.4 Repeat Example 8.1 with a stoichiometric flame surrounding the droplet.

Solution:
The flame temperature is about 2,300 K
The mean temperature $\bar{T} = (2300K + 351K)/2 \cong 1350K$

$$k_{air}(\text{at } 1300K) \sim 0.0837\text{W/m} - \text{K}.$$

Let's assume that $C_2 = 0.5$ and with $h_{fg} = 836$ kJ/kg, we have

$$\beta'_0 \equiv \frac{4k(T_f - T_b)}{\rho_l h_{fg} C_2} = \frac{4 \cdot 0.0837 \text{ W/m} - \text{K} \cdot (2300K - 351K)}{789 \text{ kg/m}^3 \cdot 836 \text{ kJ/kg} \cdot 1000 \text{ J/kJ} \cdot 0.5}$$
$$= 1.98 \cdot 10^6 (\mu m)^2/s$$

Lifetime $= 5.05$ ms which is much smaller than 163 ms in the pure evaporation case.

A droplet burning in a convective flow follows the same model for evaporation, changing T_a to T_f, we have

$$D^2 = D_0^2 - 2C_2\beta'_0 t - \beta' t, \tag{8.15}$$

where

$$\beta' \equiv \frac{1.6 k \bar{Re}_D^{1/2} \text{Pr}^{1/3} (T_f - T_b)}{\rho_l h_{fg}}.$$

Table 8.1 Equations for droplet evaporation and combustion under different conditions

Droplet condition	q_s''	$D^2(t)$	Parameter
Evaporation in quiescent air	$k\dfrac{T_a - T_b}{C_1 D}$	$D^2 = D_0^2 - \beta_0 t$	$\beta_0 \equiv \dfrac{4k(T_a - T_b)}{\rho_l h_{fg} C_1}$
Evaporation in convective air	$\left(2 + 0.4 \cdot \text{Re}_D^{1/2}\text{Pr}^{1/3}\right)$ $\dfrac{k(T_a - T_b)}{D}$	$D^2 = D_0^2 - 2C_1\beta_0 t - \beta t$	$\beta \equiv \dfrac{1.6k\bar{R}e_D^{1/2}\text{Pr}^{1/3}(T_a - T_b)}{\rho_l h_{fg}}$
Combustion in quiescent air	$k\dfrac{T_f - T_b}{C_2 D}$	$D^2 = D_0^2 - \beta_0' t$	$\beta_0' \equiv \dfrac{4k(T_f - T_b)}{\rho_l h_{fg} C_2}$
Combustion in convective air	$\left(2 + 0.4 \cdot \text{Re}_D^{1/2}\text{Pr}^{1/3}\right)$ $\dfrac{k(T_f - T_b)}{D}$	$D^2 = D_0^2 - 2C_2\beta_0' t - \beta' t$	$\beta' \equiv \dfrac{1.6k\bar{R}e_D^{1/2}\text{Pr}^{1/3}(T_f - T_b)}{\rho_l h_{fg}}$

The D^2 law governing droplet evaporation in quiescent, convective, burning, and non-burning scenarios implies that by reducing the initial droplet size by half, the droplet lifetime can be decreased by a factor of 4. It is therefore worthwhile to decrease the droplet size when a shorter lifetime is desired. The results of the preceding derivations are summarized in Table 8.1.

8.3 Initial Heating of a Droplet

To provide an estimate of the amount of time required to heat a droplet from T_0 to T_b, several assumptions are made: (1) the droplet density is constant, (2) the heat capacity is constant, (3) there is no vaporization, and (4) the same heat transfer model is applied throughout the process. Assumptions (1) and (3) also imply that the diameter of droplet is unchanged. Considering an energy balance for the droplet, one can derive the following equation

$$\frac{\pi D_0^3}{6}\rho_l c_{p,l}\frac{dT}{dt} = \pi D_0^2 \cdot q_s''. \tag{8.16}$$

Let's consider the heat transfer in quiescent air first. The heat flux at the surface is modeled as $q_s'' = k(T_a - T)/(C_1 D_0)$ and integration of Eq. 8.16 gives

$$t = \frac{\rho_l c_{p,l} C_1 D_0^2}{6k} \cdot \ln\left(\frac{T_a - T_0}{T_a - T}\right) \tag{8.17}$$

Note that Eq. 8.17 is applicable only when $T \leqslant T_b$ and the heating time required for a droplet in quiescent air to reach T_b is

$$t_{heating} = \frac{\rho_l c_{p,l} C_1 D_0^2}{6k} \cdot \ln\left(\frac{T_a - T_0}{T_a - T_b}\right) \qquad (8.18)$$

For droplet heating in a convective flow, we follow the same analysis as Eq. 8.16 by replacing the right hand with

$$q_s'' = \left(2 + 0.4 \cdot Re_D^{1/2} Pr^{1/3}\right) \frac{k(T_a - T)}{D}$$

The result is

$$t_{heating} = \frac{\rho_l c_{p,l} D^2}{6k\left(2 + 0.4 \cdot Re_D^{1/2} Pr^{1/3}\right)} \cdot \ln\left(\frac{T_a - T_0}{T_a - T_b}\right). \qquad (8.19)$$

For droplet flames, we simply replace T_a by T_f in Eqs. 8.18 and 8.19.

Example 8.5 Estimate the time required to heat the ethanol droplet considered in Example 8.1 with initial temperature at 300 K, $T_a = 500$ K, and $D_0 = 100$ μm under two conditions:

(a) quiescent air , (b) air with a relative velocity $u_d = 1$ m/s.

Solution:
Let's estimate properties at the average temperature for $T_0 = 300$ K, $\bar{T} = (300K + 351K)/2 \approx 325$ K, $\rho_l = 773 kg/m^3$, $c_{p,l} = 2.5$ kJ/kg $- K$, $C_1 = 0.5$, using air properties, $k = 0.01865$ W/m$-$K.

(a) Using Eq. 11.18 we have

$$t_{heating} = \frac{\rho_l c_{p,l} C_1 D^2}{6k} \cdot \ln\left(\frac{T_a - T_0}{T_a - T_b}\right)$$

$$= \frac{773 kg/m^3 \cdot 2.5 kJ/kg - K \cdot 0.5 \cdot (10^{-4})^2 m^2}{6 \cdot 0.01865 \ W/m - K \cdot 10^{-3} \ kJ/J} \ln\left(\frac{500 - 300}{500 - 351}\right)$$

$$= 2.6 ms$$

Note that this is small ($\sim 1.3\%$) compared to the evaporation time (186 ms).
(b) Next with $u_d = 1$ m/s
With $u_d = 1$ m/s, $Re_D = 3.85$, $Pr = 0.788$

$$t_{heating} = \frac{\rho_l c_{p,l} D^2}{6k\left(2 + 0.4 \cdot Re_D^{1/2} Pr^{1/3}\right)} \cdot \ln\left(\frac{T_a - T_0}{T_a - T_b}\right)$$

$$= \frac{773\,kg/m^3 \cdot 2.5\,kJ/kg - K \cdot (10^{-4})^2 m^2}{6 \cdot 0.01865\,W/m - K \cdot 10^{-3}\,kJ/J\left(2 + 0.4(3.85)^{1/2}(0.788)^{1/3}\right)} \ln\left(\frac{500 - 300}{500 - 351}\right)$$

$$= 1.91\,ms$$

(with $u_d = 10$ m/s, $t_{heating} = 1.18$ ms)

8.3.1 Effect of Air Temperature and Pressure

The effect of pressure and temperature on droplet evaporation/combustion is reflected in the relation between saturation temperature and saturation pressure. During evaporation, the droplet temperature will approach the saturation temperature as illustrated in Fig. 8.7.

As the air temperature increases, the temperature differences $(T_a - T_b)$ and $(T_f - T_b)$ become larger (the flame temperature also increases). These changes lead to shorter droplet lifetimes. Figure 8.8 presents the predicted time evolution of ethanol droplet size for varying temperatures of air with a relative velocity of 1 m/s. As expected, the lifetimes of droplets decrease with increasing air temperature.

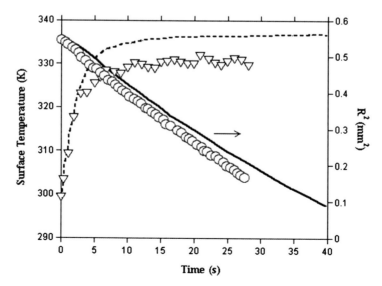

Fig. 8.7 Model predictions (*lines*) are compared to experimental data (*points*) for decane during evaporation. The droplet is heated up to saturation temperature in a short period of time (Reprinted with permission from Torres et al. [1])

Fig. 8.8 *Top*: Effect of air
temperature on ethanol
droplet evaporation.
Conditions: Air $P = 1$ atm,
$u_d = 1$ m/s, initial droplet
diameter = 100 µm. *Bottom*:
evaporation time versus air
temperature

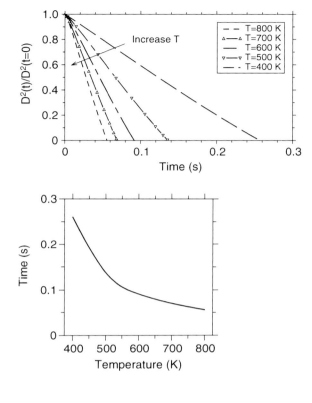

Fig. 8.9 Relations between
saturation pressure and
saturation temperature

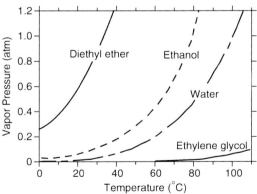

When the air pressure increases, the corresponding saturation temperature
increases and thus T_b increases. Typical relations between P_{sat} and T_{sat} are shown
in Fig. 8.9. The effect of pressure on droplet evaporation is more complex as it
also impacts many parameters through temperature, such as conductivity, heat
of vaporization, and density. As sketched in Fig. 8.10, the heat of vaporization

Fig. 8.10 Pressure-enthalpy
diagram showing the
saturation dome. Lines denote
constant temperature
contours, with A'' the lowest
temperature and A' the
highest. The *heat of
vaporization*, h_{fg}, is the
amount of enthalpy required
to bring the fluid from liquid
phase (A) to gas phase (B) at
constant temperature. h_{fg}
decreases as temperature
increases

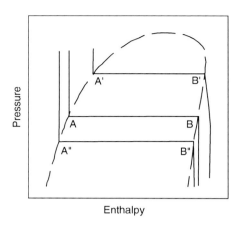

h_{fg} decreases with temperature in a nonlinear manner and drops to zero when the
critical point is reached. At and above the critical point, there is no distinct phase
change. Additionally, pressure can also affect the Reynolds number nearly linearly
through the density change. Table 8.2 lists properties of n-butanol for a range of
saturation temperatures. Additional data for other fuels can be found in Appendix 9.
If properties are not available, one can estimate the saturation temperature using the
Clausius-Clapeyron equation:

$$\frac{dP_{sat}}{P_{sat}} = \frac{h_{fg}}{R_m}\frac{dT_{sat}}{T_{sat}^2} \text{ where } R_m = \frac{\hat{R}_u}{M_f},$$
$$\text{or } d(\ln P_{sat}) = -\frac{h_{fg}}{R_m}d\left(\frac{1}{T_{sat}}\right)$$

(8.20)

where M_f is the molecular mass of fuel vapor, $\hat{R}_u$ is the universal gas constant,
and h_{fg} is the heat of vaporization that is also function of temperature. If we
approximate h_{fg} by an average value between two temperatures, T_{sat1}, T_{sat2},
Eq. 8.20 gives

$$\ln\left(\frac{P_{sat2}}{P_{sat1}}\right) \simeq \frac{h_{fg}(\bar{T})}{R_m}\left(\frac{1}{T_{sat1}} - \frac{1}{T_{sat2}}\right)$$

(8.21)

Let's consider the effect of pressure on droplet evaporation under quiescent air at
a fixed temperature, T_a. When pressure increases, T_b increases, leading to smaller
$(T_a - T_b)$ and smaller h_{fg} (the change of h_{fg} with pressure is not large until near the
critical point.) Depending on the relative magnitude of changes between $(T_a - T_b)$
and h_{fg} with pressure, the net effect could cause $\beta_0 \equiv \frac{4k(T_a - T_b)}{\rho_l h_{fg} C_l}$ to decrease or
increase. Therefore, the droplet lifetime could increase or decrease with pressure.

Table 8.2 Properties of n-butanol as function of saturation temperature/pressure[a]

n-Butanol
Critical temperature: 561.15 K
Chemical formula: $C_2H_5CH_2CH_2OH$
Critical pressure: 4,960 kPa
Molecular weight: 74.12
Critical density: 270.5 kg/m^3

T_{sat} (K)	390.65	410.2	429.2	446.5	469.5	485.2	508.3	530.2	545.5	558.9
P_{sat} (kPa)	101.3	182	327	482	759	1,190	1,830	2,530	3,210	4,030
ρ_l (kg/m^3)	712	688	664	640	606	581	538	487	440	364
ρ_v (kg/m^3)	2.30	4.10	7.9	12.5	23.8	27.8	48.2	74.0	102.3	240.2
h_{lv} (kJ/kg)	591.3	565.0	537.3	509.7	468.8	437.2	382.5	315.1	248.4	143.0
c_{pl} (kJ/kg-K)	3.20	3.54	3.95	4.42	5.15	5.74	6.71	7.76		
c_{pv} (kJ/kg-K)	1.87	1.95	2.03	2.14	2.24	2.37	2.69	3.05	3.97	
μ_l (μNs/m^2)	403.8	346.1	278.8	230.8	188.5	144.2	130.8	115.4	111.5	105.8
μ_v (μNs/m^2)	9.29	10.3	10.7	11.4	12.1	12.7	13.9	15.4	17.1	28.3
k_l (mW/m-K)	127.1	122.3	117.5	112.6	105.4	101.4	91.7	82.9	74.0	62.8
k_v (mW/m-K)	21.7	24.2	26.7	28.2	31.3	33.1	36.9	40.2	43.6	51.5
Pr_l	10.3	9.86	9.17	8.64	10.2	8.10	8.67	9.08		
Pr_v	0.81	0.83	0.81	0.86	0.87	0.91	1.01	1.17	1.56	
σ (mN/m)	17.1	15.6	13.9	12.3	10.2	7.50	6.44	4.23	2.11	0.96

[a] *Nomenclature:* ρ_l liquid density, h_{lv} heat of vaporization, ρ_v vapor density; k_l liquid conductivity, k_v vapor conductivity, σ surface tension

Figure 8.11 presents the predicted evolution of ethanol droplet sizes versus time showing an increase in lifetime with pressure.

When the droplet is injected with a relative velocity, u_d, the pressure now can impact two parts: (1) it can decrease $(T_a - T_b)$ and h_{fg} as discussed above; (2) it can increase Re_D through density changes, leading to an increase in

$$\beta \equiv \frac{1.6k\bar{Re}_D^{1/2}\mathrm{Pr}^{1/3}(T_a - T_b)}{\rho_l h_{fg}}.$$

The net effect on the droplet lifetime $(D_0^2/(2C_1\beta_0 + \beta))$ depends on the relative changes in β_0 and β. Figure 8.12 presents the effect of pressure on droplet size evolution versus time with $u_d = 1$ m/s showing a decrease in lifetime with pressure.

Fig. 8.11 Pressure increases ethanol droplet lifetime with $u_d = 0$ m/s, $T_a = 500$ K, $D_0 = 100$ μm

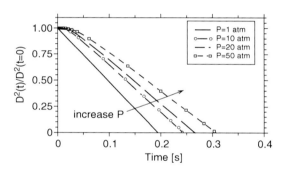

Fig. 8.12 Pressure decreases ethanol droplet lifetime with $u_d = 1$ m/s, $T_a = 500$ K, $D_0 = 100$ μm

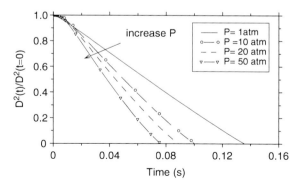

Example 8.6 Using Eq. 8.21 and Table 8.2 for n-butanol, estimate T_{sat} at $P_{sat} = 1,090$ kPa based on $T_{sat,1} = 390.65$ K, $P_{sat,1} = 101.3$ kPa.

Solution:
Since T_{sat} is not known, the average temperature is first set to 390 K. We will improve this result by iterations. With $\bar{T} = 390K$,

$$R_m = \frac{8.314\,\text{kJ/kmol} - \text{K}}{74.12\,\text{kg/kmol}} = 0.112\,\text{kJ/kg} - \text{K}$$

$$\ln\left(\frac{P_{sat2}}{P_{sat1}}\right) \cong \frac{h_{fg}(\bar{T})}{R_m}\left(\frac{1}{T_{sat1}} - \frac{1}{T_{sat2}}\right)$$

$$\ln\left(\frac{1090\,\text{kPa}}{101.3\,\text{kPa}}\right) \cong \frac{591.3\,\text{kJ/kg}}{0.112\,\text{kJ/kg} - \text{K}}\left(\frac{1}{390.65} - \frac{1}{T_{sat2}}\right)$$

solving for $T_{sat,2} = 474.09$ K

Compared to 485.2 K given in Table 8.2, the above estimate has an error of about 2.2% which usually is good for engineering purposes. To improve this, a second estimate is conducted with $\bar{T} = (390K + 474K)/2 = 432K$ with $h_{fg}(\bar{T}) = 531.9\,\text{kJ/kg}$. This gives

$$\ln\left(\frac{1090kPa}{101.3kPa}\right) \cong \frac{531.9\,\text{kJ/kg}}{0.112\,\text{kJ/kg} - \text{K}}\left(\frac{1}{390.65} - \frac{1}{T_{sat2}}\right)$$

$$\rightarrow T_{sat,2} = 485.5 \text{ K}$$

which is nearly identical to 485.2 K given in Table 8.2.

8.4 Droplet Distribution

Figure 8.13 shows a spray in a typical port injection gasoline spark ignition engine. The spray breaks up into small droplets and statistical methods are used to describe various properties of these droplets. The droplet number distribution, $\Delta N(d_i),$[1] is defined as the fraction of droplets whose sizes fall between $d_i \pm \Delta d/2$, as

$$\Delta N(d_i) = \frac{\text{number of droplets with sized such that } d_i - \Delta d/2 < d < d_i + \Delta d/2}{\text{Total number of droplets } (N_d)}, \quad (8.22)$$

---

[1] For clarity, the droplet diameter in a spray is denoted by d_i to differentiate from D used for single droplet.

Fig. 8.13 Detailed image of
a gasoline spray (Used with
permission from Dr. Chih-Yu
Wu at Kao Yuan University,
Taiwan)

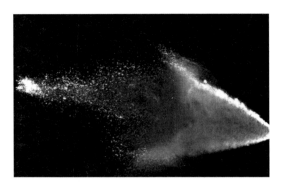

where Δd is size of bins used to sort the droplets according to their sizes and N_d is the total number of droplets in a spray. One of the most important parameters is the average droplet size. There are several ways to define a mean droplet diameter as listed in the following:

$$d_1 \equiv \text{MD (Mean Diameter)} = \sum_{i=1}^{\infty} \Delta N(d_i) \cdot d_i$$

$$d_2 \equiv \text{AMD(Area Mean Diameter)} = \left(\sum_{i=1}^{\infty} \Delta N(d_i) \cdot d_i^2 \right)^{1/2} \qquad (8.23)$$

$$d_3 \equiv \text{VMD(Volume Mean Diameter)} = \left(\sum_{i=1}^{\infty} \Delta N(d_i) \cdot d_i^3 \right)^{1/3}$$

The total area and volume are related to d_2 and d_3 as

$$\text{Total surface area of droplets} = \pi \cdot (\text{AMD})^2$$
$$\text{Total voulme occupied by droplets} = \frac{\pi}{6} \cdot (\text{VMD})^3 \qquad (8.24)$$

In most applications, the Sauter Mean Diameter (SMD) is used to quantify the average size of droplets in a spray. It is defined as

$$d_{32} \equiv \text{SMD(Sauter Mean Diameter)} = \frac{\displaystyle\sum_{i=1}^{\infty} \Delta N(d_i) \cdot d_i^3}{\displaystyle\sum_{i=1}^{\infty} \Delta N(d_i) \cdot d_i^2} = \frac{d_3^3}{d_2^2} \qquad (8.25)$$

There are two related parameters to quantify droplet distributions. The first one is the cumulative number function (or distribution) (CNF) defined as

$$\text{CNF}(d_j) = \frac{\sum\limits_{i=1}^{j} \Delta N(d_i) \cdot d_i}{\sum\limits_{i=1}^{\infty} \Delta N(d_i) \cdot d_i} = \frac{1}{MD} \sum\limits_{i=1}^{j} \Delta N(d_i) \cdot d_i \qquad (8.26)$$

One can imagine lining up all droplets according to their sizes. The product of $\sum\limits_{i=1}^{\infty} \Delta N(d_i) \cdot d_i$ and the total number of droplets, N_d, is the total distance occupied by the droplets. Similarly, the product of $\sum\limits_{i=1}^{j} \Delta N(d_i) \cdot d_i$ and N_d, represents the total distance occupied by the droplets with sizes smaller than or equal to d_i. Hence, CNF (d_j), represents the fraction of distance occupied by droplets with sizes $\leq d_j$. The second method used to quantify the droplet distribution is by volume, which is perhaps more meaningful than distance. The cumulative volume function (or distribution) (CVF) is defined as

$$\text{CVF}(d_j) = \frac{\sum\limits_{i=1}^{j} \Delta N(d_i) \cdot d_i^3}{\sum\limits_{i=1}^{\infty} \Delta N(d_i) \cdot d_i^3} \qquad (8.27)$$

Similar to CNF(d_j), CVF(d_j) represents the fraction of *volume* occupied by droplets with sizes $\leq d_j$. These two parameters and their relations to the droplet number distribution, $\Delta N(d_i)$, are sketched in Fig. 8.14.

Empirical relations to quantify the SMD for a specific spray injector are expressed in terms of several parameters:

$$\text{SMD} = \text{function of} \left\{ \begin{array}{l} \text{fluid properties}(\ \sigma, v, ..), \text{ injetction parameters } (\Delta P, \dot{m}, ..), \\ \text{swirling air properties if air blast is used},... \end{array} \right\}$$

$$(8.28)$$

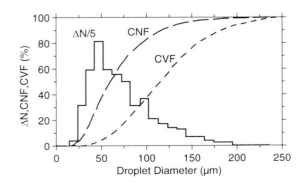

Fig. 8.14 Relations among three commonly used parameters, $\Delta N(d_i)$, CNF(d_j), and CVF(d_j) for describing droplet distribution versus droplet size

For instance, the following form has been proposed for pressure-swirl atomizers in steady flows:

$$\text{SMD} = 7.3 \cdot 10^6 \cdot \sigma_l^{0.6} v_l^{0.2} \dot{m}^{0.25} \Delta P^{-0.4} (\mu\text{m}) \tag{8.29}$$

with other properties in SI units.

Example 8.7 Estimate the SMD using Eq. 8.29 for a diesel injector.

Solution:

$\Delta P = 689$ kPa (100 psi) (pressure drop across the injector)
$\sigma_l = 0.03$ N/m (liquid surface tension)
$v_l = 2.82 \cdot 10^{-6}$ m^2/s (liquid viscosity)
$\dot{m} = 9 \cdot 10^{-3}$ kg/s

$$\text{SMD} = 7.3 \cdot 10^6 \cdot (0.03)^{0.6}(2.82 \cdot 10^{-6})^{0.2}(9 \cdot 10^{-3})^{0.25}(689 \cdot 10^3)^{-0.4}$$
$$= 98.25\mu\text{m}$$

Exercises

8.1 Consider a droplet of methanol with an initial diameter of 80 μm. It is injected into a chamber with an ambient temperature of 750 K and ambient pressure of 1 atm. Calculate the lifetime of the droplet. An additive is reported to reduce the boiling temperature of methanol by 40 K without affecting the heat of vaporization. Unfortunately, the additive also causes the initial droplet size to increase to 95 μm. Calculate the new droplet lifetime. Describe briefly how droplet lifetime would be affected if the ambient pressure were increased to 10 atm.

8.2 A turbojet flies at 250 m/s. Liquid n-heptane (C_7H_{16}) is injected in the direction of the air flow into the front of the 2.5 m long combustor where it completely combusts. Neglecting droplet breakup and drag effects, estimate the maximum allowable initial size of n-heptane droplets. Use the following information:

   (a) Air temperature and pressure inside the combustion chamber is 1,000 K and 1 atm
   (b) Droplets are injected into the combustor with a velocity 20 m/s faster than the air
   (c) The combustion chemistry process takes 1 ms after droplets are completely vaporized.
   (d) Properties of liquid n-heptane: density $= 684$ kg/m^3, boiling temperature $= 283$ K, heat of vaporization $= 317$ kJ/kg.

8.3 In a combustion chamber, fine droplets of octane with diameter 500 μm are injected into an atmosphere of air at 500°C and 1 atm. It is observed that some droplets are evaporating and others are burning. It is also observed that some of the droplets are moving with the same velocity as the air and others have significant velocities relative to the air.

(a) Calculate the lifetime of the evaporating droplets that are moving at the same velocity as the air (quiescent environment).
(b) Calculate the life time of the evaporating droplets that are moving with a velocity of 10 m/s relative to the air.
(c) Calculate the lifetime of the burning droplets that are moving at the same velocity as the air (quiescent environment).

NOTE: Assume that the thermal layer thickness and the flame stand-off distance are both equal to half the droplet diameter.

8.4 Using the data below, determine the evaporation time (droplet life time) for an n-butanol droplet of 100 μm diameter in hot air under the following conditions:

(a) $T_{air} = 900$ K, zero slip velocity ($U_d = 0$) between droplet and air, $P = 101.3$ kPa,
(b) Repeat (a) but with $U_d = 1$ m/s,
(c) $T_{air} = 900$ K, $U_d = 0$, $P = 3,210$ kPa, (same as (a) except at high pressure)
(d) Repeat (c) with a flame around the droplet with flame temperature of 2,200 K.

Note: Use the *air* property data for estimate of conductivity at $T_{ave} = (T_{air} + T_{droplet})/2$ and $T_{ave} = (T_{flame} + T_{droplet})/2$.

8.5 Estimate the evaporation lifetime of a diesel droplet (500 μm) surrounded by quiescent air at 500°C. Assume that the thickness of the thermal layer surrounding the droplet is half the droplet diameter. Compare with the combustion lifetime if the flame standoff distance is also half the droplet diameter with flame temperature of 2,305 K. If you cannot find all the needed diesel properties then use properties of n-heptane.

8.6 The droplet size data in a spray have been experimentally determined and are shown in Table 8.3. Determine the cumulative volume distribution for $d = 60$ μm, i.e., CVF ($d_j = 60$ μm).

Table 8.3 Exercise 8.6

Bin range (μm)	Number of droplets
0–10	60
10–30	100
30–40	120
40–60	300
60–80	200
80–100	20
100–130	0
130–170	0
Total droplets	800

Reference

1. Torres DJ, O'Rourke PJ, Amsden AA (2003) Efficient multicomponent fuel algorithm. Combustion Theory Modelling 7:66–86.

Chapter 9
Emissions

Emissions from combustion of fossil fuels are of great concern due to their impact on the environment and public health. The primary combustion products, carbon dioxide (CO_2) and water (H_2O), affect the environment through greenhouse effects and potential localized fog. Both products are inherent to the combustion of fossil fuels and their emission can only be reduced through modifications in the fuel or by exhaust treatment. The other major pollutants from combustion are secondary products and include carbon monoxide (CO), unburned hydrocarbons (HC), soot, nitric oxides (NO_x), sulfur oxides (SO_x), and oxides of metals. Pollutants cause health problems in humans and animals and can contribute to acid rain.

9.1 Negative Effects of Combustion Products

Combustion products cause harm at a wide range of scales. Carbon monoxide, soot, oxides of nitrogen, and unburned hydrocarbons directly harm the health of organisms that inhale the emissions. Nitrogen oxides, unburned hydrocarbons, and sulfur oxides negatively affect the environment of cities and counties. On a global scale, increased atmospheric carbon dioxide concentrations contribute to global warming through enhancement of the greenhouse effect.

Unburned hydrocarbons and soot cause respiratory problems and are known to be carcinogenic. Carbon monoxide fixes to hemoglobin in the blood so that the blood loses its ability to carry oxygen. Hemoglobin's binding affinity for CO is 200 times greater than for oxygen, meaning that small amounts of CO dramatically reduce hemoglobin's ability to transport oxygen. When hemoglobin combines with CO, it forms a bright red compound called carboxyhemoglobin. When air containing CO levels as low as 0.02% is breathed, it causes headache and nausea; if the CO concentration is increased to 0.1%, unconsciousness will follow. Fortunately, most negative symptoms of CO inhalation will disappear when an affected person decreases exposure to the pollutant. However, CO inhalation is one of the major causes of deaths in fires.

NO_x refers to the total content of NO and NO_2. These oxides of nitrogen are produced primarily from the nitrogen contained in the air. Similar to CO,

S. McAllister et al., *Fundamentals of Combustion Processes*,
Mechanical Engineering Series, DOI 10.1007/978-1-4419-7943-8_9,
© Springer Science+Business Media, LLC 2011

NO fixes to hemoglobin in the blood and threatens life if inhaled in excess. Most importantly, NO_x is the main cause of smog[1] and acid rain. Smog is produced by photochemical effects caused by the irradiation of NOx with ultravoilet light from the sun. Photochemical smog is composed of different noxious gases that cause breathing problems and allergies. Acid rain is any form of precipitation that is unusually acidic. It has harmful effects on plants, aquatic animals, and buildings. Acid rain is mostly caused by emissions of sulfur and nitrogen compounds (SO_x and NO_x) that form acids when they react with water in the atmosphere. In recent years, many governments have introduced laws to reduce these emissions.

Carbon dioxide is considered a major contributor to global warming through its role in the greenhouse effect. Though mostly transparent to incoming solar radiation, carbon dioxide absorbs and reemits the thermal-infrared radiation emitted by the earth. As a consequence, energy from solar radiation is trapped in the atmosphere. Natural occurrence of this greenhouse effect sustains Earth's temperatures at habitable levels. Unfortunately, CO_2 is formed whenever a fuel containing carbon is burned with air, and carbon dioxide's chemical stability causes it to stay trapped in the atmosphere for long periods of time once emitted. Consequently, measurements of recent global temperature increase have been linked to the rise in atmospheric carbon dioxide concentration that has accompanied the proliferation of fossil fuel combustion over recent centuries. Curbing this undesired enhancement of greenhouse warming effects requires that carbon dioxide be sequestered and emissions reduced so that atmospheric carbon dioxide concentrations can be stabilized. Carbon dioxide reduction and sequestration are currently subjects of great research interest.

9.2   Pollution Formation

The complex interaction between the chemical kinetic system and the fluid dynamics of combustion gases complicates the prediction of pollutant formation in a combustion system. Such a chemical system may involve thousands of unique chemical reactions producing and consuming hundreds of intermediate chemical species. Significant effort has been made in understanding the chemical mechanisms of flames. Global parameters such as the ignition delay time of gaseous mixtures, flame velocities, or the strain rate necessary to extinguish diffusion flames[2] can be calculated for a number of fuels in reasonable agreement with experiments. It is also

---

[1] Originally, Dr. Henry Antoine Des Voeux in his 1905 paper called the air pollution in cities as "Fog and Smoke." In the 1950s a new type of smog, known as photochemical smog, was first described. This forms when sunlight hits various pollutants in the air and forms a mix of inimical chemicals that can be very dangerous.

[2] The effect of aerodynamics on a flame is quantified as strain rate. For a one-dimensional flame, the strain rate can be defined as $|du/dx|$ with the unit of $1/s$, where u is local velocity and x is the physical coordinate. Flames under high strain rate can lead to extinction.

possible to predict the concentration profiles of fuel, oxidizer, intermediate products, and the main products (CO_2, N_2, and H_2O) of the combustion processes with reasonable accuracy. The most important chemical pathways leading to the formation of air pollutants, CO, NO_x, soot, and dioxins are reasonably known today. However, the standard of knowledge about the basic combustion processes demanded to calculate these pathways accurately is much higher than what is needed for the calculation of the global combustion parameters. The demanded accuracy is even higher if concentrations of pollutants are low, as is the case in the exhaust of modern, highly optimized combustion devices. The formation of soot is the most complex chemical system in flames. Soot particles, containing thousands of carbon atoms, are formed from simple fuel molecules within a few microseconds. The large number of molecules and particles of different forms and sizes involved by this process cannot be easily quantified. A statistical description of the chemical kinetic system is therefore necessary. It is still very difficult to accurately predict the amount of soot formation in combustion processes.

9.2.1 Parameters Controlling Formation of Pollutants

Temperature and residence time, τ_{res}, are two important parameters influencing the formation of pollutants. Temperature affects the onset of certain chemical reactions and consequently, the formation of certain chemical species. Since combustion temperature is a strong function of mixture composition, i.e. equivalence ratio, pollutant formation can be influenced by controlling reactant mixture composition. In order to complete chemical reactions in a combustion device, sufficient time must be provided for the reactants to react, i.e. the reactants must remain in the combustor longer than the time they need to react. The amount of time that reactants reside inside the combustor is called the 'residence' time. In industrial gas turbines for example, typical residence times are about 5–10 ms. The amount of time that the reactants need to react is called the chemical time, τ_{chem}. Chemical time is inversely proportional to reaction rate ($\tau_{chem} \sim 1/\dot{r}$), which depends on temperature and mixture composition as explained in Chap. 3. Typical values of chemical time are of the order of milliseconds.

The importance of the relative magnitudes of residence time and chemical time is illustrated in Fig. 9.1. The top figure sketches the variation of the concentrations of fuel (CH_4), CO, CO_2 versus time at $T = 1{,}600$ K as fuel reacts with air in a combustor. The bottom shows the same information at $T = 1{,}530$ K. Since the temperature is lower in the second case, the reaction rate is also lower and the resulting chemical time longer. The solid vertical lines denote the chemical times, τ_{chem}, at which combustion is completed, i.e., most CO is oxidized to form CO_2. The dashed lines correspond to the residence times (or the physical times) imposed by the device geometry. When $\tau_{res} \geq \tau_{chem}$ as in the case of top sketch, there will be low CO emission at the exit of the combustor. In contrast, $\tau_{res} < \tau_{chem}$ occurs in the bottom sketch and CO emissions will be high. Unburned hydrocarbons (HC) have a very similar trend as CO since both are intermediate species during combustion processes.

Fig. 9.1 Computed time
evolution of chemical species
concentration for two
scenarios of atmospheric
stoichiometric methane-air
combustion at constant
temperature. *Top*: Enough
residence time is provided, so
CO emission is low. *Bottom*:
Insufficient residence causes
high levels of CO emission

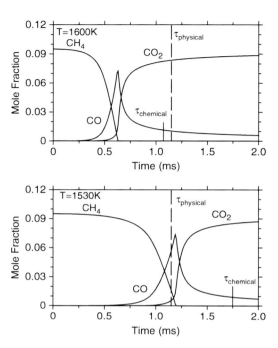

Temperature is the most important parameter in combustion processes because of the reaction rate's exponential dependence on temperature. In Fig. 9.2, the top sketch shows flame temperature versus equivalence ratio, ϕ. The flame temperature peaks slightly on the rich side. The bottom sketch illustrates trends of emissions versus equivalence ratio. As detailed in Sect. 9.2.3.1, NO formation is strongly temperature dependent, and tends to peak at slightly lean conditions where the temperature is high and there is available O_2. Therefore, the trend of NO versus ϕ closely follows the trend of temperature versus ϕ. Both CO and HC have an inverted bell shape and their levels become large at very rich and lean mixtures. In rich mixtures, insufficient oxidizer results in incomplete combustion and high levels of CO and HC. In very lean mixtures, the temperature is too low for oxidation of CO and HC. Furthermore, since the temperature is low, the chemical time is long and the reactants may not have enough time to react in the reactor.

From Fig. 9.2 it can be concluded that the most effective way to reduce pollutants emissions is by operating the combustors in a lean combustion mode. Unfortunately, it is difficult to run lean combustors reliably. Figure 9.3 presents the potential difficulties in achieving lean combustion. Flame stability becomes an issue when combustion temperature is low. As such, the interactions between acoustics and the flame may become strong in some systems and lead to pressure coupling effects that can disrupt the flame or damage the combustor. This is especially troublesome in current 'dry' low NO_x lean industrial gas turbine burners.

Fig. 9.2 *Top*: Temperature peaks near the stoichiometric equivalence ratio, $\phi = 1$. *Bottom*: Trends of emission versus ϕ show that lean combustion can achieve low emissions

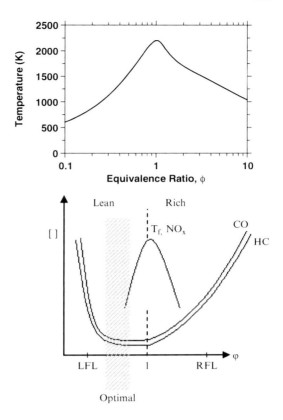

Fig. 9.3 Emissions reductions and potential difficulties in achieving lean combustion: flame stability becomes an issue when combustion temperature is low

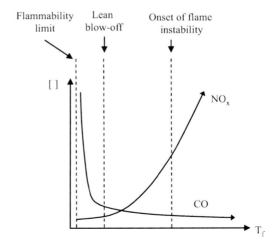

9.2.2 CO Oxidation

In hydrocarbon combustion, CO is the major intermediate species before CO_2 is formed. As such, a substantial amount of CO is formed once the fuel and intermediate hydrocarbon fragments are consumed. Oxidation of CO to CO_2 occurs in the late stages of a combustion process, and it produces a large amount of heat. In most practical systems, the oxidation of CO to CO_2 proceeds through the so-called "wet" route because OH radicals provide the primary reaction as

$$CO + OH \rightarrow CO_2 + H \quad k_f = 1.51 \cdot 10^7 \cdot T^{1.3} \cdot \exp(381/T) \qquad (R1)$$

The rate constant of (R1) has a slightly negative activation temperature but the term $T^{1.3}$ gives the reaction (R1) a weak positive temperature dependence as sketched on the right of Fig. 9.4. The radical OH is thus the determining factor in CO oxidation via step (R1). OH is produced mainly through the chain branching step $H + O_2 \rightarrow OH + O$, which practically stops when temperature drops below 1,100 K.

The other step called the "dry" route involves the following reaction step

$$CO + O_2 \rightarrow CO_2 + O \qquad (R2)$$

Other steps in the oxidation of CO may include

$$CO + O + M \rightarrow CO_2 + M \qquad (R3)$$

Reactions (R1) and (R2) require high temperature (above 1,100 K). The complete conversion of CO to CO_2 also depends on the availability of O_2. In rich mixtures, a large amount of CO remains unconverted due to low temperatures and inadequate O_2. In very lean mixtures, CO remains unconverted because of low

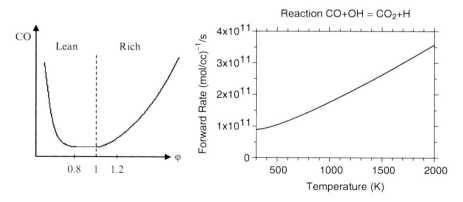

Fig. 9.4 *Left*: CO emissions are minimized at slightly lean equivalence ratios; *Right*: The reaction rate constant of the "wet" CO oxidation route increases with temperature but not a strong function of temperature

temperatures and long chemical times. As such, CO emission has a U-shaped dependence on equivalence ratio as sketched in the left plot of Fig. 9.4.

Reduction of CO emissions can be achieved by increasing combustion temperature or by burning lean. In spark ignition engines or gas turbines, lean burn is used with some success, but unsteady flame dynamics are a common issue in gas turbines due to their acoustic interactions. Since CO oxidation is significantly decreased when the flame temperature drops below 1,100 K, general rules of thumb are to keep combustion gas temperatures above this level and to avoid cold spots. A small amount of water addition to the burn out zone can help by creating more OH, leading to more complete CO oxidation.

Example 9.1 Is 1% CO emission significant in terms of percent of fuel unburned when burning a stoichiometric mixture of C_8H_{18} with air?

Solution:
The stoichiometric combustion is described by

$$C_8H_{18} + 12.5 \cdot (O_2 + 3.76N_2) \rightarrow 8CO_2 + 9H_2O + 12.5 \cdot 3.76N_2 + \text{emissions}$$

The total number of moles in the exhaust per mole of fuel is about $8 + 9 + 12.5 \cdot 3.76 = 64$ mol. 1% CO in the exhaust means 0.64 mol of CO per mol of exhaust gas. Since 1 mol of fuel produces 8 mol of C atoms, 0.64 mol of CO means that $0.64/8 = 0.08$ mol $C_8H_{18} = 8\%$ of fuel is not fully reacted.

9.2.3 Mechanisms for NO Formation

Nitrogen oxides (NO_x) consist of nitric oxide (NO) and nitrogen dioxide (NO_2). NO is formed in combustion processes, and part of the NO can be converted into NO_2 in the cold regions of a combustor. Four major routes of NO formation have been identified in combustion processes: thermal NO (Zeldovich Mechanism), prompt NO (Fenimore), N_2O route, and Fuel Bound Nitrogen (FBN).

9.2.3.1 Thermal NO

Three major steps are responsible for thermal NO formation. The first two form the basis of the well-known Zeldovich mechanism, named after the Russian scientist who proposed it in 1939.

$$N_2 + O \rightarrow NO + N \quad k_1 = 1.8 \cdot 10^{14} \exp(-38,370/T) \tag{R4}$$

$$N + O_2 \rightarrow NO + O \quad k_2 = 1.8 \cdot 10^{10}T \exp(-4,680/T) \tag{R5}$$

$$N + OH \rightarrow NO + H \quad k_3 = 7.1 \cdot 10^{13} \exp(-450/T) \tag{R6}$$

The first reaction is the rate limiting step due to its high activation temperature of about 38,000 K. The high activation energy is caused by the need to break the triple bond in N_2. Once an N atom is formed via reaction (R4), N is consumed immediately by reaction (R5). Reaction (R6) is important in rich parts of flames. One can assume that the N atom is in *quasi-steady state*, i.e., the production rate $\approx$ consumption rate.

$$\frac{d[N]}{dt} = k_{f1}[N_2][O] - k_{f2}[N][O_2] - k_{f3}[N][OH] \approx 0 \qquad (9.1)$$

Equation. 9.1 leads to

$$[N] = \frac{k_{f1}[N_2][O]}{k_{f2}[O_2] + k_{f3}[OH]}$$

With this approximation, the NO production rate becomes[3]

$$\frac{d[NO]}{dt} = k_{f1}[N_2][O] + k_{f2}[N][O2] \cong 2k_{f1}[N_2][O] \qquad (9.2)$$

Since *NO* is formed only at high temperatures ($T > 1,800$ K), the radical O can be assumed to be in the partial equilibrium state via

$$O_2 \leftrightarrow 2O \qquad (9.3)$$

The O concentration is determined by $P_o = \sqrt{K_p}P_{O_2}{}^{1/2}$ where P_o and P_{o2} are the partial pressures of O and O_2. The equilibrium constant K_p depends only on temperature and is determined by

$$K_p = \exp\left(\frac{\hat{g}^o_{O2}}{\hat{R}_u T} - 2\frac{\hat{g}^o_O}{\hat{R}_u T}\right), \qquad (9.4)$$

where $\hat{g}^o$ is the Gibbs free energy at the standard pressure. With $[C] = P_c/\hat{R}_u T$, the concentration of O is related to that of O_2 as

$$[O] = \sqrt{K_c}[O_2]^{1/2} = \sqrt{K_p/\hat{R}_u T}[O_2]^{1/2} \qquad (9.5)$$

Note that K_c depends on temperature only and is approximated by

$$\sqrt{K_c} = 4.1 \cdot \exp\left(-\frac{29,150}{T}\right) \qquad (9.6)$$

---

[3] Same result is obtained if (R6) $N + OH \rightarrow NO + H$ is not included in the analysis.

Using Eq. 9.6, the O radical concentration is determined as

$$[O] = \sqrt{K_c}[O_2]^{1/2} = [O_2]^{1/2} \cdot 4.1 \exp\left(-\frac{29,150}{T}\right) \tag{9.7}$$

and the NO formation rate can be obtained as

$$\frac{d[NO]}{dt} \cong 2k_{f1}[N_2][O]$$

$$\cong 1.476 \cdot 10^{15}[N_2][O_2]^{1/2} \exp\left(-\frac{67,520}{T}\right)(\text{mol/cc} - \text{s}) \tag{9.8}$$

The formation of NO is often expressed in terms of ppm per second (ppm/s). In terms of mole fractions of species, the equation can be written as

$$\frac{dx_{NO}}{dt} \cong 1.476 \cdot 10^{21} x_{N_2} x_{O_2}^{1/2} \exp\left(-\frac{67,520}{T}\right)\left(\frac{P}{\hat{R}_u T}\right)^{1/2}(\text{ppm/s}) \tag{9.9}$$

The above analysis is applicable when the NO level is low. When NO is formed to a sufficiently high level, reverse reactions become important and the net formation of NO is decreased. Figure 9.5 shows a comparison of [O] obtained using Eq. 9.7 against those obtained from equilibrium calculations with good agreement. One can get a nearly perfect match if

$$\sqrt{K_c} = 3.8 \cdot \exp\left(-\frac{29,150}{T}\right)$$

is used. Note that the O atom concentration peaks at the lean side of the flame. Figure 9.6 presents the predicted NO formation rates versus equivalence ratio for hydrogen-air combustion.

Fig. 9.5 Comparison of analytic mole fractions of O atom obtained from the equilibrium expression Eq. 9.7 (*solid dots*) and numerical equilibrium calculations (*line*) showing good agreement for atmospheric hydrogen-air combustion

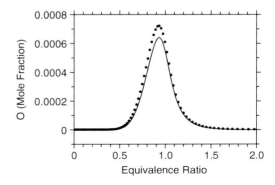

Fig. 9.6 Comparison of predicted NO formation rate using Eq. 9.8 (*solid dots*) against the numerical equilibrium calculations (*lines*)

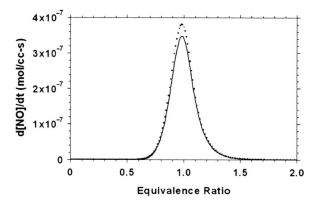

Fig. 9.7 Predicted NO formation rate versus temperature showing little NO formation when temperature is below 1,800 K. The right branch is for rich mixtures (equivalence ratio >1) and the left is for lean combustion

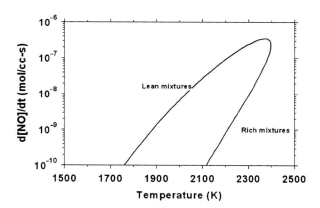

The corresponding NO formation rate is plotted versus temperature for a wide range of equivalence ratios in Fig. 9.7 showing that little NO is formed when temperature is below 1,800 K (note the logarithm scale on the y-axis).

Example 9.2 Estimate the concentration of [O], [N], and $d[NO]/dt$ in a flame at $T = 2,000$ K, $x_{N2} = 0.6$, $x_{O2} = 0.03$ (mole fractions) at 1 atm. Repeat the estimate for $T = 2,100$ K.

Solution:
Using the ideal gas law

$$[C] = \frac{P}{\hat{R}_u T} = 6.1 \cdot 10^{-6} \text{mol/cc},$$

$$[O_2] = 0.03 \cdot (6.1 \cdot 10^{-6}) = 1.83 \cdot 10^{-7} \text{mol/cc},$$

$$[N_2] = 0.6 \cdot (6.1 \cdot 10^{-6}) = 3.66 \cdot 10^{-6} \text{mol/cc}$$

By Eq. 9.7,

$$[O] = \sqrt{K_c} [O_2]^{1/2} = [O_2]^{1/2} \cdot 4.1 \exp\left(-\frac{29,150}{T}\right) = 8.2 \cdot 10^{-10} \text{mol/cc}$$

$$\frac{d[NO]}{dt} \cong 2k_{f1} [N_2][O] \cong 1.476 \cdot 10^{15} [N_2][O_2]^{1/2} \exp\left(-\frac{67,520}{T}\right)$$

$$= 5.035 \cdot 10^{-9} \text{mol/cc} - \text{s}$$

or

$$\frac{dx_{NO}}{dt} \cong 1.476 \cdot 10^{21} x_{N_2} x_{o_2}^{1/2} \exp\left(-\frac{67,520}{T}\right) \left(\frac{P}{\hat{R}_u T}\right)^{1/2}$$

$$= 825 \text{ppm/s}$$

For [N] we will explore two methods:

(a) Using a similar partial equilibrium approach as for [O], we get $K_P = 8 \times 10^{-19}$, and

$$[N] = \sqrt{K_p/\hat{R}_u T} [N_2]^{1/2} = 4.20 \cdot 10^{-15} \text{mol/cc},$$

which is very small in comparison to [O].

(b) Using the quasi-steady state approach and $k_{f1} = 8.382 \times 10^5$, $k_{f2} = 3.468 \times 10^{12}$,

$$[N] = \frac{k_{f1}[N_2][O]}{k_{f2}[O_2]} = 3.96 \cdot 10^{-15} \text{mol/cc},$$

which is in good agreement with the estimate from the partial equilibrium approach.

At $T = 2,100$ K, $[C] = 5.81 \cdot 10^{-6}$ mol/cc, $[O_2] = 1.74 \cdot 10^{-7}$ mol/cc, $[N_2] = 3.49 \cdot 10^{-6}$ mol/cc,

$$\frac{d[NO]}{dt} \cong 2k_{f1}[N_2][O] \cong 1.476 \cdot 10^{15} [N_2][O_2]^{1/2} \exp\left(-\frac{67,520}{T}\right)$$

$$= 2.64 \cdot 10^{-8} \text{mol/cc} - \text{s}$$

or

$$\frac{dx_{NO}}{dt} \cong 1.476 \cdot 10^{21} x_{N_2} x_{O_2}{}^{1/2} \exp\left(-\frac{67,520}{T}\right) \left(\frac{P}{\hat{R}_u T}\right)^{1/2}$$

$$= 4020.2 \, \text{ppm/s}$$

Note that the NO production rate increases more than fourfold when temperature increases by merely 100 K.

9.2.3.2 Prompt NO (Fenimore NO_x)

Oxides of nitrogen can be produced promptly at the flame front by the presence of CH radicals, an intermediate species produced only at the flame front at relatively low temperature. NO generated via this route is named "prompt NOx" as proposed by Fenimore [2]. CH radicals react with nitrogen molecules with the following sequence of reaction steps

$$CH + N_2 \rightarrow HCN + N \tag{R7}$$

$$HCN + N \rightarrow \cdots \rightarrow NO \tag{R8}$$

N atoms generated from (R7) can react with O_2 to produce NO or can react further with HCN leading to NO via a series of intermediate steps. The activation temperature of (R7) is about 9,020 K. In contrast to thermal mechanisms that have an activation temperature about 38,000 K from (R4), prompt NO can be produced starting at low temperatures around 1,000 K. Note that in hydrogen flames, there is no prompt NO as there are no CH radicals.

9.2.3.3 N_2O Route

Under high pressures, the following three-body recombination reaction can produce N_2O through

$$N_2 + O + M \rightarrow N_2O + M \tag{R9}$$

Due to the nature of three-body reactions, the importance of (R9) increases with pressures. Once N_2O is formed, it reacts with O to form NO via

$$N_2O + O \rightarrow NO + NO \tag{R10}$$

Reaction (R10) has an activation temperature around 11,670 K and therefore NO can be formed at low temperatures of around 1,200 K.

9.2.3.4 Fuel-Bound Nitrogen (FBN)

NO_x can be formed directly from fuels, such as coal, containing nitrogen compounds such as NH_3 or pyridine (C_5NH_5). In coal combustion, these compounds evaporate during gasification and react to produce NO_x in the gas phase. This type of NO_x formation can exceed 50% of the total NO_x in coal combustion. FBN is also significant in the combustion of biologically-derived fuels since they typically contain more nitrogen than their petroleum-based counterparts.

9.2.4 Controlling NO Formation

Since the formation of thermal NO is highly sensitive to temperature, reduction in peak flame temperature is the primary mechanism for decreasing NO emissions. When the flame temperature exceeds 1,800 K, a decrease of 30–70 K in peak flame temperature can decrease NO formation by half. As such, reducing peak flame temperature provides an effective means of NO_x reduction. For instance, Fig. 9.8 presents measured NO_x emissions (corrected for 15% O_2) versus equivalence ratio from gas turbines with various flame stabilization devices. Cleary the flame temperature is the dominant controlling parameter in thermal NO_x emissions.

As indicated in the top of Fig. 9.2, lower flame temperatures can be achieved by burning either rich or lean. If a rich mixture is burned, this mixture needs to mix with additional air in order to complete combustion. Due to the difficulty of quickly mixing rich-burned mixtures with air, NO_x can be formed when the mixture passes the stoichiometric point. Such a combustion scheme is called Rich-burn, Quick-mix, Lean-burn (RQL) combustion and it is considered as a potential means to reduce NO_x in various combustion systems including furnaces, aircraft turbines and other internal combustion engines. In land-based practical devices, burning lean is more feasible than burning rich, but flame stability is a challenging issue.

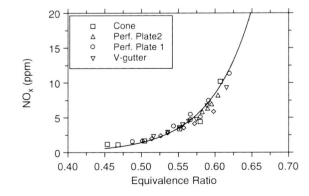

Fig. 9.8 Emission of NO_x (corrected for 15% O_2) from gas turbines with various flame stabilization devices (reprinted with permission from [1])

Injection of water has been practiced in industrial gas turbines to reduce NO by reducing the temperature. However, the water needs to be purified to remove minerals before injection, otherwise minerals in the water will deposit on the combustor liners as well as on downstream turbines. New industrial gas turbines for power generation are run with very lean mixtures, with equivalence ratios of about 0.5. Such turbines are said to run with 'dry' low-NOx burners as water injection is not needed. NO_x levels of 15 ppm corrected to 15% O_2 are now achieved with such technology. The challenges of dry low NO_x technology lie in abatement of the interactions between acoustics and flames. Pressure waves generated by such acoustic-flame interactions can reach 10–20 kPa, which can cause premature fatigue in combustor parts such as transition zones between the burner and turbine inlet.

For automobiles, exhaust gases can be reintroduced into the intake as inert gases to reduce the peak flame temperature. This method is referred to as "exhaust gas recirculation" (EGR). This method has been also effectively applied to furnaces and boilers where it is called "flue gas recirculation" (FGR). Staged combustion that avoids high temperature regions has been shown to reduce NO.

In non-premixed combustion systems, diffusion flames react at near stoichiometric conditions, resulting in near-maximum flame temperatures and consequent production of large amounts of NO_x. Thus, from a NO_x reduction point of view, one should always avoid non-premixed flames when possible. However, non-premixed flames are far more stable than lean premixed flames. For instance, in aviation gas turbines, only non-premixed flames are used for safety reasons. A potential approach to reducing NO_x formation is to induce turbulence so that the fuel burns in a partially premixed way and the flame temperature is reduced.

9.2.5   Soot Formation

Flickering candle lights, fires, and combustors produce soot. Formation of soot means a loss of usable energy. Deposits of soot vitiate the thermal and mechanical properties of an engine. The distribution of soot directly affects the heat radiation and the temperature field of a flame. In boilers, one may want to increase soot formation to enhance radiative heat transfer to the water. The exhaust gas of diesel engines contains fine soot particles which are suspected to cause cancer.

Soot consists of agglomerates with diameters of up to several hundred nanometers. These have a fine structure of spherical primary particles. Soot formation starts with the pyrolysis of fuel molecules in the rich part of flames and the formation of polycyclic aromatic hydrocarbons (PAH). The most important precursor of the formation of higher hydrocarbons is acetylene (C_2H_2). Two-dimensional condensation processes follow. Finally, a rearrangement produces spherical primary particles that continue growing at their surface. Three distinct steps are used to model soot in flames: nucleation, agglomeration, and oxidation.

The black soot clouds of the diesel engines prior to the 1980s are gone, as industry uses high pressure injectors to decrease the size of soot. However, the remaining invisible fine particles are a severe toxicological problem, as they can penetrate deeper into human tissues. These fine particles likely cause asthma and cardiac infarctions. Soot formation in engines driven by hydrocarbons, especially diesel engines and aircraft turbines, are the focus of current research. However, while the formation of nitric oxides in internal combustion engines is well understood, formation of soot is by far more complicated and difficult to examine. The formation of soot particles in diesel sprays is so fast and complex that it is not sufficiently understood yet. Practical approaches to trap the soot particles at the exhaust of the diesel engine are currently being implemented successfully.

9.2.6   Relation Between NO_x and Soot Formation

The top plot in Fig. 9.9 illustrates the relation between soot and NO_x with equivalence ratio and temperature as two independent parameters. For spark ignition engines running with a stoichiometric mixture, the reaction pathway for the combustible mixture is represented by the horizontal line at $\phi = 1$. As the flame temperature can reach 2,500–2,600 K in an internal combustion engine, a large amount of NO_x is formed with exhaust levels reaching 1,000 ppm. Since the mixture is premixed, there is basically no soot formed during the combustion process. In contrast, a diesel engine operates with injection of fuel into hot compressed air near top dead center. The arrows sketched in Fig. 9.9 represent a desirable pathway of the fuel mixture from rich toward lean during the rich flame premixed zone followed by non-premixed flames. The goal is to modulate the injection of fuel to avoid both soot and NO_x formation. Experience shows that since NO_x and soot are formed in different regions in the (ϕ, T) map, NO_x and soot often exhibit a trade-off relation as shown in the bottom plot in Fig. 9.9. A small soot production is at the expense of large NO_x formation and vice versa.

9.2.7   Oxides of Sulfur (SO_x)

Oxides of sulfur from combustion processes may consist of SO, SO_2, and SO_3. Among these species, SO_3 has great affinity for water. At low temperatures, it creates sulfuric acid (H_2SO_4) via

$$SO_3 + H_2O \rightarrow H_2SO_4 \qquad \text{(R11)}$$

In a combustion system with a fuel containing elemental sulfur or a sulfur-bearing compound, the predominant product is SO_2. However, the concentration of SO_3 is generally larger than that expected from the equilibrium value for the reaction

Fig. 9.9 *Top*: Soot and NO_x
relation in terms of
equivalence ratio and
temperature. *Bottom*: tradeoff
between NO_x and soot as
function of injection timing

See p 234

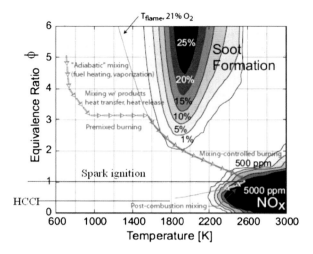

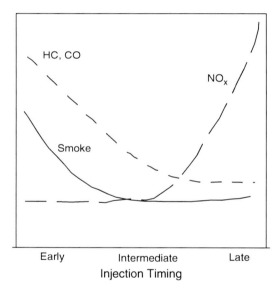

$$SO_2 + \frac{1}{2}O_2 \rightarrow SO_3 \qquad \qquad \text{(R12)}$$

Under fuel rich conditions, the stable products are sulfur dioxide (SO_2), hydrogen
sulfide (H_2S), carbon disulfide (CS_2), and disulfur (S_2). The radical sulfur monoxide
(SO) is an intermediate species that is highly reactive with O at high temperatures to
form SO_2 via

$$SO + O \rightarrow SO_2 \qquad \qquad \text{(R13)}$$

Sulfur trioxide (SO_3) is important because of the production of H_2SO_4 via (R11).
As indicated in reaction step (R12), SO_3 production is very sensitive to the initial

O_2 concentration. There is practically no SO_3 formed under fuel rich conditions even close to the stoichiometric point. However, if there is even 1% excess air, a sharp increase in SO_3 is observed. The melting point of H_2SO_4 is 10°C and formation of aerosols may occur if the temperature drops below 10°C.

9.3 Quantification of Emissions

There are many ways to quantify the emissions depending on the particular application of interest. One generic way to define the level of emission is called the "emission index" (*EI*). The EI_i for a certain chemical species is defined as the ratio of the mass of the pollutant species i to the mass of fuel burned as

$$EI_i \equiv \frac{m_{i,emitted}}{m_{f,burned}} \tag{9.10}$$

Since *EI* is a dimensionless quantity, the units are conventionally expressed as g/kg. Measurements of exhaust gases can be used to estimate *EI*. For instance, measurements of CO_2, O_2, CO, NO_x, and HC can be made by using a sampling probe and gas analyzers. Results are expressed in term of dry mole fractions, as water vapor needs to be removed before the exhaust gas is sent to the gas analyzer. Otherwise, water will condense inside the gas analyzer and cause the analyzer to malfunction. The unburned hydrocarbons are measured as equivalent to a certain hydrocarbon species, such as C_3H_8 or C_6H_{14}. Assuming that CO_2, CO, and unburned hydrocarbons are the major combustion products and all other species are negligible, EI_i can be determined for general hydrocarbon fuels by

$$\begin{aligned} EI_i &\equiv \frac{m_{i,emitted}}{m_{f,burned}} = \frac{x_{i,emitted} M_i}{(x_{CO_2} + x_{CO} + 3x_{C_3H_8})/\alpha \cdot M_f} \\ &\approx \frac{x_{i,emitted} M_i}{(x_{CO_2} + x_{CO})/\alpha \cdot M_f} \end{aligned} \tag{9.11}$$

Here $\alpha = N_c/N_f$ is the number of moles of carbon in 1 mol of fuel. M_i denotes the molecular mass of *i-th* species and x_i is its mole fraction. The last approximation of Eq. 9.11 is reasonable if the concentration of unburned HCs is small (<1,000 ppm).

Although *EI* is general, it may not be the best representative for all applications. For instance, in automobile or power generation applications, mass species emission (MSE, units g/(kW-h)) is used and its definition is

$$MSE \equiv \frac{\text{mass flow of pollutant species}}{\text{brake power produced}} \tag{9.12}$$

Using the definition of *EI*, *MSE* can be expressed in terms of *EI* as

$$MSE \equiv \frac{\dot{m}_f EI}{\dot{W}}, \tag{9.13}$$

where $\dot{m}_f$ is the fuel mass flow rate (kg/h), $\dot{W}$ is the brake power generated (kW), and EI is the emission index (g/kg). For furnace applications, the level of pollution is often expressed as

$$\frac{\text{mass of pollutant}}{\text{heat of combustion}} = \frac{EI}{\text{LHV or HHV}}. \tag{9.14}$$

The natural gas industry uses

$$\frac{\text{mass of pollutant}}{10^6 \text{m}^3 \text{ of natural gas}}.$$

Example 9.3 A spark-ignition engine runs with isooctane. The measured exhaust data indicates: $CO_2 = 12.47\%$, $CO = 0.12\%$, $O_2 = 2.3\%$, $HC = 367$ ppm, and $NO = 76$ ppm. All concentrations are expressed by volume (mole fractions) on a "dry" basis, i.e. water vapor is removed before measurements are taken. For this analysis, HC is assumed to be composed exclusively of C_6H_{14}. Determine emission indices for NO, CO, and HC.

Solution:

$$C_8H_{18} + (1/\phi) \cdot 12.5 \cdot (O_2 + 3.76N_2) \rightarrow \text{exhaust gases}$$

Using Eq. 9.11

$$EI_i \approx \frac{x_{i,emitted}M_i}{(x_{CO2} + x_{CO})/\alpha \cdot M_f},$$

since the ratio x_{dry} between x_{wet} is the same for all species, we can use dry values (see note below).

$$\begin{aligned}
EI_{CO} &\approx \frac{x_{CO}M_{co}}{(x_{CO_2} + x_{CO})/\alpha \cdot M_f} \\
&= \frac{0.0012 \cdot 28}{(0.1247 + 0.0012)/8 \cdot 114.2} = 0.0187 \text{ kg/kg} \\
&= 18.7 \text{ g/kg}
\end{aligned}$$

Similarly, $EI_{NO} = 1.27$ g/kg, $EI_{HC} = 17.63$ g/kg

Note that when a "dry" mixture is measured, the relation between real (wet) mole fraction and dry fraction can be derived for lean combustion $\phi \leqslant 1$ as follows:

$$C_\alpha H_\beta O_\gamma + \frac{1}{\phi}\left(\alpha + \frac{\beta}{4} - \frac{\gamma}{2}\right)(O_2 + 3.76N_2) \rightarrow$$

$$\alpha CO_2 + \frac{\beta}{2}H_2O + \frac{3.76}{\phi}\left(\alpha + \frac{\beta}{4} - \frac{\gamma}{2}\right)N_2 + \left(\alpha + \frac{\beta}{4} - \frac{\gamma}{2}\right)\left(\frac{1}{\phi} - 1\right)O_2 +$$

trace of pollutant species

$$\frac{x_{i,dry}}{x_{i,wet}} = \frac{\alpha + \frac{\beta}{2} + \frac{3.76}{\phi}\left(\alpha + \frac{\beta}{4} - \frac{\gamma}{2}\right) + \left(\alpha + \frac{\beta}{4} - \frac{\gamma}{2}\right)\left(\frac{1}{\phi} - 1\right)}{\alpha + \frac{3.76}{\phi}\left(\alpha + \frac{\beta}{4} - \frac{\gamma}{2}\right) + \left(\alpha + \frac{\beta}{4} - \frac{\gamma}{2}\right)\left(\frac{1}{\phi} - 1\right)} = \frac{\alpha + \frac{\beta}{2} + \left(\frac{4.76}{\phi} - 1\right)\left(\alpha + \frac{\beta}{4} - \frac{\gamma}{2}\right)}{\alpha + \left(\frac{4.76}{\phi} - 1\right)\left(\alpha + \frac{\beta}{4} - \frac{\gamma}{2}\right)}$$

Example 9.4 In the gas turbine industry, NO and CO emissions are quantified as ppm of NO_x and % of CO in the exhaust stream corrected to 15% O_2. Considering lean methane-air combustion, find the equivalence ratio at which the exhaust gases contain 15% of O_2. For a lean combustion system, the exhaust gases contain 12.47% CO_2, 0.12% CO, 2.3% O_2, 76 ppm of NO, and 367 ppm of HC equivalent to C_6H_{14}. Determine the corresponding emissions corrected at 15% O_2.

Solution:

1. For lean combustion, we have

$$C_\alpha H_\beta O_\gamma + \frac{1}{\phi}\left(\alpha + \frac{\beta}{4} - \frac{\gamma}{2}\right)(O_2 + 3.76N_2) \rightarrow$$

$$\alpha CO_2 + \frac{\beta}{2}H_2O + \frac{3.76}{\phi}\left(\alpha + \frac{\beta}{4} - \frac{\gamma}{2}\right)N_2 + \left(\alpha + \frac{\beta}{4} - \frac{\gamma}{2}\right)\left(\frac{1}{\phi} - 1\right)O_2$$

$$+ \text{ trace of pollutant species}$$

For methane,

$$\alpha + \frac{\beta}{4} - \frac{\gamma}{2} = 2,$$

so that x_{O2} is

$$x_{O_2} = \frac{2\left(\frac{1}{\phi} - 1\right)}{1 + 2 + \frac{3.76}{\phi}2 + 2\left(\frac{1}{\phi} - 1\right)} = \frac{2\left(\frac{1}{\phi} - 1\right)}{1 + 2\frac{4.76}{\phi}}$$

For the exhaust to contain 15% O_2, $\phi = 0.266$.

2. Let's denote the amount of extra air to be added to the exhaust mixture as X mol for 1 mol of exhaust gas. The resulting mole fraction of O_2 is

$$x_{O_2} = \frac{0.023 + X \cdot 0.21}{1 + X} = 0.15$$

Solving for X, $X = 2.117$.
The emission levels corrected to 15% O_2 are:

$$CO: 0.12\%/(1 + X) = 0.038\%$$

$$NO: 76 \text{ ppm}/(1 + X) = 24.38 \text{ ppm}$$

Exercises

9.1. A mixture of gases containing 3% O_2 and 60% N_2 by volume at room temperature is suddenly heated to 2,000 K at 1 atm pressure. Assume that the fractions of O_2 and N_2 are unchanged by the sudden gas heating process. Find the initial rate of formation of NO (ppm/s) after the mixture is heated. Indicate whether the NO formation rate will then increase or decrease as time progresses. Use the Zeldovich mechanism and assume there is no hydrogen in the mixture. Repeat the same analysis at a temperature of 1,200 K.

EPA Pollution Standards

Pollutant	Criteria pollutants		
	Description	Sources	Health effects
Carbon monoxide (CO)	An odorless, tasteless, colorless gas which is emitted primarily from any form of combustion	Mobile sources (autos, trucks, buses), wood stoves, open burning, industrial combustion sources	Deprives the body of oxygen by reducing the blood's capacity to carry oxygen; causes headaches, dizziness, nausea, listlessness and in high doses, may cause death*
Hydrocarbons (HC)	Unburned, partially burnt fuel	Mobile sources (autos, trucks, buses), formed by the incomplete combustion of fuel	When combined with sun light produces photo chemical (smog)
Lead (Pb)	A widely used metal, which may accumulate in the body	Leaded gasoline, smelting, battery manufacturing and recycling	Affects motor function and reflexes and learning; causes damage to the central nervous system, kidneys and brain. Children are affected more than adults
Ozone (O_3)	Formed when nitrogen oxides and volatile organic compounds react with one another in the presence of sunlight and warm temperatures. A component of smog	Mobile sources, industry, power plants, gasoline storage and transfer, paint	Irritates eyes, nose, throat and respiratory system; especially bad for those with chronic heart and lung disease, as well as the very young and old, and pregnant women
Nitrogen dioxide (NO_2)	A poisonous gas produced when nitrogen oxide is a by-product of	Fossil fuel power, mobile sources, industry, explosives	Harmful to lungs, irritates bronchial and respiratory systems; increases

Pollutant	Criteria pollutants		
	Description	Sources	Health effects
	sufficiently high burning temperatures	manufacturing, fertilizer manufacturing	symptoms in asthmatic patients
Particulate Matter $PM_{10}PM_{2.5}$	Particles of soot, dust, and unburned fuel suspended in the air	Wood stoves, industry, dust, construction, street sand application, open burning	Aggravates ailments such as bronchitis and emphysema; especially bad for those with chronic heart and lung disease, as well as the very young and old, and pregnant women
Sulfur dioxide (SO_2)	A gas or liquid resulting from the burning of sulfur-containing fuel	Fossil fuel power plants, non-ferrous smelters, kraft pulp production	Increases symptoms in asthmatic patients; irritates respiratory system

Air Quality Standards

New standards for particulate matter smaller than 2.5 μm in size ($PM_{2.5}$) and ozone were adopted by EPA in 1997.

Pollutant	National		Washington state
	Primary	Secondary	
Carbon monoxide (CO)			
8 h average	9 ppm	9 ppm	9 ppm
1 h average	35 ppm	35 ppm	35 ppm
Lead (Pb)			
Quarterly average	1.5 $\mu g/m^3$	1.5 $\mu g/m^3$	No standard
Nitrogen dioxide (NO_2)			
Annual average	0.053 ppm	0.053 ppm	0.05 ppm
Ozone (O_3)			
1 h average	0.12 ppm	0.12 ppm	0.12 ppm
8 h average[b]	0.08 ppm	0.08 ppm	No standard
Particulate Matter (PM_{10})			
Annual arithmetic mean	50 $\mu g/m^3$	50 $\mu g/m^3$	50 $\mu g/m^3$
24 h average	150 $\mu g/m^3$	150 $\mu g/m^3$	150 $\mu g/m^3$
Particulate Matter (PM2.5)			
Annual arithmetic mean	15 $\mu g/m^3$	15 $\mu g/m^3$	No standard
24 h	65 $\mu g/m^3$	65 $\mu g/m^3$	No standard
Sulfur dioxide (SO_2)			
Annual average	0.030 ppm	No Standard	0.02 ppm
24 h average	0.14 ppm	No Standard	0.10 ppm
3 h average	No standard	0.5 ppm	No standard
1 h average	No standard	No standard	0.40 ppm[b]

Pollutant	National		Washington state
	Primary	Secondary	
Total suspended particulates			
Annual geometric mean	No standard	No standard	60 $\mu g/m^3$
24 h average	No standard	No standard	150 $\mu g/m^3$

Primary standards are listed in this table as they appear in the federal regulations. Ambient concentrations are rounded using the next higher decimal place to determine whether a standard has been exceeded. The data charts in this report are shown with these un-rounded numbers.

Details of the national standards are available in 40 CFR Part 50

ppm parts per million, $\mu g/m^3$ micrograms per cubic meter

[a] 0.25 not to be exceeded more than two times in any 7 consecutive days

[b] Eight hour ozone standard went into effect on September 16, 1997, but implementation is limited.

References

1. Correa, SM (1993) "A review of NO$_x$ Formation under gas-turbine combustion conditions," Comb. Sci. Technol. 87:329-362.
2. Fenimore CP (1971) Formation of nitric oxide in premixed hydrocarbon flames. Symposium (International) on Combustion 13(1):373-380.

Chapter 10
Premixed Piston IC Engines

Internal combustion (IC) engines have been moving the industrial world for over three centuries. Huygens and Papin's first proposal of a gunpowder-powered engine in the 1680s started a revolution for the new industrial world. For the next 50 years, numerous types of engines (mainly steam engines) were invented and produced. Many failed to meet the commercial needs of the time, but others prevailed. An example of the first successful IC engine was Lenoir's single-cylinder, two-stroke gas engine in 1860. By the early nineteenth century, liquid fuels were made increasingly available from oil wells in the United States. The convenience of liquid fuels and their high energy density compared to gaseous fuels promoted the rise of internal combustion engines. Otto patented his first four-stroke IC engine in 1876. Otto claimed that his engine was more quiet and efficient than steam engines. Many others such as Daimler followed in Otto's footsteps. Descendants of Otto's engine, the modern spark-ignited (SI) engines can be found in every corner of the globe. Because of the high power density, low cost of production, and the vast infrastructure for gasoline, SI engines are ideal power platforms for passenger cars, small trucks, motorcycles, lawn mowers, and small electrical power generators. SI engines are robust and capable of producing high levels of power at wide speed ranges. However, SI engines usually require throttling to control the power output, which increases the engine's pumping losses and decreases overall efficiency. Current opportunities for internal combustion engine research include efficiency improvement, novel fuel implementation, and pollution reduction.

10.1 Principles of SI Engines

The premixed piston SI engine is an engine in which premixed fuel and oxidizer are introduced into the combustion chamber through an intake manifold. The combustible mixture is compressed by a piston to reach a high temperature and pressure. When the piston is near the top of the compression stroke (top dead center or TDC), combustion is initiated by a spark plug, and a premixed flame develops and propagates through the cylinder, creating gases with even higher temperature and pressure.

S. McAllister et al., *Fundamentals of Combustion Processes*,
Mechanical Engineering Series, DOI 10.1007/978-1-4419-7943-8_10,
© Springer Science+Business Media, LLC 2011

Expansion of these gases produces direct force on the piston, thereby producing useful mechanical work. Because of combustion stability problems, the spark-ignited engine requires the use of near-stoichiometric air/fuel mixtures to ensure a successful ignition event and subsequent flame propagation. As we learned in Chap. 2, a mixture (often referred to as "charge") at stoichiometric conditions produces the highest flame temperature possible and consequently the highest power output. Unfortunately, the high temperatures also generate high levels of nitric oxide (NO_x) emissions.

The thermal efficiency of a SI engine is strongly dependent on the compression ratio of the engine, thus one might attempt to improve efficiency by increasing the compression ratio. However, the amount that the compression ratio can be increased is limited by the onset of a phenomenon known as engine knock, which is the autoignition of the gases ahead of the propagating flame front in the combustion chamber. This autoignition, or knocking, is a result of compression heating of the unburned mixture by the expanding burned gases. A rapid pressure rise occurs upon autoignition of the unburned "end gas," initiating propagation of a strong pressure wave across the combustion chamber that can "scrape off" the boundary layer, exposing the piston surface to the core gas temperature. In time, piston damage may result. The high peak pressures can also damage the spark plug and head gasket. Spark-ignited engines are also notorious for cyclic variation in performance. Cyclic variation can result in loss of engine efficiency as well as increased engine emissions. Figure 10.1 shows a set of cylinder pressure traces obtained from a Pontiac 1.6 L SI engine. Note that the peak cylinder pressure varies between each cycle. The main cause of cyclic variation in SI engines is ignition lag, which is the time required for initiating a flame kernel following the passage of a spark.

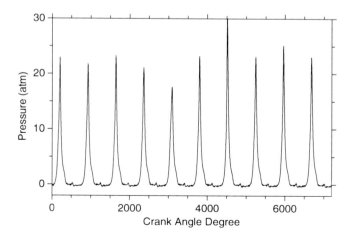

Fig. 10.1 Typical SI engine pressure traces (ten consecutive cycles)

10.2 Thermodynamic Analysis

The thermodynamic cycle that describes the SI engine is the Otto cycle. Thus, thermodynamic efficiency of a SI engine under idealized conditions (standard air assumption[1]) is given by

$$\eta = 1 - \frac{1}{CR^{\gamma-1}}, \qquad (10.1)$$

where γ is the ratio of specific heats c_p/c_v, and CR is the compression ratio, V_{max}/V_{min}. It is interesting that the thermal efficiency depends only on compression ratio and γ. The temperature after the isentropic compression stroke is $T_2 = T_1 \cdot CR^{\gamma-1}$. Higher compression ratios lead to higher flame temperatures and therefore one anticipates an increase in thermodynamic efficiency. For the same reason, at a given CR, η increases with γ as shown in Table 10.1 for $CR = 8.5$.

Therefore, it is desirable to use a working media with a large γ value. The highest compression ratio that can be used in an IC engine is limited by autoignition during combustion (engine knock). The relation between the critical pressure and temperature discussed in Chap. 5 plays a vital role. As shown in Fig. 10.2 below, at the

Table 10.1 Dependence of theoretical thermal efficiency on ratio of specific heats

Working media	$\gamma = c_p/c_v$	Efficiency, η (%)
Air	1.4	57.5
CO_2	1.288	46.0
Ar	1.667	76.0
He	1.667	76.0

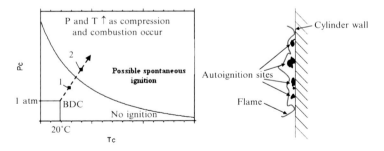

Fig. 10.2 *Left:* Autoignition can occur when critical pressure and temperature are exceeded in the engine *Right:* Engine knock occurs when unburned gases autoignite

---

[1] The standard air assumptions are that the mixture is entirely air that behaves as an ideal gas, all processes are internally reversible, and that the combustion and exhaust processes are heat addition/rejection processes.

beginning of the compression stroke, the mixture of fuel and air is at a temperature and pressure well below the critical pressure-temperature curve. As compression occurs, the mixture temperature and pressure increase. The mixture is typically ignited with a spark plug before the compression stroke ends (TDC). The flame then propagates through the mixture, the hot combustion products expand, and the unburned gases are further heated and compressed. Ideally, the flame will propagate through the entire mixture before the unburned gases reach the critical pressure and temperature (for instance, point 1 in Fig. 10.2). If this doesn't occur and the unburned mixture reaches state 2, autoignition can occur causing the engine to knock. Increasing the compression ratio increases both temperature and pressure at the end of the compression stroke and therefore increases the likelihood of auto-ignition. In addition, any hot spot in the combustion chamber can also promote autoignition.

A fuel's ability to resist knock is quantified by its octane number (for more detail, see Sect. 10.4). Increasing a fuel's octane number shifts the critical pressure-temperature curve seen in Fig. 10.2 upward, so that a higher temperature and pressure, and thus compression ratio, can be reached without autoignition. Different octane number gasolines are produced through the crude oil distillation process and by addition of chemical components. Figure 10.3 sketches power output and required octane number as function of compression ratio for a typical gasoline engine. When the compression ratio is increased, engine output increases, but a higher octane number fuel is needed to prevent autoignition. As will be discussed in Chap. 11, diesel engines operate on a different principle: autoignition of the fuel mixture is desired. In this case, the engine operates so that the temperature and pressure of the mixture at the end of the compression stroke is well above the autoignition curve, at say point 2 in Fig. 10.2.

Example 10.1 You are given a new biofuel and need to figure out if it will cause your spark ignition engine to knock. At the beginning of the compression stroke, the stoichiometric fuel/air mixture is at 25°C and 101.3 kPa. The mixture is then isentropically compressed with a volumetric compression ratio of 10. If the engine cooling system provides a convective heat transfer coefficient of 100 W/m^2−K, does the mixture autoignite? Assume that the surface area to volume ratio is 0.05 m^{-1} and the engine coolant is at 97°C. The properties of the fuel are:

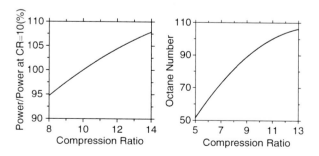

Fig. 10.3 Increasing compression ratio increases power but requires higher octane fuel

$$E_a/R = 20,000K \quad \hat{Q}_c = 1.81 \text{ MJ/mol fuel} \quad a = 0.25 \quad b = 1.5$$

$$A_0 = 2.1 \cdot 10^9 \quad \text{Stoichiometric relation: Fuel} + 6.5 \cdot \text{Air} \rightarrow \text{Products}$$

Solution:
We first need to calculate the temperature and pressure of the mixture at the end of the compression stroke. Using the isentropic relations assuming the mixture is mostly air:

$$\left(\frac{T_2}{T_1}\right) = \left(\frac{V_1}{V_2}\right)^{k-1} = (10)^{1.387-1} \rightarrow T_2 = (25^{\circ}C + 273)(10)^{0.387} = 726.5K$$

$$\left(\frac{P_2}{P_1}\right) = \left(\frac{V_1}{V_2}\right)^{k} = (10)^{1.387} \rightarrow P_2 = 101.3\text{kPa} \cdot (10)^{1.387} = 2469.5\text{kPa}$$

The minimum condition for autoignition is when the heat losses balance the heat generation. Because the temperature and pressure increase during the compression stroke, the amount of heat generated by the combustible mixture will also increase so that autoignition is most likely going to occur at the end of the compression stroke. To determine whether autoignition will occur, we must evaluate the heat generated and the heat lost at the top of the compression stroke:

$$\dot{q}'''_{loss} = h\frac{A}{V}(T - T_\infty) = \left(100\,\frac{\text{kW}}{\text{m}^2\text{K}}\right)\left(0.05\,\frac{1}{\text{m}}\right)(726.5 - 370K)$$

$$= 1783\,\frac{\text{W}}{\text{m}^3}$$

$$\dot{q}'''_{gen} = \hat{r}\hat{Q}_c = A_0 \exp\left(\frac{-E_a}{RT}\right)x_f x_o \left(\frac{P}{RT}\right)^{a+b}\hat{Q}_c$$

$$x_f = \frac{1}{1 + 6.5 * 4.76} = 0.0313 \qquad x_o = \frac{6.5}{1 + 6.5 * 4.76} = 0.2035$$

$$\hat{r} = (2.1 \cdot 10^9) \exp\left(\frac{-20,000K}{726.4K}\right)(0.0313)^{0.25}(0.2035)^{1.5}$$

$$\times \left[\frac{2469.5\text{kPa} \cdot 1000\,\frac{\text{Pa}}{\text{kPa}}}{\left(8.314\,\frac{\text{Pa}\cdot\text{m}^3}{\text{mol}-\text{K}}\right)\left(100^3\,\frac{\text{cm}^3}{\text{m}^3}\right)726.4K}\right]^{0.25+1.5}$$

$$\hat{r} = 1.05 \cdot 10^{-10}\,\frac{\text{mol}}{\text{cc} - \text{s}}$$

$$\dot{q}'''_{gen} = \left(1.05 \cdot 10^{-10}\,\frac{\text{mol}}{\text{cc} - \text{s}}\right)\left(100^3\,\frac{\text{cc}}{\text{m}^3}\right)\left(1.81 \cdot 10^6\,\frac{\text{J}}{\text{mol}}\right) = 191\,\frac{\text{W}}{\text{m}^3}$$

Fig. 10.4 Pressure-volume
trace from a typical IC engine

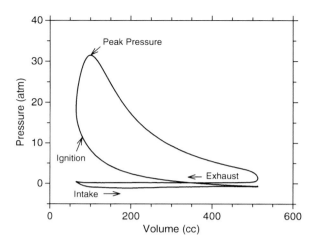

Because the heat lost $(1,783 \text{ W/m}^3)$ is greater than the heat generated (191 W/m^3) the mixture will not autoignite due to the compression.

The $P\text{-}V$ diagram of actual engines differs somewhat from the ideal Otto cycle diagram due to heat losses, friction, and the finite amount of time required for release of the fuel energy. Figure 10.4 sketches a typical pressure trace versus volume. The volume of the combustion chamber is a function of the rotational position of the crankshaft (θ), which can be measured with units of crank angle degrees (CAD) using a shaft encoder mounted on the crankshaft. With knowledge of crankshaft position and engine geometry, the engine cylinder volume can be determined by using the slider-crank formula [2].

$$V = V + \frac{\pi B^2}{4}(l + a - s) \tag{10.2}$$

where Vc is the clearance volume (volume at TDC), B is the bore (cylinder diameter), l is the connecting rod length (rod between crankshaft and piston), a is the crankshaft radius, and s is the distance between the center of the crankshaft and the piston and is given by

$$s = a\cos(\theta) + (l^2 - a^2 \sin^2 \theta)^{\frac{1}{2}}$$

The indicated work done by a piston engine between crank angle degree θ_1 and crank angle degree θ_2 can be calculated by integrating the cylinder pressure data as

$$Work = \int_{\theta_1}^{\theta_2} PdV \approx \sum_{\theta_1}^{\theta_2} P(\theta)\frac{dV}{d\theta}(\theta)\Delta\theta \tag{10.3}$$

The Indicated Mean Effective Pressure (IMEP) is defined as

$$IMEP = \frac{Work}{Swept \cdot Volume} \tag{10.4}$$

Brake mean effective pressure (BMEP) is the measured mean effective pressure from dynamometer testing of the engine. Brake specific fuel consumption (BSFC)

Table 10.2 Lower Heating Values (LHV) of some commonly used fuels

Fuel	(MJ/kmol)	(MJ/kg)
Methane (CH_4)	802.64	50.031
Propane (C_3H_8)	2043.15	46.334
Butane (C_4H_{10})	2652.34	45.73
Methanol (CH_3OH)	676.22	21.104
Iso-Octane (C_8H_{18})	5100.50	44.651

Table 10.3 Current design and operation of IC engines

IC	Operation	CR	Max. RPM	BMEP (atm)	BSFC (g/kW-h)
Small	2 S[a], 4 S[a]	6–11	4,500–7,500	4–10	350
Cars	4 S	8–10	4,500–6,500	7–10	270
Trucks	4 S	8–12	3,600–5,000	6.5–7	300
Large gas engines	2 S, 4 S	8–12	300–900	6.8–12	200
Wankel engines	4 S	9	6,000–8,000	9.5–10.5	300

[a] 2 S: 2-stroke; 4 S: 4-stroke

is a measure of an engine's efficiency. It is the rate of fuel consumption divided by power production. The indicated efficiency (η_i) is defined as followed:

$$\eta_i = \frac{Power_i}{\dot{m}_f \, LHV},$$
(10.5)

where the indicated power is measured[2] in kW, $\dot{m}_f$ is the mass flow rate of fuel and LHV is the lower heating value of the fuel in MJ/kg. Table 10.2 shows the LHVs of several commonly used fuels. This definition is a thermodynamic measurement only and neglects mechanical losses such as driveline losses and oil /coolant pump losses.

Due to fluid-dynamic losses during intake and exhaust gas exchanges, the transfer of gases through combustion chamber valves is not perfect. Volumetric efficiency (η_v) is an indication of the engine's intake and exhaust performance compared to the ideal situation without any loss. Volumetric efficiency is defined as

$$\eta_v = \frac{\dot{V}_a}{V_s \cdot N},$$
(10.6)

where $\dot{V}_a$ is the actual *volumetric flow rate* at standard temperature and pressure (STP) for engines without boost pressure (turbocharging or supercharging), V_s is the cylinder swept volume, and N is half the number of revolutions per second for 4-stroke engines. Since the combustible mixture is introduced into the cylinder through a relatively small opening between the intake valve and engine block, volumetric efficiency decreases with engine speed [3]. With proper tuning of an intake manifold (sometimes with the help of an acoustic box), the volumetric efficiency can be extended to higher engine speeds before it starts to decrease. Table 10.3 summarizes the typical design and operation of IC engines.

---

[2] When the engine is connected to a dynamometer, the power produced by an engine can be determined by $Power(\text{kw}) = \frac{2\pi * Torque(NM) * RPM}{60,000}$.

Combustion efficiency (η_c) is a measure of how completely a mixture combusts in the engine cylinder and is defined as follows:

$$\eta_c = \frac{(\dot{m}_f\, h_f) + (\dot{m}_a\, h_a) - (\dot{m}_e\, h_e)}{(\dot{m}_f\, LHV)}, \tag{10.7}$$

where $\dot{m}_f$ is the mass flow rate of fuel into the engine, $\dot{m}_a$ is the mass flow rate of air into the engine, $\dot{m}_e$ is the mass flow rate of exhaust flowing out of the engine, and h_f, h_a, and h_e are the enthalpy of fuel, air, and exhaust gas, respectively. A combustion efficiency, η_c, of about 90% is considered as a successful combustion event. In most IC engines, about 10% of the inducted mass leaves the engine unburned due to cold boundary layers near cylinder walls and crevices.

10.3   Relationship between Pressure Trace and Heat Release

Heat release data can provide valuable information useful for better understanding engine performance. Though direct measurement of heat release rates in an engine would be difficult, heat release rate can be deduced from time histories of cylinder pressure and volume. In-cylinder pressure can be measured using a pressure transducer. Again, with knowledge of crankshaft position (CAD, θ) and engine geometry, the engine cylinder volume can be determined by using the slider-crank formula. With this information, the relation between heat release rate and pressure changes is deduced in the following:

The first law of thermodynamics gives

$$\delta Q = dE + \delta W + \delta Q_{loss} \tag{10.8}$$

The internal energy is $E = mc_vT$. For the period from compression stroke to expansion stroke, let us assume that the mass inside the cylinder is constant ($m =$ constant) and c_v is constant.

$$\delta Q = mc_v dT + PdV + \delta Q_{loss} \tag{10.9}$$

Using the ideal gas law

$$PV = mRT \tag{10.10}$$

we get $dT = d(PV)/mR$. Eq. (10.9) becomes

$$\delta Q = \frac{c_v}{R} d(PV) + PdV + \delta Q_{loss}$$
$$\delta Q - \delta Q_{loss} = \left(\frac{c_v}{R} + 1\right)PdV + \frac{c_v}{R} VdP \tag{10.11}$$

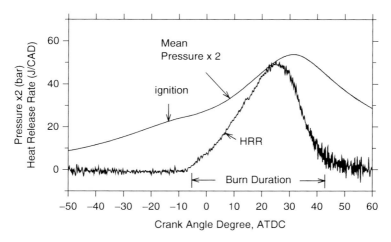

Fig. 10.5 Pressure trace and heat release rate versus CAD for a research engine with a fuel mixture of 70% isooctane and 30% n-heptane

With $c_p/c_v = \gamma$ and $c_p\text{-}c_v = R$, the net heat release rate $(\delta Q - \delta Q_{loss})$ i CAD (θ) can be calculated by the following equation

$$\frac{dQ_{net}}{d\theta} = \frac{\gamma}{\gamma - 1} P \frac{dV}{d\theta} + \frac{1}{\gamma - 1} V \frac{dP}{d\theta}, \tag{10.12}$$

where $dQ_{net}/d\theta$ is the net (gross heat production minus heat losses to wall) heat release rate, γ = ratio of gas heat capacities, P = cylinder pressure, $dV/d\theta$ = the rate of change in cylinder volume with crank angle, V = cylinder volume, and $dP/d\theta$ = the rate of change in cylinder pressure with crank angle. The cylinder gas temperature can be estimated using the equation of state. Once the temperature is found, R and γ can be calculated. The rate of change of cylinder volume, $dV/d\theta$ can be calculated from the slider-crank formula. Figure 10.5 shows typical profiles of pressure and heat release rate deduced from Eq. (10.12) versus CAD for a typical engine.

10.4 Octane Number

10.4.1 Definition of Octane Rating

The octane number is a quantity for gauging the autoignition resistance of fuels used in spark-ignition internal combustion engines. The octane rating is evaluated on the basis of the knock resistance compared to a mixture of isooctane (2,2,4-Trimethylpentane) and normal heptane (n-heptane); these two fuels are referred to as the Primary Reference Fuels (PRF). By definition, isooctane is assigned an octane rating of 100 and n-heptane is assigned an octane rating of zero. An octane number is expressed as the percentage of isooctane by volume in a mixture of isooctane and normal heptane (n-heptane) that would have the same anti-knocking capacity as the tested fuel. For example, 87-octane gasoline possesses the same anti-knock rating of a mixture of 87% (by volume) isooctane and 13% (by volume) n-heptane. However, this does not mean that the gasoline actually contains these hydrocarbons in these proportions. It simply means that the fuel has the same autoignition resistance as the described mixture of primary reference fuels. A fuel that has high tendency to autoignite is undesirable in a spark ignition engine but desirable in a diesel engine. Such a fuel would have low octane numbers. The standard for the combustion quality of diesel fuel is the cetane number to be discussed in Chap. 11.

10.4.2 Measurement Methods

The most common type of octane rating worldwide is the Research Octane Number (RON). RON is determined by running a stoichiometric fuel-air mixture through a specific variable-compression-ratio test engine, the "Co-operative Fuel Research engine" (CFR). Results obtained using the test fuel under controlled conditions are compared to results obtained for mixtures of isooctane and n-heptane. There is a second type of octane rating, called Motor Octane Number (MON) that is a better measure of how the fuel behaves when under load. MON testing uses a similar test engine to that used in RON testing, but with a preheated fuel mixture, a higher engine speed, and variable ignition timing to further stress the fuel's knock resistance. Table 10.4 lists the engine conditions of a typical CFR engine used for determining RON and MON. With mixtures of PRF containing a range of isooctane content, a reference relation between RON and compression ratio at the onset of knocking is established as shown in Fig. 10.6. The RON of a test fuel is determined by running this fuel under the same engine settings. The compression ratio at the onset of knocking is determined, say 6.75 in Fig. 10.6, and then cross-referenced to give RON = 91.6. Depending on the composition of the fuel, the MON of a modern gasoline will be about 8–10 points lower than the RON. Some example values of RON and MON are listed in Table 10.5. Normally fuel specifications require both a minimum RON and a minimum MON.

Table 10.4 Test conditions of RON and MON	MON	RON
Engine speed (rpm)	900	600
Intake temperature (°C)	149	52
Intake pressure (bar)	1	1
Ignition time (degrees BTDC)	19–26 (Varies with compression ratio)	13

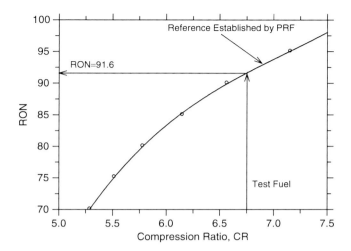

Fig. 10.6 Determination of octane number based on relation between RON and compression ratio established by mixtures of isooctane and n-heptane (PRF)

Table 10.5 Examples of octane numbers (RON and MON)

Fuel	RON	MON
n-Octane	−10	
n-Heptane	0	0
2-Methylheptane	23	
n-Hexane	25	26
2-Methylhexane	44	
1-Heptene	60	
n-Pentane	62	
1-Pentene	84	
n-Butane	91	71
Cyclohexane	97	
Isooctane	100	100
Benzene	101	
Methane	107	
Ethane	108	
Methanol	133	105
Ethanol	129	116
E85 Ethanol	105	
Toluene	114	95
Xylene	117	

In the United States, Canada, and some other countries, the headline number on the pump is the average of the RON and the MON, sometimes called the Anti-Knock Index (AKI). In many other countries, including all of Europe and Australia, the octane number on the pump is simply the RON. Because of the 8–10 point difference noted above, this means that the octane number shown on the pumps in the United States will be about 4–5 points lower than the same fuel elsewhere. For instance, 87 octane fuel, the regular gasoline in the US and Canada, would be 91–92 in Europe.

It is possible for a fuel to have a RON greater than 100, because isooctane is not the most knock-resistant substance available. Racing fuels, straight ethanol, and Liquified Petroleum Gas (LPG) typically have octane ratings of 110 or significantly higher – ethanol's RON is 129 (MON 116, AKI 122). High octane number fuels can be used as octane booster additives.

10.5   Fuel Preparation

Premixing of fuel with air is an important step for premixed IC engines. The quality of the fuel-air mixture can greatly affect engine performance. Before the advent of electronic fuel injection, carburetors were used to mix fuel with air using a *venturi*. The top and middle of Fig. 10.7 shows how a carburetor prepares of the fuel and air mixture. The quality of fuel/air mixtures from a carburetor is not precise enough to use with three-way catalysts (see Sect. 10.9), so port fuel injection (PFI) is now widely used as sketched in the bottom of Fig. 10.7. Fuel is sprayed at the intake valve stem area when the intake valve is closed. The fuel spray usually splashes on the stem, breaking up the droplets to form a gaseous fuel-air mixture. The fuel is first pressurized by a pump to about $300 - 500$ kPa, so the amount of fuel injected is controlled by the injection duration and managed by an on-board computer. However, due to the higher cost of electronic systems, carburetors are still used on small engines such as lawnmowers. The quality of fuel/air mixture can influence engine torque, with a typical relationship shown in Fig. 10.8.

Example 10.2   Estimate the power from a typical 4-cylinder 1.6 Liter 4-stroke gasoline engine at 6,000 rpm with an overall thermal efficiency of 25% and volumetric efficiency of 90%. Also determine the energy needed to vaporize the fuel and compare it to the total power produced.

Solution:
We will use isooctane as a representative fuel for cars. The stoichiometric relation is

$$C_8H_{18} + 12.5(O_2 + 3.76N_2) \rightarrow 8CO_2 + 9H_2O + 12.5 \cdot 3.76N_2$$

The power produced under standard conditions is

$$\dot{W} = \eta \cdot LHV \cdot \dot{m}_f$$
$$\dot{m}_f = \eta_v \cdot \dot{n}_{air} \cdot (FAR)_{mole} \cdot M_f$$
$$\dot{n}_{air} = \frac{V_d \cdot rpm \cdot Stroke/2}{60} \frac{1}{22.4}$$

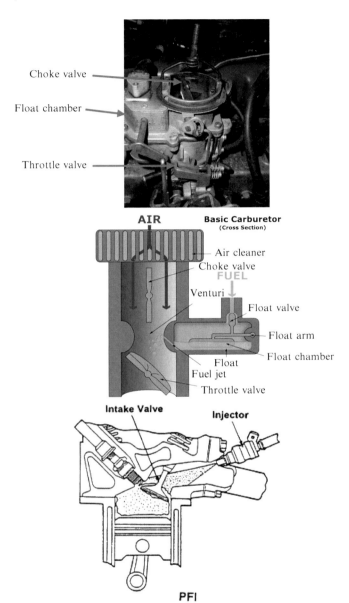

Fig. 10.7 *Top*: Carburetor and its operation within an IC engine (K. Aainsqatsi, under license CC-BY-SA-2.5). *Bottom*: Port fuel injection (Reproduced with permission from Zhao et al. [4])

Fig. 10.8 Engine torque versus air fuel ratio (AFR)

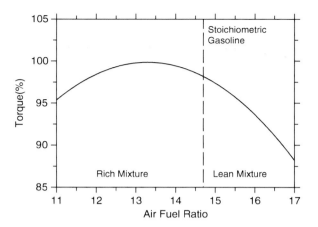

In 4-stroke engines, every two revolutions finish one thermodynamic cycle. The total volume entering the engine is $6,000/2 \times 1.6 = 4,800$ Liter/min $= 80$ Liter/s. This corresponds to 3.57 moles of air per second that requires 0.06 moles of fuel per second (6.827 g/s) to run at stoichiometric.

$$\dot{W} = \eta \cdot LHV \cdot \dot{m}_f = 0.25 \cdot 44.65 \text{ kJ/g} \cdot 6.827 \text{ g/s} = 76 \text{ kW}(\sim 100 \text{ hp})$$

$$\text{Energy for vaporization} = h_{fg} \cdot \dot{m}_f = 283 \text{J/g} \cdot 6.827 \text{g/s} = 1.932 \text{kW}$$

which is about 0.6% of the power produced.

Example 10.3 Consider gasoline having a chemical composition of $C_{8.26}H_{15.5}$. Determine the mole fractions of CO_2 and O_2 in the exhaust for an IC engine with normalized air/fuel ratio $\lambda = 1.2$.

Solution:
Since the overall equivalence ratio, $\phi = 1/\lambda = 1/1.2 = 0.83$, the mixture is lean. Using Eq. (2.14)

$$C_\alpha H_\beta O_\gamma + \frac{1}{\phi}\left(\alpha + \frac{\beta}{4} - \frac{\gamma}{2}\right)(O_2 + 3.76N_2)$$
$$\rightarrow \alpha CO_2 + \frac{\beta}{2}H_2O + \frac{3.76}{\phi}\left(\alpha + \frac{\beta}{4} - \frac{\gamma}{2}\right)N_2 + \left(\alpha + \frac{\beta}{4} - \frac{\gamma}{2}\right)\left(\frac{1}{\phi} - 1\right)O_2$$

with $\alpha = 8.26$, $\beta = 15.5$, and $\gamma = 0$, we have

$$C_{8.26} H_{15.5} + \frac{12.135}{\phi} \cdot (O_2 + 3.76 N_2)$$

$$\rightarrow 8.26 \cdot CO_2 + 7.75 \cdot H_2O + \frac{42.63}{\phi} N_2 + 12.135 \cdot \left(\frac{1}{\phi} - 1\right) O_2$$

The mole fractions of CO_2 and O_2 are

$$x_{CO_2} = \frac{8.26}{8.26 + 7.75 + \frac{42.63}{\phi} + 12.135(\frac{1}{\phi} - 1)}$$

$$= \frac{8.26}{8.26 + 7.75 + 42.63 \cdot \lambda + 12.135(\lambda - 1)} = 0.119$$

$$x_{O_2} = \frac{12.135(1/\phi - 1)}{8.26 + 7.75 + \frac{42.63}{\phi} + 12.135(\frac{1}{\phi} - 1)}$$

$$= \frac{12.135(\lambda - 1)}{8.26 + 7.75 + 42.63 \cdot \lambda + 12.135(\lambda - 1)} = 0.035$$

Note that the dry-based mole fractions are slightly higher due to the removal of water as

$$\text{dry} - based\, x_{CO_2} = \frac{8.26}{8.26 + \frac{42.63}{\phi} + 12.135(\frac{1}{\phi} - 1)}$$

$$= \frac{8.26}{8.26 + 42.63 \cdot \lambda + 12.135(\lambda - 1)} = 0.134$$

$$\text{dry} - based\, x_{O_2} = \frac{12.135(1/\phi - 1)}{8.26 + \frac{42.63}{\phi} + 12.135(\frac{1}{\phi} - 1)}$$

$$= \frac{12.135(\lambda - 1)}{8.26 + 42.63 \cdot \lambda + 12.135(\lambda - 1)} = 0.039$$

10.6 Ignition Timing

Spark ignition timing has a significant impact on the performance of an SI engine. The finite speed of turbulent flames requires that the mixture be ignited before the piston reaches top dead center in order to achieve maximum output and assure complete combustion before the exhaust valves open. Typically, ignition timing is tuned to give the best performance in terms of engine torque and pollutant emissions. To produce the maximum torque for a given rpm, the best timing is found when the peak pressure occurs around 5–10 CAD after TDC. This optimal timing is referred to as the maximum brake torque (MBT) timing as sketched in Fig. 10.9.

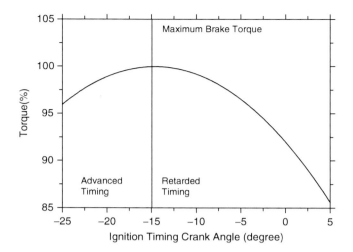

Fig. 10.9 Torque versus timing for a typical engine

Since combustion chemistry takes a certain amount of time to complete, ignition timing needs to be changed according to engine speed. When the engine speed increases, timing is advanced to achieve the best thermal efficiency. If timing is advanced too early, an engine may experience knocking. In modern engines, a knock sensor is used to detect such occurrences to protect the engine from damage. When knocking is detected, the timing of the engine is retarded slightly until knocking ceases.

10.7 Flame Propagation in SI Engines

Once the spark ignites the combustible mixture, a flame kernel develops. After a short period of time, a turbulent flame starts to form and propagate into the unburned mixture. The left picture of Fig. 10.10 was taken from the top of an optical engine showing the propagation of a turbulent flame inside a typical SI engine. Because the unburned mixture is subject to continuous compression and heating, it may autoignite, causing knocking. The pressure waves due to knock are shown on the right plot of Fig. 10.10 from a Co-operative Fuel Research Engine (CFR) for a fuel with 70% isooctane and 30% n-heptane (by volume) at CR ~ 6.0. Tremendous effort has been made to design engines that can achieve high thermo-dynamic efficiency by running at the highest possible compression ratios without knocking. Increasing turbulent flame speed is an effective method to increase an engine's maximum allowable compression ratio, as the residence time of the unburned mixture can be decreased, thereby reducing the chance of autoignition.

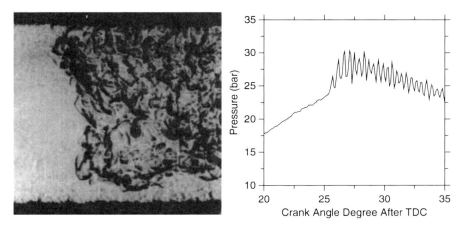

Fig. 10.10 *Left*: Picture of turbulent flame propagation inside a spark ignition engine (Reproduced with permission from Gatowski et al. [1]); *Right*: The pressure trace of an IC engine experiencing knocking shows unsteady waves

10.8 Modeling of Combustion Processes in IC Engines

Numerical models are useful tools for studying combustion processes inside an engine as well as for assisting in the design of advanced engines. Figure 10.11 presents the various physical models needed for simulation of IC engines. Due to the complexity of interactions among the different processes involved in an engine, a detailed model may demand impractically large CPU time to compute. Advancements in both Computational Fluid Dynamics (CFD) and various submodels have been made in the last two decades, and large-scale simulations using parallel computers are now run. In the foreseeable future, CFD will increase its role as an engine design tool.

The amount of CPU time required to calculate detailed chemistry can be quite severe. Figure 10.12 presents an estimate of required CPU times showing that the CPU time scales with the total number of grid cells used in CFD. In engine CFD, grid cells are used to resolve the details of the flow field, with each cell storing values of local temperature, velocity, pressure, and chemical composition. In a typical 3-D simulation, the total number of grid cells is on the order of millions. With simplified combustion chemistry, such a simulation would take a few days to a few weeks depending on the complexity of the engine geometry. Evidently from Fig. 10.12, the inclusion of detailed chemical kinetics into a detailed CFD for modeling practical engines is not practical unless a massively parallel computing facility is used. This may not be economically feasible even for a large car designer.

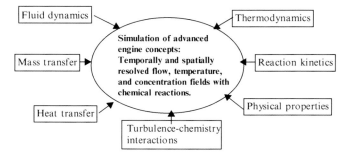

Fig. 10.11 Various physical models needed for simulations of combustion in an IC engine

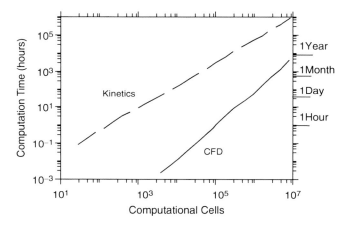

Fig. 10.12 Estimates of CPU time versus number of cells with and without combustion chemistry (From Lawrence Livermore National Laboratory)

10.8.1 A Simplified Two-Zone Model of Engine Combustion

Simplified models are often used to gain understanding of certain aspects of combustion in IC engines. The simplest model for a spark ignition (SI) engine consists of two zones, one for the burned gases and one for the unburned gases. Such a model may be used to assess overall heat release and perhaps predict the onset of knocking when an empirical model for the turbulent burning rate is properly tuned. The turbulent flame is modeled by a spherical flame front with its center located at the spark. In a more general model, the turbulent flame front can be modeled by a wrinkled front as sketched in Fig. 10.13.

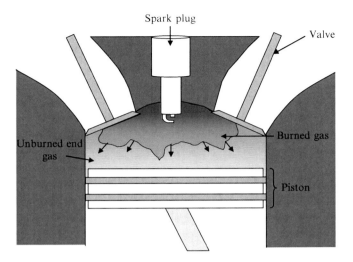

Fig. 10.13 A two-zone model for SI engine combustion with a turbulent flame front propagating from the burned zone into the unburned zoned

In most engines, experimental data indicate that the turbulent flame falls into the laminar flamelet regime[3]. Under this regime, turbulent flame speed is reasonably correlated with laminar flame speed. For engineering purposes, the turbulent propagation flame velocity is represented by an empirical model that depends on several parameters

$$\frac{S_T}{S_L} = f(u'/S_L, P/P_m, \theta_{ign}), \tag{10.13}$$

where S_T is the turbulent flame speed, S_L is the laminar flame speed, u' is the characteristic turbulent fluctuation velocity, P is cylinder pressure, P_m is the motoring pressure, θ_{ign} is the ignition timing in terms of CAD before TDC. As sketched in Fig. 10.14, the ratio S_T/S_L for general turbulent flames increases slowly with u'/S_L at low values and then increases rapidly when turbulence is intensified. For IC engines, data suggest that S_T/S_L also depends on P/P_m and ignition timing. For instance, the following empirical relation has been used in modeling engine combustion:

$$\frac{S_T}{S_L} = 1 + 1.21 \frac{u'}{S_L} \left(\frac{P}{P_m}\right)^{0.82} \left(1 + 0.05 \cdot \theta_{ign}^{0.4}\right) \tag{10.14}$$

---

[3] Under certain regimes of turbulence-chemistry interactions, the turbulent flames consist of an ensemble of laminar flames that are merely wrinkled by turbulence. These flames are called flamelets.

Fig. 10.14 Correlation
between turbulent flame
speed normalized by laminar
flame speed versus turbulent
fluctuation velocity
normalized by laminar flame
speed

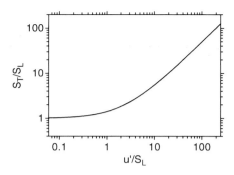

The governing equations for the two-zone model include those for energy conservation, mass conservation, and two ideal gas equations:

$$\frac{d(m_u u_u)}{dt} = h_u \frac{dm_u}{dt} - P\frac{dV_u}{dt} - \dot{q}_{u,L} \tag{10.15}$$

$$\frac{d(m_b u_b)}{dt} = h_b \frac{dm_b}{dt} - P\frac{dV_b}{dt} - \dot{q}_{b,L} \tag{10.16}$$

$$\begin{aligned} m_u + m_b &= m \\ V_u + V_b &= V \end{aligned} \tag{10.17}$$

where m_u and m_b denote the masses of unburned and burned mixtures respectively, h_u and h_b are the respective enthalpies, and V_u and V_b are the corresponding volumes. Heat transfer rates to engine walls, $\dot{q}_{u,L}$ and $\dot{q}_{b,L}$, are modeled by empirical correlations. The pressure is assumed to be uniform. Using the two ideal gas equations, we have

$$P = \frac{m_u R_{un} T_u}{V_u} = \frac{m_b R_{bn} T_b}{V_b} \tag{10.18}$$

The overall mass burning rate inside an IC engine is computed by

$$\frac{dm_b}{dt} = -\rho_u \cdot A_f \cdot S_T, \tag{10.19}$$

where ρ_u and A_f are the unburned density and flame surface area respectively. One may consider the two-zone model as an extremely simplified CFD model with two grid cells. As such, detailed chemistry may be incorporated into such a simplified model. Figure 10.15 presents a typical predicted pressure trace of an IC engine running at 600 RPM using a detailed isooctane combustion mechanism

Fig. 10.15 Predicted
pressure trace using a two-
zone model coupled with
detailed chemistry for
isooctane. Ignition is initiated
at 13 CAD before TDC as
shown by the first pressure
jump. The second pressure
jump near 38 CAD after TDC
indicates the onset of
knocking

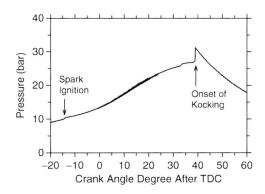

(856 species, 3,660 steps). Ignition is initiated at 13 CAD before TDC as shown by the first pressure jump. The compression ratio is varied to predict the onset of knocking as indicated by the small jump in the pressure trace near 38 CAD after TDC.

10.9   Emissions and Their Control

The most common emissions from a typical spark-ignition engine are summarized in Table 10.6. Most engines run with near-stoichiometric mixtures, causing high NO_x emissions in the range of 1,000 ppm. Levels of unburned hydrocarbons and CO, present primarily because of reaction quenching in the cylinder walls and crevices[4], are also high. Untreated exhaust gases can pose a severe challenge to the environment because there are so many cars on the road. The environmental impact of various exhaust species is summarized in Fig. 10.16.

Four basic methods can be used to decrease engine emissions:

1. Engineering of the combustion process
2. Optimizing the choice of the operating parameters
3. Using after-treatment devices in the exhaust system
4. Using reformulated fuels

As it was explained in Chap. 9, lean combustion is the most effective way to reduce emissions of HC, CO, and NO_x. Unfortunately, combustion instabilities in

---

[4] Crevices are narrow volumes present around the surface of the combustion chamber, having high surface-to-volume ratio into which flame will not propagate. They are present between the piston crown and cylinder liner, along the gasket joints between cylinder head and block, along the seats of the intake and exhaust valves, space around the plug center electrode and between spark plug threads.

Table 10.6 Typical engine emissions without treatment	HC	750 ppm[a]	CO_2	13.5 vol-%
	NO_x	1,050 ppm	O_2	0.51 vol-%
	CO	0.68 vol-%	H_2O	12.5 vol-%
	H_2	0.23 vol-%	N_2	72.5 vol-%

[a]Based on C3

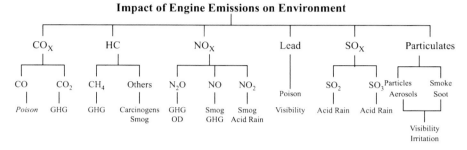

Fig. 10.16 Impact of engine emissions on the environment. GHG: Green House Gases; Ozone Depletion: OD

the cylinder limit the use of this technique in SI engines. A considerable amount of research currently attempts to improve the use of lean combustion in engines and combustors. Staged combustion – rich burning followed by lean burning – has also been used in SI engines with some success, but the accompanying reduction in power has deterred its wide implementation. Reformulated fuels, such as oxygenated gasoline in winter to reduce CO and low volatility gasoline in summer to reduce evaporative HC, are often used. Advancements in fuel injector design, oxygen sensors, on-board computers, and catalysts have lead to significant emissions reductions in SI engines in the past decades.

10.9.1 Three-Way Catalyst

Figure 10.17 gives a schematic of a three-way catalytic converter used for emission control. A three-way catalytic converter simultaneously performs three main tasks:

1. Reduction of nitrogen oxides to nitrogen and oxygen:

$$2NO_x \rightarrow xO_2 + N_2$$

2. Oxidation of carbon monoxide to carbon dioxide:

$$2CO + O_2 \rightarrow 2CO_2$$

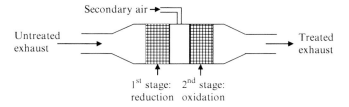

Fig. 10.17 Three-way catalytic converter with interiors exposed

3. Oxidation of unburned hydrocarbons (HC) to carbon dioxide and water:

$$2C_xH_y + (2x + y/2)O_2 \rightarrow 2xCO_2 + yH_2O$$

The catalysts used are usually a platinum/rhodium blend for the reducing reactions and a platinum/palladium blend for the oxidizing reactions. The catalytic reactions occur on the surface of the catalyst so the metals are often coated onto either a ceramic honeycomb or ceramic beads to increase the available catalyst surface area. These three reactions occur most efficiently when the catalytic converter receives exhaust from an engine running slightly lean. Typically gasoline SI engines are run with an air-to-fuel ratio between 14.8 and 14.9 (by weight), which corresponds to an equivalence ratio of 0.993–0.987. Figure 10.18 presents the transformation effectiveness of a three-way catalyst as function of product mixture. When there is more oxygen than required, the system is said to be running lean, and the system is in an oxidizing condition. In that case, the converter's two oxidizing reactions (oxidation of CO and hydrocarbons) are favored at the expense of the reducing reaction. When there is excessive fuel, the engine is running rich. The reduction of NO_x is favored at the expense of CO and HC oxidation. To compensate, additional air is often supplied to the catalytic converter in between the reducing and oxidizing stages.

In most automotive applications, an oxygen sensor (also called lambda sensor) installed in the exhaust monitors the O_2 level. The signal is used for feedback control of fuel injection duration such that the overall equivalence ratio is maintained near stoichiometric for maximum conversion of all emissions. Figure 10.19 shows the typical placement of an oxygen sensor and its voltage signal as a function of λ.

10.10 Gasoline Direct Injection (GDI) Engines

At a fixed engine speed, the amount of work produced by SI engines is controlled by a throttle plate upstream of intake manifold. When this throttling plate is partially closed, it restricts the amount of air flow, in turn restricting the amount of combustible mixture flowing into the engine. As such, for a partial load, the work required to bring combustible mixture into the cylinder increases. This loss is called

Fig. 10.18 Effectiveness of a three-way catalyst versus deviation from stoichiometric mixture

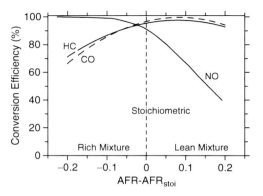

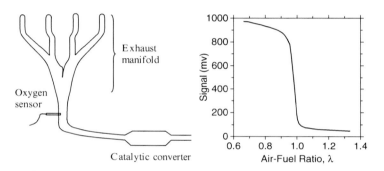

Fig. 10.19 *Left*: oxygen sensor and its typical installation in the exhaust pipe. *Right*: signal (voltage) from an oxygen sensor as function of normalized air/fuel ratio, λ

'pumping loss.' One potential means for reducing pumping loss is to manage the load by direct injection of fuel into the cylinder, similar to what is done in a diesel engine. This eliminates the need for a throttling valve and the losses associated with pulling air past the restriction. Such an engine is called a gasoline direct injection (GDI) engine and is sketched in Fig. 10.20. In principle, a throttle plate is not required in GDI engines, but in practice it is often used as a safety device. The potential benefits of GDI engines over the traditional premixed spark ignition engines with a throttling valve are: enhanced fuel economy, improved transient response, and reduced cold-start hydrocarbon emissions.

Due to the lack of a throttle plate, operating a GDI engine is more complex than a traditional gasoline engine. Figure 10.21 depicts the operation map of a typical GDI engine with three distinct modes noted. At high load (shown in the top region), a GDI engine operates similar to a traditional engine with the throttle wide open. The only difference is that fuel is injected directly into the cylinder. Injection of fuel takes place during the intake stroke of the engine to ideally generate a homogeneous mixture. To achieve a homogeneous mixture, the fuel should be injected as early as possible to allow sufficient time for vaporization of the liquid fuel as well as

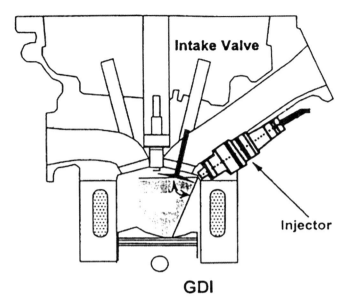

Fig. 10.20 Sketch of a gasoline direct injection engine (Reproduced with permission from Zhao et al. [4])

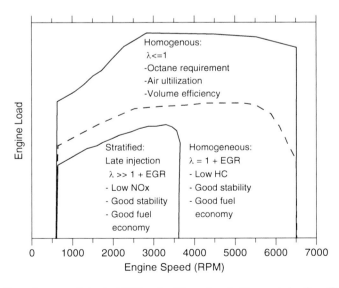

Fig. 10.21 Operation map of a typical GDI engine (Reproduced with permission from Zhao et al. [4])

subsequent mixing with air. Due to the presence of the piston near TDC, injection of fuel right after the opening of the intake valve may lead to impingement of fuel on the piston. Any wetting of interior metal surfaces inside the cylinder is undesirable, as the liquid film of fuel will not vaporize, causing large amounts of unburned hydrocarbon emission. A lean burn mode is often employed for enhancing fuel economy and lowering NO_x.

In the intermediate load regime, a homogeneous stoichiometric mixture as shown on the right is used for good running stability and fuel economy. Exhaust gas recirculation (EGR) is often used for reducing NO_x as well as for load control. Next, under the low load conditions, the amount of air taken in by the engine exceeds that required by combustion. If a homogeneous mixture is prepared by early injection, the mixture becomes too lean for flame propagation. The current method for overcoming this difficulty is to inject the fuel late in the compression stroke so that a stratified fuel-air mixture is created. Ideally, the mixture near the spark plug is near stoichiometric, making flame propagation feasible. There are two main drawbacks to such a mode: (1) In the stoichiometric region, high temperatures create high NO_x emission levels; (2) Since the mixture is stratified, a region exists where the fuel-containing mixture becomes too lean to burn; thus high levels of unburned hydrocarbons remain in the emissions. These two issues require further improvements in current GDI engines before they can be widely used in countries with strict emission laws.

Exercises

10.1 Using 87 octane gasoline, a spark-ignited internal combustion engine is designed to run at an equivalence ratio of 0.7 and a compression ratio of 9. Do you anticipate any potential problems if the engine is modified to run at a compression ratio of 12 while still running 87 octane gasoline? What about a compression ratio of 12 with 93 octane gasoline?

10.2 Assuming the spark plugs usually fire at 15 crank angle degrees before top dead center, how would the power output and emissions change if the engine was modified so that the spark plugs fire at 30 crank angle degrees before top dead center?

10.3 In a single-cylinder gasoline spark ignition premixed engine, the following data are given:

Engine geometry: bore (cylinder diameter) = 6 cm, displacement volume = 400 cm^3, compression ratio = 8

Laminar flame speed: $S_L = 70$ cm/s (constant throughout combustion)

Turbulence fluctuation: $V' = 120$ cm/s (constant throughout combustion)

Spark timing: 15 CAD BTDC ($\theta = 15$)

The following empirical formulation is used for the ratio of turbulent flame speed (S_t) to laminar flame speed (S_L)

$$\frac{S_t}{S_L} = 2 + 5 * \frac{V'}{S_L}(1 + 0.05\sqrt{\theta})$$

Estimate the total burn duration in terms of CAD at 1,000 rpm.

10.4 Considering internal combustion engines, answer the following questions.

(a) What is the purpose of an intake throttle plate commonly used in a spark ignition (SI) engine?

(b) Consider an SI engine with a volumetric efficiency of 0.85 at 2,000 rpm. How much can power be increased if the volumetric efficiency is increased to 0.95 at the same operating condition (in terms of % of power at $\eta_v = 0.85$)?

(c) The CO emissions measured in the tailpipe of a SI engine are 2,000 ppm. The calculated chemical equilibrium concentration of CO at the tailpipe conditions is 2 ppm. How is it possible that 2,000 ppm CO levels are measured in the tailpipe?

(d) What is the main purpose of a turbocharger?

10.5 In a gasoline spark ignition premixed engine running with a stoichiometric mixture, perform an analysis to determine whether or not engine knocking will occur with the following information:

Assumptions:

1. the turbulent flame propagates at a constant speed.
2. the turbulent premixed flame has a spherical shape.

Conditions, engine data, and simplifications:

1. Engine geometry: bore (cylinder diameter) = 6 cm, displacement volume = 400 cm³, compression ratio = 8
2. Laminar flame speed: $S_L = 70$ cm/s (constant throughout combustion)
3. Turbulence fluctuation: $V' = 120$ cm/s (constant throughout combustion)
4. Spark timing: 15 CAD BTDC ($\theta = 15$)
5. Spark plug location: top center of engine cylinder
6. Unburned gas temperature = 1,650 K (constant throughout combustion)
7. Unburned gas pressure = 0.5 MPa (constant throughout combustion)

Empirical formulas:

1. The following empirical formulation is used for the ratio of turbulent flame speed (S_t) to laminar flame speed (S_L)

$$\frac{S_t}{S_L} = 2 + 5 * \frac{V'}{S_L}(1 + 0.05\sqrt{\theta[\text{CAD BTDC}]})$$

2. Empirical relation for autoignition delay of a stoichiometric gasoline-air mixture

$$\tau_{\text{ignitondelay}}[\text{ms}] = 0.08 \cdot \frac{1}{P^{1.5}[\text{MPa}]} \exp\left(\frac{3800}{T[\text{K}]}\right)$$

The units are expressed inside [ ].

10.6 From an internal combustion engine, measurements of the *exhaust* gases show that accelerating the engine speed (rpm) above a certain value increases the concentration (emission) of CO but decreases the concentration of NO. These measurements are taken right at the exhaust port before the catalyst.

(a) Explain the main reason for the emission trend *vs.* rpm.
(b) How would the emissions of pollutants change if the engine were cold or hot? Why?
(c) At a certain RPM, measurement of some (not all) exhaust species indicate: $CO_2 = 12\%$, $CO = 0.2\%$, $O_2 = 2.3\%$, and $NO = 70$ ppm. Using isooctane as the fuel, determine the NO emission index.
(d) Sketch the conversion efficiencies of an automobile catalyst for CO, HC, and NO versus equivalence ratio.

References

1. Gatowski JA, Heywood JB, Deleplace C (1984) Flame Photographs in a Spark–Ignition Engine. Combustion and Flame 56:71–81.
2. Heywood, JB (1988) Internal Combustion Engine Fundamentals. McGraw-Hill Book Company, New–York.
3. Lumley JL (1999) Engines, an introduction. Cambridge University Press, Cambridge.
4. Zhao F, Lai MC, Harrington DL (1999) Automotive spark-ignited direct-injection gasoline engines. Progress in Energy and Combustion 25:437–562.

Chapter 11
Diesel Engines

The term "diesel" derives from the name of the German engineer, Dr. Rudolph Diesel, who is widely credited for the development of compression ignition (CI) engines. Modern compression-ignition engines (diesel engines) have evolved from the 3:1 compression ratio engine that Rudolph Diesel built in 1890 to compression ratios up to 20:1 with high-pressure fuel injection systems, outputting up to 10,000 hp. CI engines are merited with high engine efficiency (up to 45%) because of (1) higher compression ratios, (2) no throttling, (3) lower running speed than SI engines, therefore less friction losses, and (4) lean air/fuel mixture. At most load ranges, CI engines are more fuel efficient than SI engines. However, these engines are heavier than spark ignition engines because of the need to support higher internal pressures in the cylinders. They are also noisier because of the spontaneous ignition of the charge. CI engines are generally found on heavy-duty trucks, construction vehicles/equipment, stationary power generators, trains, and large ships because of the higher power output required.

The concerns of greenhouse gases demand improvement of vehicle mileage and reduction of pollutant emissions. Diesel engines have high fuel economy and thus the highest CO_2 reduction potential among all other thermal engines due to their superior thermal efficiency. However, particulate matter (PM) and nitrogen oxides (NO_x) emissions from diesel engines are comparatively higher than those emitted from modern SI gasoline engines. PM consists of tiny particles of solid or liquid suspended in a gas or liquid. Increased levels of fine particles in the air are linked to health hazards such as heart disease, altered lung function, and lung cancer. Therefore, reduction of diesel emitted pollutants, especially PM and NO_x, without an increase of the specific fuel consumption is a challenging problem requiring immediate action. This chapter provides the fundamental background on the physical processes occurring in typical diesel engines.

11.1 Overall Comparisons to SI Engines

Unlike SI engines, where the amount of air allowed into the cylinder is controlled, only the amount of fuel injected needs to be controlled to regulate the power of a CI engine. This eliminates the need for throttling and the associated loss of

S. McAllister et al., *Fundamentals of Combustion Processes*,
Mechanical Engineering Series, DOI 10.1007/978-1-4419-7943-8_11,
© Springer Science+Business Media, LLC 2011

efficiency. Since CI engines do not use a premixed charge like an SI engine, the motion of the air and the injected fuel inside the combustion chamber must be designed to obtain the best performance possible. A certain swirl (ordered rotation of air about the cylinder axis) is needed to ensure mixing of the fuel and air and proper combustion. A Direct-Injected (DI) system, where the fuel is injected directly into the cylinder, requires masked inlet valves and a powerful fuel injection system for this purpose. An Indirect-Injection system (IDI) is sometimes used instead of the DI system. IDI systems have pre-chambers where fuel is evaporated by a heated element before flowing into the main cylinder for better mixing with the air. The use of IDI systems lowers the requirement for powerful fuel injection systems. However, the size of the pre-chambers must be increased significantly as the cylinder size increases.

Combustion timing of a CI Direct Injected (CIDI) engine is controlled not by spark but by autoignition of the injected fuel. Fuel is injected at high pressure as a spray into the engine cylinder late in the compression stroke. After the fuel evaporates and mixes with air, local autoignition occurs, and a non-premixed flame emerges between the fuel and air along the outer region of the spray. This flame burns at stoichiometric conditions, giving the highest flame temperatures possible and thus producing significant amounts of NO_x and soot. Nitric oxide formation takes place in the lean regions of the mixture and particulate matter (PM) formation takes place in the rich regions.

11.1.1 Advantages of Diesel Engines as Compared to SI Engines

(a) Compression ratio (CR) is higher, leading to higher thermal efficiency.
(b) Since no throttling valve is needed, intake losses are lower, thus efficiency is higher.
(c) Overall equivalence ratio is lean ($\phi \sim 0.7$–0.8), so less unburned hydrocarbons and CO are leftover from the gas phase combustion.
(d) Walls and crevices contain air only during the compression stroke, so in principle, no hydrocarbons and CO go unburned due to quenching in the crevices.

11.1.2 Disadvantages of Diesel Engines as Compared to SI Engines

(a) The liquid spray flame burns in diffusion flame mode, causing high temperatures that result in high NO_x.
(b) At high loads, soot/particles are formed.
(c) Cost of diesel engines is high due to the high-pressure injection system.

(d) Engines must be heavier to withstand the higher pressures.
(e) Maximum operable engine speed (RPM) is lower than in SI engines, so peak power output is lower.

11.2 Thermodynamics of Diesel Engines

Figure 11.1 (left) presents the ideal Diesel cycle in terms of a P-V diagram. The dashed lines denote the corresponding Otto cycle with the same compression ratio, CR, i.e., CR $= V_1/V_2$. The thermal efficiency of the standard diesel engine is

$$\eta_D = 1 - \frac{1}{(CR)^{\gamma-1}} \left[\frac{r_c^\gamma - 1}{\gamma(r_c - 1)} \right] \tag{11.1}$$

where γ is the ratio of specific heats, c_p/c_v and $r_c = V_3/V_2$ is the cut-off ratio.

For the same CR and γ, the only difference between the Diesel and Otto cycle efficiencies (Eqs. 11.1 and 10.1) lies in the terms in the bracket

$$\left[\frac{r_c^\gamma - 1}{\gamma(r_c - 1)} \right] \tag{11.2}$$

which is greater than one when $r_c > 1$, and equal to 1 when $r_c = 1$. Therefore, the Otto cycle is more efficient than the Diesel cycle if CR and γ are kept the same as presented in the right plot of Fig. 11.1. The difference in thermal efficiencies can be understood in the left plot of Fig. 11.1 where dashed lines denote the corresponding Otto cycle. Since the Otto cycle assumes heat addition at constant volume, extra work is produced in comparison to the Diesel cycle. For instance, with $CR = 18$, $\gamma = 1.4$, the thermal efficiency of the Otto cycle is 0.685, in contrast to 0.632 from the Diesel cycle. In reality, unwanted autoignition would occur at high

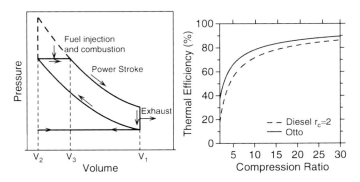

Fig. 11.1 *Left*: P-V diagram of an ideal Diesel cycle *Right*: efficiency versus compression ratio showing that Diesel cycle is less efficient for a given compression ratio

compression in the Otto cycle with gasoline, so the compression ratio in practice is limited to about 10, resulting in thermal efficiencies in the range of 0.3–0.35. For diesel engines, the compression ratio can range from 18 to 25 with thermal efficiencies in the range of 0.45–0.5.

11.3   Diesel Spray and Combustion

Diesel spray consists of three distinct zones as sketched in Fig. 11.2: (1) spray evaporation, (2) mixing with surrounding hot air, and (3) combustion. Although this breakdown is somewhat oversimplified, it gives an estimate of the total physical time required to complete the entire spray combustion process in a diesel engine as

$$t_{totalphysicaltime} = t_{evap} + t_{mix} + t_{comb} \tag{11.3}$$

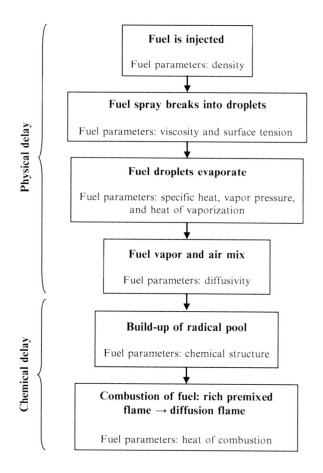

Fig. 11.2 Diesel spray consists of several processes in sequence including evaporation, mixing with air, and ultimately combustion

The total physical time places an upper limit on how fast the engine can run. Usually the injection timing is set around $30°$ Before Top Dead Center (BTDC) with a total burn duration of $70°$ Crank Angle Degrees (CAD). When the engine is run at 3,000 rpm, the total time available for spray combustion is about 3.9 ms. For reference, droplets of size of 10 μm can be vaporized at 900 K and 4 MPa (40 bar) within 0.5 ms.

It is desirable to decrease the total physical time necessary to burn the fuel and generate work through combustion, enhancing the available power output of a diesel engine. The evaporation rate can be increased by using several methods to decrease droplet sizes. For instance, the following empirical correlation has been proposed for the Sauter Mean Diameter (SMD) by El-Kotb (1982) [3] for diesel injectors:

$$SMD = 3.08 \cdot 10^6 \sigma_l^{0.737} v_l^{0.385} \rho_l^{0.737} \rho_{air}^{0.06} \Delta P^{-0.54} (\mu m) \qquad (11.4)$$

where the subscript 'l' denotes the properties of the liquid fuel. To reduce the SMD, Eq. 11.4 suggests using an increase in pressure drop across the fuel injector (ΔP) and a decrease in surface tension (σ) and viscosity (v) of the fuel.

Past measurements of a liquid spray have established two important parameters for quantifying the spray: (1) spray cone angle (Φ) and, (2) tip penetration distance (L_p). The spray cone angle is usually correlated to the details of injector geometry and is obtained from experiments. The spray penetration distance is a useful quantity for estimating possible spray impingement on engine cylinder walls. Experimental observations of liquid spray into stagnant air reveal that for a short period of time after injection ($t < t_b$), the tip of the spray travels linearly with time as [1]

$$L_p = 0.55142 \cdot \sqrt{\frac{\Delta P}{\rho_l}} \cdot t \qquad (11.5)$$

where

$$t_b = 28.65 \sqrt{\frac{\rho_l}{\rho_a}} \frac{d_0}{\sqrt{\Delta P / \rho_l}},$$

d_0 is the injector diameter, ΔP is the pressure drop across the injector into stagnation air, and ρ is the density.

After $t > t_b$, the penetration distance increases with pressure drop to the ¼ power as

$$L_p = 2.95 \cdot \sqrt{d_0 t} \cdot \left(\frac{\Delta P}{\rho_a} \right)^{1/4} \qquad (11.6)$$

Therefore, increasing the pressure drop also increases the total length of a jet spray, scaling roughly with $\Delta P^{0.25}$. This could lead to impingement of diesel fuel on the cylinder walls. When such impingement occurs, the lubricant oil on the cylinder walls is displaced by diesel fuel leading to early wear of the engine. Additionally, the evaporation of the fuel is delayed. By carefully designing the piston shape and the orientation of injection, one can prevent the spray from hitting walls as illustrated in Fig. 11.3.

Preheating the fuel is now a common practice to decrease both surface tension and viscosity of the fuel before injection. For instance, Fig. 11.4 shows the surface tension of n-heptane decreasing as temperature increases. Note that surface tension vanishes when temperature exceeds the critical point where phase transition disappears. Unfortunatley, only empirical formulations exist to describe such temperature dependence. One potential problem of preheating the diesel fuel is the coking of the fuel that forms carbonaceous particles that may clog the fuel injectors. In winter, additives to diesel fuel such as ethanol help by reducing the viscosity of the fuel for easier cold start and for better combustion due to smaller droplet size.

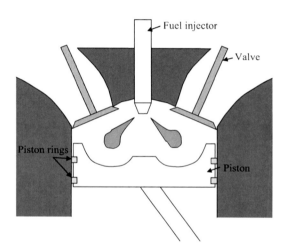

Fig. 11.3 A bowl shaped piston design prevents the diesel fuel spray from hitting the surface

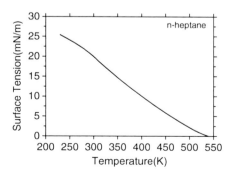

Fig. 11.4 Like water, the surface tension of n-heptane decreases with temperature

Fig. 11.5 Diesel engine
piston designed to provide
squish flow in the bowl when
the piston is moving upward
near top dead center

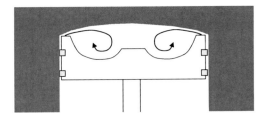

Increasing turbulence reduces the mixing time, t_{mix}, which quantifies the time for droplet vapor and air to mix into a combustible mixture. Introduction of tumbling motions via intake valve arrangement is commonly used to promote turbulence. In addition, designing engine shape to create a "squish" flow provides another way to increase turbulence as illustrated in Fig. 11.5.

The time required for combustion (t_{comb}) is due to the delay in autoignition ($t_{ignition}$) as well as the time required for the chemical reactions to occur (t_{chem}). The autoignition delay time, $t_{ignition}$, can be decreased by increasing temperature and pressure. Often, the autoignition delay can be correlated to temperature, equivalence ratio, and pressure as

$$t_{\text{ignition}} = C_e \left(\frac{P}{P_0}\right)^a \phi^b \exp\left(\frac{E}{\bar{R}_u T}\right)$$

or

$$t_{\text{ignition}} = C_e \left(\frac{P}{P_0}\right)^a [Fuel]^b [Oxidizer]^c \exp\left(\frac{E}{\bar{R}_u T}\right) \tag{11.7}$$

where C_e is an empirical constant; P_0 is a reference pressure; $-1.9 < a < -0.8$ and $-1.9 < b, c < -1.6$ are empirically determined exponents; and [ ] denotes reactant concentration. Due to the high activation energy of combustion chemistry, temperature has the most profound effect on ignition delay followed by pressure and equivalence ratio. Similar to SI engines, once the flame is ignited by autoignition, the combustion process is strongly influenced by the overall temperature and pressure. The overall combustion can be greatly enhanced by turbulence.

After autoignition, the combustion process in a typical diesel spray usually takes place in two stages: (a) rich premixed flame, and (b) subsequent diffusion flame as illustrated in Fig. 11.6.

As such, the heat release rate in a typical diesel engine exhibits two peaks as shown in Fig. 11.7. Two other combustion modes (spark ignition and homogeneous charge compression ignition (HCCI)) are also shown for comparison. The first peak corresponds to the rich premixed flame and the second to the diffusion flame. Empirical correlations are often used to describe both heat release rates for use in analytical models.

Fig. 11.6 *Top*: Spray
combustion (graphic courtesy
of Dr. John Dec, Sandia
National Laboratories, from
[2] and [4]). *Bottom*: Sketch
of combustion processes in
a typical diesel engine.
Residence time, which is
influenced by the physical
path, must also be considered

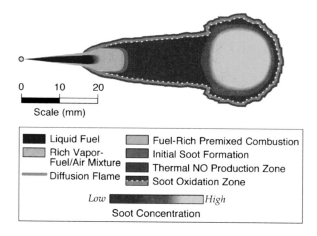

See p 192

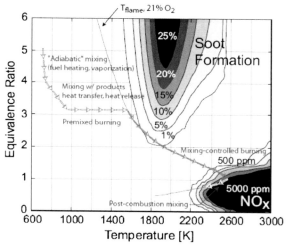

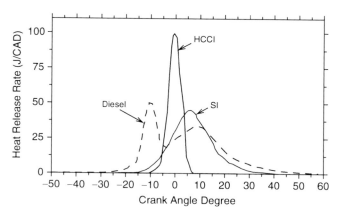

Fig. 11.7 Comparison of heat release rates between three different types of engine combustion:
spark ignition, direct diesel, and homogeneous charge compression ignition (HCCI)

Example 11.1 Estimate and plot the penetration distance versus time for a liquid fuel spray into stagnant air for the time period of $0 < t < 2$ ms after injection with the following data: single hole injector nozzle 0.2 mm diameter, air density 25 kg/m^3, pressure drop across the injector $= 10$ MPa, fuel density $= 850$ kg/m^3.

Solution:

First let's determine

$$t_b = 28.65 \sqrt{\frac{\rho_l}{\rho_a}} \frac{d_0}{\sqrt{\Delta P/\rho_l}}$$

$$= 28.65 \sqrt{\frac{850}{25}} \frac{0.2 \cdot 10^{-3}}{\sqrt{10^7/850}}$$

$$= 3.08 \cdot 10^{-4} s = 0.308 \text{ ms}$$

Second, for $t < t_b$, we have

$$L_p = 0.55142 \cdot \sqrt{\frac{\Delta P}{\rho_l}} \cdot t = 59.8 \cdot t \quad m = 5.98 \cdot 10^4 \cdot t \quad \text{mm}$$

for $t \geq t_b$,

$$L_p = 2.95 \cdot \sqrt{d_0 t} \cdot \left(\frac{\Delta P}{\rho_a}\right)^{1/4} = 1.049 \cdot \sqrt{t} \text{ m} = 1.049 \cdot 10^3 \cdot \sqrt{t} \text{ mm}$$

The above results are plotted against time up to 2 ms in Fig. 11.8. Note that the above formulas have a discontinuity in slope at $t = t_b \sim 0.3$ ms.

11.4   Cetane Number

For diesel fuels, the most important feature is the autoignition delay time under high pressure and temperature (700–900 K). The cetane number measures the ignition quality of a diesel fuel. The cetane number of a fuel is determined similarly

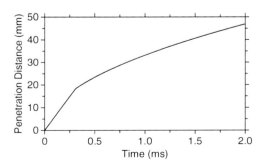

Fig. 11.8 Example 11.1

to the octane number by comparing its autoignition delay to a mixture of two referenced fuels (n-cetane and iso-cetane) as

$$CN = (\% \text{ n - cetane}) + 0.15(\% \text{ of } iso \text{ - cetane}) \tag{11.8}$$

Cetane number is the % volume of cetane (n-hexadecane, cetane number = 100) in alpha methyl naphthalene (cetane number = 0), that provides the specified standard of 13° (crankshaft angle) ignition delay at the identical compression ratio to that of the fuel sample. These days, heptamethyl nonane - with a cetane number of 15 – is used in place of alpha methyl naphthalene because it is a more stable reference compound.

Cetane number is measured in a special ASTM variable compression ratio test engine that is closely controlled with regard to temperatures (coolant 100°C, intake air 65.6°C), injection pressure (~100 atm or 1,500 psi), injection timing (13° BTDC), and speed (900 rpm). The compression ratio is adjusted until combustion occurs at TDC (the ignition delay is 13°). The test is then repeated with reference fuels with five cetane numbers difference, until two of them have compression ratios that bracket the sample. The cetane number is then determined by interpolation. Higher cetane numbers mean the delay between injection and ignition is shorter. If the fuel is pure hydrocarbons (does not contain cetane number improving agents like alkyl or amyl nitrates) then the cetane number can be predicted fairly well using some physical properties, such as boiling point and aniline point.

There is a negative correlation between octane and cetane number; that is a fuel with high octane number is more resistant to autoignition and therefore it has low cetane number and vice versa. It's obvious from the above that a higher cetane number (100 = normal alkane, 15 = iso-alkane) signifies a lower octane number (100 = iso-alkane, 0 = normal alkane). This is because the desirable property of gasoline is the ability to resist autoignition to prevent knock, whereas for diesel, the desirable property is autoignition. The octane number of normal alkanes decreases as carbon chain length increases, whereas the cetane number increases as the carbon chain length increases. Many other factors also affect the cetane number, and around 0.5% by volume of cetane number improvers will increase the cetane number by ten units. Cetane number improvers can be alkyl nitrates, primary amyl nitrates, nitrites, or peroxides.

Typically, engines are designed to use fuels with cetane numbers of 40–55, because below 38 there is a more rapid increase in ignition delay. Most engines show an increase in ignition delay time when the cetane number is decreased from around 50 to 40, with an increase of 2° being typical, and minimal advantages accrue if lower CN fuels are used. The significance of the cetane number increases with the speed of the engine. Large, low speed diesel engines (marine usages) often only specify viscosity, combustion, and contaminant level requirements, as the cetane number requirement of the engine is met by most distillate and residual fuels that have the appropriate properties. High speed diesel engines in cars and small trucks are almost all designed to accept fuels around 50 cetane numbers, with higher numbers being a waste.

Cetane number is only one important property of diesel fuels. There are three others that must be considered. Firstly, the viscosity is important because many injection systems rely on the lubricity of the fuel for lubrication. Secondly, the cold weather properties are important. Remember that normal alkanes are desirable, but diesel alkanes have melting points above $0°C$ temperature, so special flow-enhancing additives and changes to the hydrocarbon profiles occur seasonally. Thirdly, diesel in many countries has a legal minimum flash point. In all cases it's usually well above ambient ($60°C+$, kerosene is $37°C+$, whereas gasoline is typically below $-30°C$), and mixing a lower flash point fraction with diesel usually voids all insurance and warranties on the vehicle.

11.5   Diesel Emissions

Due to the overall lean combustion and ideally having only air (not fuel) in contact with the cylinder walls, unburned hydrocarbons and CO are not a major problem in diesel engines. Particulates (soot) and NO_x are the two main issues. As illustrated in Fig. 11.6, soot formation starts in the rich premixed flame and it is oxidized near the stoichiometric and lean regions of combustion. Similar to CO emissions, soot oxidation is heavily dependent on radicals, such as OH, and the temperature must be high enough. In addition, soot oxidation requires sufficient residence time. NO_x is formed primarily at the diffusion flame front. Figure 11.9 presents the NO_x and soot engine exhaust emissions as a function of injection timing, showing a tradeoff between NO_x and soot. For early injection timing, more NO_x is formed due to higher pressures and temperatures in the cylinder and the longer combustion time. Soot emissions have the opposite trend as NO_x emissions. Lower soot emissions are found with early injection because the soot is burned prior to the exhaust valve opening due to the higher temperatures and longer combustion times.

Similar to SI engines, NO_x production can be decreased by using Exhaust Gas Recirculation (EGR) to lower the peak flame temperature. Various after treatments to reduce emissions from diesel engines are also available. An oxidation catalyst can further reduce unburned hydrocarbons and CO.

Diesel PM filters (DPF) are now available to decrease soot emissions. Urea or ammonia injection in the exhaust together with an oxidation catalyst is now

Fig. 11.9 *Left*: Emissions of NO_x decrease but soot emissions increase with injection timing delay. *Right*: NOx-soot trade-off diagram (*right*)

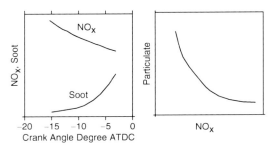

being used to reduce HC, CO and NO_x. Various lean NO_x catalysts are now being researched for conversion of NO_x into N_2 and O_2. For example, one exhaust treatment scheme injects diesel fuel into a platinum catalyst to reduce NO_x. Recently, silver catalysts were found capable of converting up to 90% of NO_x with injection of ethanol. Further research is needed to better understand various issues, such as emission of formaldehyde from the catalyst.

11.6 Homogeneous Charge Compression Ignition (HCCI)

11.6.1 HCCI Overview

HCCI technology offers major advancements in high efficiency and low emissions from engines. HCCI approaches the high fuel efficiency of diesel engines by using a high compression ratio. Similar to SI engines, the charge is a well-mixed fuel and air mixture that is lean and introduced in the cylinder prior to compression. Combustion is initiated by autoignition that occurs almost simultaneously throughout the engine cylinder near the end of the compression stroke. Such a combustion event causes a sudden pressure jump exactly like engine knock experienced by SI engine. HCCI is thus a combustion process that combines the ignition process of a compression ignition (CI) engine with the premixed nature of the spark ignition (SI) engine. In HCCI, very lean mixtures ($\phi = 0.1$–0.4) are used such that the peak flame temperature is below 1,800 K to prevent large amount of thermal NO_x formation. The lean premixed charge helps minimize particulate emissions.

The HCCI engine platform is nearly the same as the traditional CI engine. This mode of engine operation takes with it many of the advantages of the CI engine, but at the same time, brings with it some serious challenges. HCCI engines may produce diesel-like efficiency due to high compression ratios and very rapid heat release, while maintaining low nitrogen oxide (NO_x), particulate matter (PM), and soot emissions. Also, HCCI engines are fuel flexible and may be cost competitive to manufacture since a high-pressure fuel injection system is not required. In addition, unlike SI engines, HCCI has very low cyclic variation, resulting in steady engine performance and emission characteristics. Figure 11.10 shows that peak pressure remains fairly constant from cycle to cycle. This is due to the lack of the ignition lag typical of SI combustion.

11.6.2 HCCI Emissions

HCCI combustion is a multi-point premixed auto-ignition process, with little or no flame propagation. This is similar to the autoignition process (commonly

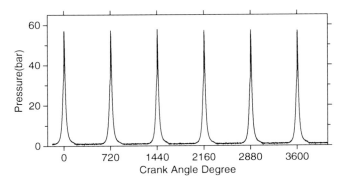

Fig. 11.10 HCCI engines typically have steady cylinder pressure traces

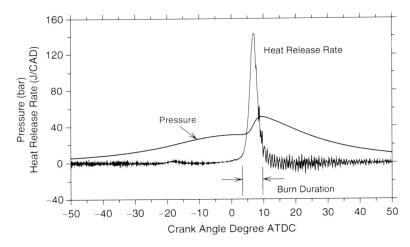

Fig. 11.11 Experimental data of average pressure and heat release for a HCCI engine

known as "knocking") that is undesirable in an SI engine. The distinction that can be made between SI "knocking" and HCCI autoignition is that for HCCI the autoignition is typically contained in the core gas. Crevices and boundary layers typically have higher heat transfer and are not compressed rapidly enough to autoignite. Figure 11.11 presents experimental data taken from a HCCI engine showing the average pressure trace and the deduced heat release rate. Compared to the burn duration of SI engines (30–40 CAD), HCCI combusts in a relatively short period of time. The use of lean mixtures (lower flame temperature) and the lack of flame propagation (eliminating local heterogeneity) greatly reduce NO_x formation. In a HCCI engine, thermal NO_x formation is typically minimal due to the low combustion temperature (below 1,700 K). However, a small amount of NO_x is still formed by the prompt and to a lesser extent, the N_2O mechanism.

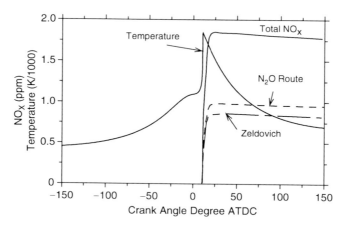

Fig. 11.12 Predicted temperature, NO_x and two pathways (N_2O and thermal NO) are plotted versus CAD showing N_2O becomes equally important as thermal NO

It is important to note that HCCI combustion is controlled largely by chemical kinetics, with fluid dynamics playing a less significant role. Figure 11.12 shows that the predicted NO_x level from HCCI engines is very low (~2 ppm) and the N_2O pathway becomes an equally important route (>50%) as the thermal NO pathway.

11.6.3 Challenges with HCCI

Several technical challenges must be overcome before HCCI can be widely used. These include (1) controlling the combustion autoignition and phasing, (2) expanding the load regime into high load application (limited by rapid pressure rise), (3) noise, (4) cold start, and (5) reducing the amount of unburned hydrocarbons and CO.

Maintaining optimal combustion phasing is a challenging task. Combustion is not initiated by a spark or an injection of fuel. Instead, the autoignition is controlled by chemical kinetics, which are sensitive to temperature, pressure, mixture composition, and EGR. There are many possibilities for HCCI engine control: variable compression ratio, variable valve timing, operation with multiple fuels, and thermal control. Out of these options, thermal control is inexpensive to implement and purely based on technologies familiar to manufactures. It therefore may be adopted if demonstrated satisfactory.

Using EGR can alleviate the rapid pressure rise by slowing down combustion chemistry at the expensive of reducing power density (power per unit weight). To increase power density, boosted pressure can be used with additional cost.

It is difficult to cold start in the HCCI mode because preheating of the intake charge is typically required. A proposed solution is to start the engine in the regular

CI or SI mode and transition to the HCCI mode once the required operating conditions are met.

Noise is a disadvantage of the HCCI engine. This is a direct result of the explosive nature of autoignition combustion. At low to medium loads, an HCCI engine has a noise level that is comparable to a CI engine of the same class. However, at high load, the noise level of an HCCI engine can be painful to human ears.

Unlike diesel engines, emission of unburned hydrocarbons and CO are high from HCCI engines due to the premixed charge reaching the crevices and the cooler boundary layers of the engine walls. Reducing the crevices is limited by the piston ring properties, so after-treatment is necessary for CO and HC removal. Without after-treatment, HC and CO emissions likely will not meet the current automotive emission standards; the use of an oxidation catalyst is thus called for. However, due to the high efficiency of HCCI, the exhaust temperature is low which may require the development of a low-temperature oxidation catalyst.

As HCCI is a promising technology for increasing engine performance while achieving low emissions, research is currently being conducted all over the world to advance HCCI technology.

References

1. Arai M, Tabata M, Hiroyasu H (1984) Disintegration Process and Spray Characterization of Fuel Jet Injected by a Diesel Nozzle. SAE Paper no. 840275.
2. Dec J (1997) A conceptual model of DI diesel combustion based on lased-sheet imaging. SAE paper 970873.
3. El-Kotb MM (1982) Fuel Atomization for Spray Modeling. Prog. Energy Comb. Sci. 8:61–91.
4. Flynn PF, Durrett RP, Hunter GL, Loye AZ, Akinyemi OC, Dec JE, Westbrook CK (1999) Diesel Combustion: An Integrated View Combining Laser Diagnostics, Chemical Kinetics, and Empirical Validation. SAE paper 1999-01-0509. Presented at International Congress & Exposition, March 1999, Detroit, MI, USA, Session: Diesel Engine Combustion Processes (Part C&D) Future Transportation Technology Conference & Exposition, August 1999, Costa Mesa, CA, USA.

Appendix 1
Properties of Fuels

S. McAllister et al., *Fundamentals of Combustion Processes*,
Mechanical Engineering Series, DOI 10.1007/978-1-4419-7943-8,
© Springer Science+Business Media, LLC 2011

Formula	Fuel	M (kg/kmol)	T_b (°C)	c_{pg}^a (kJ/kg-K)	T_{ig} (°C)	HHV (MJ/kg)	LHV (MJ/kg)	h_{fg}^b (kJ/kg)	AFRs	T_f (K)^c	$\Delta\hat{h}^0$ (kJ/mol)	RON^d	MON^e
CH_4	Methane	16.04	−161	2.21	537	55.536	50.048	510	17.2	2,226	−74.4	120	120
C_2H_2	Acetylene	26.04	−84	1.60	305	49.923	48.225	–	13.2	2,540	8.7	50	50
C_2H_4	Ethylene	28.05	−104	1.54	490	50.312	47.132	–	14.7	2,380	52.4	–	–
C_2H_6	Ethane	30.07	−89	1.75	472	51.902	47.611	489	16.1	2,370	−83.8	115	99
C_3H_8	Propane	44.10	−42	1.62	470	50.322	46.330	432	15.7	2,334	−104.7	112	97
C_4H_{10}	n-Butane	58.12	−0.5	1.64	365	49.511	45.725	386	15.5	2,270	−146.6	94	90
C_4H_{10}	iso-Butane	58.12	−12	1.62	460	49.363	45.577	366	15.5	2,310	−153.5	102	98
C_5H_{12}	n-Pentane	72.15	36	1.62	284	49.003	45.343	357	15.3	2,270	−173.5	62	63
C_5H_{12}	iso-Pentane	72.15	28	1.60	420	48.909	45.249	342	15.3	2,310	−178.5	93	90
C_6H_{14}	n-Hexane	86.18	69	1.62	233	48.674	45.099	335	15.2	2,271	−198.7	25	26
C_6H_{14}	iso-Hexane	86.18	50	1.58	421	48.454	44.879	305	15.2	2,300	−207.4	104	94
C_7H_{16}	n-Heptane	100.20	99	1.61	215	48.438	44.925	317	15.2	2,273	−224.2	0	0
C_8H_{18}	n-Octane	114.23	126	1.61	206	48.254	44.786	301	15.1	2,275	−250.1	20	17
C_8H_{18}	iso-Octane	114.23	114	1.59	418	48.119	44.651	283	15.1	2,300	−259.2	100	100
C_9H_{20}	n-Nonane	128.6	151	1.61	–	48.119	44.688	295	15.1	2,274	−274.7	–	–
$C_{10}H_{22}$	n-Decane	142.28	174	1.61	210	48.002	44.599	277	15.1	2,278	−300.9	−41	−38
$C_{10}H_{22}$	iso-Decane	142.28	171	1.61	–	48.565	44.413	–	15.1	2,340	–	113	92
$C_{12}H_{26}$	n-Dodecane	170.33	216	1.61	204	47.838	44.574	256	15.0	2,276	−350.9	−88	−90
CH_4O	Methanol	32.04	65	1.37	385	22.663	19.915	1,099	6.5	2,183	−201.5	106	92
C_2H_6O	Ethanol	46.07	78	1.42	365	29.668	26.803	836	9.0	2,144	−235.1	107	89
H_2	Hydrogen	2.02	−253	14.47	400	141.72	119.96	451	34.3	2,345	0	–	–

^{a} Gas phase specific heat evaluated at 25°C
^{b} Heat of vaporization at 1 atm
^{c} Estimated equilibrium flame temperature
^{d} Research octane number
^{e} Motoring octane number

Appendix 2
Properties of Air at 1 atm

Temp (K)	Specific heat c_p (kJ/kg-K)	Specific heat c_v (kJ/kg-K)	Ratio of specific heats γ, (c_p/c_v)	Viscosity, $\mu \cdot 10^5$ (kg/m-s)	Thermal conductivity, $k \cdot 10^5$ (kW/m-K)	Prandtl number v/α	Kinematic viscosity, $v \cdot 10^5$ (m^2/s)	Density ρ (kg/m^3)
175	1.0023	0.7152	1.401	1.182	1.593	0.744	0.586	2.017
200	1.0025	0.7154	1.401	1.329	1.809	0.736	0.753	1.765
225	1.0027	0.7156	1.401	1.467	2.020	0.728	0.935	1.569
250	1.0031	0.716	1.401	1.599	2.227	0.720	1.132	1.412
275	1.0038	0.7167	1.401	1.725	2.428	0.713	1.343	1.284
300	1.0049	0.7178	1.400	1.846	2.624	0.707	1.568	1.177
325	1.0063	0.7192	1.400	1.962	2.816	0.701	1.807	1.086
350	1.0082	0.7211	1.398	2.075	3.003	0.697	2.056	1.009
375	1.0106	0.7235	1.397	2.181	3.186	0.692	2.317	0.9413
400	1.0135	0.7264	1.395	2.286	3.365	0.688	2.591	0.8824
450	1.0206	0.7335	1.391	2.486	3.710	0.684	3.168	0.7844
500	1.0295	0.7424	1.387	2.670	4.041	0.680	3.782	0.706
550	1.0398	0.7527	1.381	2.849	4.357	0.680	4.439	0.6418
600	1.0511	0.7540	1.376	3.017	4.661	0.680	5.128	0.5883
650	1.0629	0.7758	1.370	3.178	4.954	0.682	5.853	0.543
700	1.0750	0.7879	1.364	3.332	5.236	0.684	6.607	0.5043
750	1.0870	0.7999	1.359	3.482	5.509	0.687	7.399	0.4706
800	1.0987	0.8116	1.354	3.624	5.774	0.690	8.214	0.4412
850	1.1101	0.8230	1.349	3.763	6.030	0.693	9.061	0.4153
900	1.1209	0.8338	1.344	3.897	6.276	0.696	9.936	0.3922
950	1.1313	0.8442	1.34	4.026	6.520	0.699	10.83	0.3716
1,000	1.1411	0.8540	1.336	4.153	6.754	0.702	11.76	0.3530
1,050	1.1502	0.8631	1.333	4.276	6.985	0.704	12.72	0.3362
1,100	1.1589	0.8718	1.329	4.396	7.209	0.707	13.70	0.3209
1,150	1.1670	0.8799	1.326	4.511	7.427	0.709	14.07	0.3069
1,200	1.1746	0.8875	1.323	4.626	7.640	0.711	15.73	0.2941
1,250	1.1817	0.8946	1.321	4.736	7.849	0.713	16.77	0.2824
1,300	1.1884	0.9013	1.319	4.846	8.054	0.715	17.85	0.2715
1,350	1.1946	0.9075	1.316	4.952	8.253	0.717	18.94	0.2615

(continued)

Temp (K)	Specific heat c_p (kJ/kg-K)	Specific heat c_v (kJ/kg-K)	Ratio of specific heats γ, (c_p/c_v)	Viscosity, $\mu \cdot 10^5$ (kg/m-s)	Thermal conductivity, $k \cdot 10^5$ (kW/m-K)	Prandtl number ν/α	Kinematic viscosity, $\nu \cdot 10^5$ (m^2/s)	Density ρ (kg/m^3)
1,400	1.2005	0.9134	1.314	5.057	8.450	0.719	20.06	0.2521
1,500	1.2112	0.9241	1.311	5.262	8.831	0.722	22.36	0.2353
1,600	1.2207	0.9336	1.308	5.457	9.199	0.724	24.74	0.2206
1,700	1.2293	0.9422	1.305	5.646	9.554	0.726	27.20	0.2076
1,800	1.2370	0.9499	1.302	5.829	9.899	0.728	29.72	0.1961
1,900	1.2440	0.9569	1.300	6.008	10.233	0.730	32.34	0.1858

Appendix 3
Properties of Ideal Combustion Gases

Ideal gases are assumed for combustion gases. The enthalpy, $\hat{h}(T)$, of a gaseous species consists of two parts: (1) enthalpy of formation at the standard condition (25°C and 1 atm) and (2) sensible enthalpy. Enthalpy of a species is evaluated by

$$\hat{h}(T) = \Delta\hat{h}^0 + (\hat{h}_s(T) - \hat{h}_s(T = 25^\circ C)) = \Delta\hat{h}^0 + (\hat{h}_s(T) - \hat{h}_s^0)$$

This formula can be extended to include phase change from liquid to gas by including the latent heat of vaporization.

For an elementary reaction

$$aA + bB \leftrightarrow cC + dD$$

the equilibrium constant based on concentrations, $K_c = k_f/k_b$, can be determined by thermodynamics properties as

$$K_c = \frac{k_f}{k_b} = \frac{[C]_{eq}^c [D]_{eq}^d}{[A]_{eq}^a [B]_{eq}^b} = K_p(T) \left(\frac{\hat{R}_u T}{1 \text{ atm}}\right)^{a+b-c-d}$$

where $K_p(T) = \exp\left\{a\frac{\hat{g}_A^0}{\hat{R}_u T} + b\frac{\hat{g}_B^0}{\hat{R}_u T} - c\frac{\hat{g}_C^0}{\hat{R}_u T} - d\frac{\hat{g}_D^0}{\hat{R}_u T}\right\}$ is the equilibrium constant based on partial pressures and $\hat{g}_i^0(T) = \hat{h}_i(T) - T\hat{s}_i^0(T)$ is the Gibbs free energy at reference pressure (1 *atm*).

CO_2

T (K)	$\hat{c}_p$ (kJ/kmol-K)	$\hat{h} - \hat{h}^0$ (MJ/kmol)	$\hat{s}$ (kJ/kmol-K)	$g^0/\hat{R}_u T$
200	32.39	−3.42	199.87	−262.76
250	34.96	−1.74	207.37	−215.11
298	37.2	0	213.73	−184.46
300	37.28	0.07	213.96	−183.48
350	39.37	1.99	219.86	−161
400	41.27	4	225.25	−144.22
450	43	6.11	230.21	−131.24
500	44.57	8.3	234.82	−120.91

(continued)

CO_2 (continued)

T (K)	$\hat{c}_p$ (kJ/kmol-K)	$\hat{h} - \hat{h}^0$ (MJ/kmol)	$\hat{s}$ (kJ/kmol-K)	$g^0/\hat{R}_u T$
550	46	10.56	239.14	−112.51
600	47.31	12.9	243.2	−105.55
650	48.51	15.29	247.03	−99.7
700	49.61	17.75	250.67	−94.72
750	50.62	20.25	254.13	−90.43
800	51.55	22.81	257.42	−86.7
850	52.38	25.41	260.57	−83.43
900	53.13	28.05	263.59	−80.55
950	53.79	30.72	266.48	−77.99
1,000	54.36	33.42	269.25	−75.7
1,050	54.86	36.15	271.92	−73.64
1,100	55.33	38.91	274.48	−71.79
1,150	55.78	41.69	276.95	−70.11
1,200	56.2	44.49	279.33	−68.58
1,250	56.6	47.31	281.64	−67.19
1,300	56.98	50.15	283.86	−65.91
1,350	57.34	53	286.02	−64.74
1,400	57.67	55.88	288.11	−63.66
1,450	57.99	58.77	290.14	−62.67
1,500	58.29	61.68	292.11	−61.74
1,550	58.57	64.6	294.03	−60.89
1,600	58.83	67.54	295.89	−60.1
1,650	59.08	70.48	297.71	−59.36
1,700	59.31	73.44	299.47	−58.67
1,750	59.53	76.41	301.2	−58.02
1,800	59.73	79.4	302.88	−57.42
1,850	59.93	82.39	304.52	−56.86
1,900	60.11	85.39	306.12	−56.33
1,950	60.27	88.4	307.68	−55.83
2,000	60.43	91.42	309.21	−55.36
2,050	60.58	94.44	310.7	−54.92
2,100	60.71	97.47	312.16	−54.5
2,150	60.84	100.51	313.59	−54.11
2,200	60.96	103.56	314.99	−53.74
2,250	61.08	106.61	316.37	−53.39
2,300	61.18	109.66	317.71	−53.06
2,350	61.28	112.73	319.03	−52.74
2,400	61.37	115.79	320.32	−52.45
2,450	61.46	118.86	321.58	−52.16
2,500	61.55	121.94	322.83	−51.9
2,550	61.62	125.02	324.05	−51.64
2,600	61.7	128.1	325.24	−51.4
2,650	61.77	131.19	326.42	−51.17
2,700	61.84	134.28	327.57	−50.95
2,750	61.9	137.37	328.71	−50.74
2,800	61.96	140.47	329.83	−50.54

(continued)

CO$_2$ (continued)

T (K)	$\hat{c}_p$ (kJ/kmol-K)	$\hat{h} - \hat{h}^0$ (MJ/kmol)	$\hat{s}$ (kJ/kmol-K)	$g^0/\hat{R}_u T$
2,850	62.02	143.57	330.92	−50.35
2,900	62.08	146.67	332	−50.17
2,950	62.14	149.78	333.06	−50
3,000	62.19	152.88	334.11	−49.83
3,050	62.25	155.99	335.14	−49.68
3,100	62.3	159.11	336.15	−49.53
3,150	62.35	162.22	337.15	−49.38
3,200	62.4	165.34	338.13	−49.25
3,250	62.45	168.46	339.1	−49.12
3,300	62.51	171.59	340.05	−48.99
3,350	62.56	174.72	340.99	−48.87
3,400	62.61	177.84	341.92	−48.76
3,450	62.66	180.98	342.83	−48.65
3,500	62.72	184.11	343.73	−48.54

H$_2$O

T (K)	$\hat{c}_p$ (kJ/kmol-K)	$\hat{h} - \hat{h}^0$ (MJ/kmol)	$\hat{s}$ (kJ/kmol-K)	$g^0/\hat{R}_u T$
200	32.25	−3.23	175.59	−168.5
250	32.9	−1.6	182.86	−139.11
298	33.45	0	188.71	−120.26
300	33.47	0.06	188.91	−119.66
350	33.97	1.75	194.11	−105.85
400	34.44	3.46	198.68	−95.58
450	34.89	5.19	202.76	−87.64
500	35.34	6.95	206.46	−81.34
550	35.8	8.73	209.85	−76.22
600	36.29	10.53	212.98	−71.99
650	36.81	12.35	215.91	−68.43
700	37.36	14.21	218.65	−65.41
750	37.96	16.09	221.25	−62.81
800	38.59	18	223.72	−60.56
850	39.25	19.95	226.08	−58.59
900	39.93	21.93	228.34	−56.85
950	40.62	23.94	230.52	−55.31
1,000	41.31	25.99	232.62	−53.94
1,050	41.99	28.07	234.65	−52.71
1,100	42.64	30.19	236.62	−51.6
1,150	43.26	32.34	238.53	−50.6
1,200	43.87	34.52	240.39	−49.69
1,250	44.46	36.72	242.19	−48.87
1,300	45.02	38.96	243.94	−48.11

(continued)

H₂O   (continued)

T (K)	$\hat{c}_p$ (kJ/kmol-K)	$\hat{h} - \hat{h}^0$ (MJ/kmol)	$\hat{s}$ (kJ/kmol-K)	$g^0/\hat{R}_u T$
1,350	45.57	41.23	245.65	−47.42
1,400	46.1	43.52	247.32	−46.79
1,450	46.61	45.84	248.95	−46.2
1,500	47.1	48.18	250.53	−45.66
1,550	47.58	50.55	252.09	−45.16
1,600	48.03	52.94	253.6	−44.7
1,650	48.47	55.35	255.09	−44.28
1,700	48.9	57.78	256.54	−43.88
1,750	49.31	60.24	257.97	−43.51
1,800	49.7	62.71	259.36	−43.16
1,850	50.08	65.21	260.73	−42.84
1,900	50.45	67.72	262.07	−42.54
1,950	50.8	70.25	263.38	−42.26
2,000	51.14	72.8	264.67	−42
2,050	51.47	75.37	265.94	−41.75
2,100	51.78	77.95	267.19	−41.52
2,150	52.08	80.54	268.41	−41.31
2,200	52.38	83.16	269.61	−41.1
2,250	52.66	85.78	270.79	−40.91
2,300	52.92	88.42	271.95	−40.73
2,350	53.18	91.07	273.09	−40.56
2,400	53.43	93.74	274.21	−40.4
2,450	53.67	96.42	275.32	−40.25
2,500	53.9	99.11	276.4	−40.11
2,550	54.12	101.81	277.47	−39.98
2,600	54.34	104.52	278.53	−39.85
2,650	54.54	107.24	279.56	−39.73
2,700	54.74	109.97	280.58	−39.62
2,750	54.93	112.72	281.59	−39.52
2,800	55.11	115.47	282.58	−39.42
2,850	55.29	118.23	283.56	−39.32
2,900	55.46	120.99	284.52	−39.23
2,950	55.62	123.77	285.47	−39.15
3,000	55.78	126.56	286.41	−39.07
3,050	55.93	129.35	287.33	−39
3,100	56.07	132.15	288.24	−38.92
3,150	56.21	134.96	289.14	−38.86
3,200	56.35	137.77	290.03	−38.8
3,250	56.48	140.59	290.9	−38.74
3,300	56.61	143.42	291.76	−38.68
3,350	56.73	146.25	292.62	−38.63
3,400	56.85	149.09	293.46	−38.58
3,450	56.96	151.94	294.29	−38.53
3,500	57.07	154.79	295.11	−38.49

N_2

T (K)	$\hat{c}_p$ (kJ/kmol-K)	$\hat{h} - \hat{h}^0$ (MJ/kmol)	$\hat{s}$ (kJ/kmol-K)	$g^0/\hat{R}_u T$
200	28.79	−2.84	179.95	−23.35
250	28.95	−1.4	186.39	−23.09
298	29.07	0	191.5	−23.03
300	29.07	0.05	191.68	−23.03
350	29.19	1.51	196.17	−23.08
400	29.32	2.97	200.08	−23.17
450	29.46	4.44	203.54	−23.29
500	29.63	5.92	206.65	−23.43
550	29.84	7.41	209.49	−23.58
600	30.08	8.9	212.09	−23.72
650	30.37	10.42	214.51	−23.87
700	30.68	11.94	216.77	−24.02
750	31.03	13.48	218.9	−24.17
800	31.39	15.04	220.92	−24.31
850	31.76	16.62	222.83	−24.45
900	32.13	18.22	224.66	−24.59
950	32.47	19.84	226.4	−24.72
1,000	32.76	21.47	228.08	−24.85
1,050	33.01	23.11	229.68	−24.98
1,100	33.26	24.77	231.22	−25.1
1,150	33.49	26.44	232.71	−25.22
1,200	33.71	28.12	234.14	−25.34
1,250	33.91	29.81	235.52	−25.46
1,300	34.11	31.51	236.85	−25.57
1,350	34.3	33.22	238.14	−25.68
1,400	34.48	34.94	239.39	−25.79
1,450	34.64	36.67	240.6	−25.9
1,500	34.8	38.4	241.78	−26
1,550	34.95	40.15	242.92	−26.1
1,600	35.1	41.9	244.04	−26.2
1,650	35.23	43.66	245.12	−26.3
1,700	35.36	45.42	246.17	−26.4
1,750	35.48	47.19	247.2	−26.49
1,800	35.59	48.97	248.2	−26.58
1,850	35.7	50.75	249.18	−26.67
1,900	35.8	52.54	250.13	−26.76
1,950	35.9	54.33	251.06	−26.85
2,000	35.99	56.13	251.97	−26.93
2,050	36.07	57.93	252.86	−27.01
2,100	36.15	59.73	253.73	−27.1
2,150	36.23	61.54	254.58	−27.18
2,200	36.3	63.36	255.42	−27.26
2,250	36.36	65.17	256.23	−27.34
2,300	36.43	66.99	257.03	−27.41
2,350	36.49	68.82	257.82	−27.49
2,400	36.54	70.64	258.59	−27.56

(continued)

N$_2$ (continued)

T (K)	$\hat{c}_p$ (kJ/kmol-K)	$\hat{h} - \hat{h}^0$ (MJ/kmol)	$\hat{s}$ (kJ/kmol-K)	$g^0 / \hat{R}_u T$
2,450	36.59	72.47	259.34	−27.64
2,500	36.64	74.3	260.08	−27.71
2,550	36.69	76.13	260.81	−27.78
2,600	36.74	77.97	261.52	−27.85
2,650	36.78	79.81	262.22	−27.92
2,700	36.82	81.65	262.91	−27.98
2,750	36.86	83.49	263.58	−28.05
2,800	36.89	85.33	264.25	−28.12
2,850	36.93	87.18	264.9	−28.18
2,900	36.96	89.03	265.54	−28.25
2,950	36.99	90.88	266.17	−28.31
3,000	37.03	92.73	266.8	−28.37
3,050	37.06	94.58	267.41	−28.43
3,100	37.09	96.43	268.01	−28.49
3,150	37.11	98.29	268.61	−28.55
3,200	37.14	100.14	269.19	−28.61
3,250	37.17	102	269.77	−28.67
3,300	37.2	103.86	270.33	−28.73
3,350	37.22	105.72	270.89	−28.79
3,400	37.25	107.58	271.45	−28.84
3,450	37.27	109.45	271.99	−28.9
3,500	37.3	111.31	272.53	−28.95

O$_2$

T (K)	$\hat{c}_p$ (kJ/kmol-K)	$\hat{h} - \hat{h}^0$ (MJ/kmol)	$\hat{s}$ (kJ/kmol-K)	$g^0 / \hat{R}_u T$
200	28.47	−2.84	193.51	−24.98
250	28.9	−1.4	199.91	−24.72
298	29.31	0	205.03	−24.66
300	29.33	0.05	205.21	−24.66
350	29.77	1.53	209.77	−24.7
400	30.21	3.03	213.77	−24.8
450	30.66	4.55	217.36	−24.93
500	31.11	6.1	220.61	−25.07
550	31.57	7.66	223.6	−25.22
600	32.03	9.25	226.36	−25.37
650	32.48	10.87	228.94	−25.53
700	32.93	12.5	231.37	−25.68
750	33.35	14.16	233.65	−25.83
800	33.76	15.84	235.82	−25.98
850	34.13	17.53	237.88	−26.13
900	34.45	19.25	239.84	−26.28
950	34.73	20.98	241.71	−26.42

(continued)

O$_2$ (continued)

T (K)	$\hat{c}_p$ (kJ/kmol-K)	$\hat{h} - \hat{h}^0$ (MJ/kmol)	$\hat{s}$ (kJ/kmol-K)	$g^0/\hat{R}_u T$
1,000	34.93	22.72	243.49	−26.55
1,050	35.1	24.47	245.2	−26.69
1,100	35.27	26.23	246.84	−26.82
1,150	35.43	28	248.41	−26.95
1,200	35.59	29.77	249.92	−27.08
1,250	35.75	31.56	251.38	−27.2
1,300	35.9	33.35	252.78	−27.32
1,350	36.05	35.15	254.14	−27.44
1,400	36.2	36.95	255.46	−27.55
1,450	36.35	38.77	256.73	−27.66
1,500	36.49	40.59	257.96	−27.77
1,550	36.63	42.42	259.16	−27.88
1,600	36.77	44.25	260.33	−27.99
1,650	36.9	46.09	261.46	−28.09
1,700	37.03	47.94	262.56	−28.19
1,750	37.17	49.8	263.64	−28.29
1,800	37.29	51.66	264.69	−28.38
1,850	37.42	53.53	265.71	−28.48
1,900	37.54	55.4	266.71	−28.57
1,950	37.67	57.28	267.69	−28.66
2,000	37.79	59.17	268.64	−28.75
2,050	37.9	61.06	269.58	−28.84
2,100	38.02	62.96	270.49	−28.93
2,150	38.14	64.86	271.39	−29.01
2,200	38.25	66.77	272.27	−29.1
2,250	38.36	68.69	273.13	−29.18
2,300	38.47	70.61	273.97	−29.26
2,350	38.58	72.53	274.8	−29.34
2,400	38.68	74.46	275.61	−29.42
2,450	38.79	76.4	276.41	−29.5
2,500	38.89	78.34	277.2	−29.57
2,550	38.99	80.29	277.97	−29.65
2,600	39.09	82.24	278.73	−29.72
2,650	39.19	84.2	279.47	−29.79
2,700	39.29	86.16	280.2	−29.86
2,750	39.38	88.13	280.93	−29.94
2,800	39.48	90.1	281.64	−30
2,850	39.57	92.07	282.34	−30.07
2,900	39.66	94.06	283.03	−30.14
2,950	39.75	96.04	283.7	−30.21
3,000	39.84	98.03	284.37	−30.27
3,050	39.93	100.03	285.03	−30.34
3,100	40.02	102.02	285.68	−30.4
3,150	40.11	104.03	286.32	−30.47
3,200	40.19	106.03	286.96	−30.53
3,250	40.28	108.05	287.58	−30.59

(continued)

O_2 (continued)

T (K)	$\hat{c}_p$ (kJ/kmol-K)	$\hat{h} - \hat{h}^0$ (MJ/kmol)	$\hat{s}$ (kJ/kmol-K)	$g^0/\hat{R}_u T$
3,300	40.36	110.06	288.2	−30.65
3,350	40.44	112.08	288.8	−30.71
3,400	40.52	114.11	289.4	−30.77
3,450	40.6	116.13	290	−30.83
3,500	40.68	118.17	290.58	−30.89

CO

T (K)	$\hat{c}_p$ (kJ/kmol-K)	$\hat{h} - \hat{h}^0$ (MJ/kmol)	$\hat{s}$ (kJ/kmol-K)	$g^0/\hat{R}_u T$
200	28.69	−2.84	186.01	−90.55
250	28.89	−1.4	192.43	−77
298	29.07	0	197.54	−68.35
300	29.08	0.05	197.72	−68.08
350	29.25	1.51	202.21	−61.79
400	29.43	2.98	206.13	−57.14
450	29.63	4.46	209.61	−53.57
500	29.86	5.94	212.74	−50.75
550	30.11	7.44	215.6	−48.48
600	30.41	8.95	218.23	−46.61
650	30.73	10.48	220.68	−45.06
700	31.09	12.03	222.97	−43.74
750	31.47	13.59	225.13	−42.62
800	31.86	15.17	227.17	−41.66
850	32.25	16.78	229.11	−40.82
900	32.63	18.4	230.97	−40.09
950	32.97	20.04	232.74	−39.45
1,000	33.25	21.7	234.44	−38.88
1,050	33.49	23.36	236.07	−38.38
1,100	33.72	25.04	237.63	−37.93
1,150	33.94	26.74	239.13	−37.53
1,200	34.15	28.44	240.58	−37.17
1,250	34.34	30.15	241.98	−36.84
1,300	34.53	31.87	243.33	−36.55
1,350	34.7	33.6	244.64	−36.28
1,400	34.87	35.34	245.9	−36.04
1,450	35.03	37.09	247.13	−35.82
1,500	35.18	38.85	248.32	−35.62
1,550	35.32	40.61	249.48	−35.43
1,600	35.45	42.38	250.6	−35.27
1,650	35.57	44.15	251.69	−35.11
1,700	35.69	45.93	252.76	−34.97
1,750	35.8	47.72	253.79	−34.84

(continued)

CO   (continued)

T (K)	$\hat{c}_p$ (kJ/kmol-K)	$\hat{h} - \hat{h}^0$ (MJ/kmol)	$\hat{s}$ (kJ/kmol-K)	$g^0/\hat{R}_u T$
1,800	35.91	49.51	254.8	−34.72
1,850	36.01	51.31	255.79	−34.62
1,900	36.1	53.12	256.75	−34.52
1,950	36.19	54.92	257.69	−34.42
2,000	36.27	56.73	258.6	−34.34
2,050	36.35	58.55	259.5	−34.26
2,100	36.42	60.37	260.38	−34.19
2,150	36.49	62.19	261.24	−34.13
2,200	36.55	64.02	262.08	−34.07
2,250	36.61	65.85	262.9	−34.01
2,300	36.67	67.68	263.7	−33.96
2,350	36.72	69.51	264.49	−33.91
2,400	36.77	71.35	265.27	−33.87
2,450	36.82	73.19	266.02	−33.83
2,500	36.87	75.03	266.77	−33.79
2,550	36.91	76.88	267.5	−33.76
2,600	36.95	78.72	268.22	−33.73
2,650	36.99	80.57	268.92	−33.71
2,700	37.02	82.42	269.61	−33.68
2,750	37.06	84.27	270.29	−33.66
2,800	37.09	86.13	270.96	−33.64
2,850	37.12	87.98	271.62	−33.62
2,900	37.15	89.84	272.26	−33.61
2,950	37.18	91.7	272.9	−33.59
3,000	37.21	93.56	273.52	−33.58
3,050	37.24	95.42	274.14	−33.57
3,100	37.27	97.28	274.74	−33.56
3,150	37.29	99.15	275.34	−33.55
3,200	37.32	101.01	275.93	−33.55
3,250	37.35	102.88	276.51	−33.54
3,300	37.37	104.75	277.08	−33.54
3,350	37.4	106.61	277.64	−33.53
3,400	37.42	108.49	278.19	−33.53
3,450	37.45	110.36	278.74	−33.53
3,500	37.47	112.23	279.28	−33.53

C (gas)

T (K)	$\hat{c}_p$ (kJ/kmol-K)	$\hat{h} - \hat{h}^0$ (MJ/kmol)	$\hat{s}$ (kJ/kmol-K)	$g^0/\hat{R}_u T$
200	20.84	−2.05	149.68	411.79
250	20.84	−1	154.33	325.78
298	20.83	0	158	270.13
300	20.83	0.04	158.13	268.35

(continued)

C (gas) (continued)

T (K)	$\hat{c}_p$ (kJ/kmol-K)	$\hat{h} - \hat{h}^0$ (MJ/kmol)	$\hat{s}$ (kJ/kmol-K)	$g^0/\hat{R}_u T$
350	20.83	1.08	161.34	227.27
400	20.82	2.12	164.12	196.41
450	20.81	3.16	166.57	172.38
500	20.81	4.2	168.76	153.12
550	20.8	5.24	170.75	137.35
600	20.8	6.28	172.56	124.18
650	20.79	7.32	174.22	113.02
700	20.79	8.36	175.76	103.45
750	20.79	9.4	177.2	95.14
800	20.79	10.44	178.54	87.85
850	20.8	11.48	179.8	81.42
900	20.8	12.52	180.99	75.69
950	20.81	13.56	182.11	70.56
1,000	20.81	14.6	183.18	65.93
1,050	20.8	15.64	184.2	61.74
1,100	20.79	16.68	185.16	57.92
1,150	20.78	17.72	186.09	54.43
1,200	20.78	18.76	186.97	51.23
1,250	20.78	19.8	187.82	48.28
1,300	20.78	20.84	188.63	45.55
1,350	20.78	21.88	189.42	43.02
1,400	20.78	22.92	190.17	40.67
1,450	20.79	23.95	190.9	38.48
1,500	20.8	24.99	191.61	36.43
1,550	20.81	26.03	192.29	34.51
1,600	20.82	27.08	192.95	32.71
1,650	20.83	28.12	193.59	31.01
1,700	20.84	29.16	194.21	29.41
1,750	20.86	30.2	194.82	27.9
1,800	20.88	31.24	195.41	26.48
1,850	20.9	32.29	195.98	25.12
1,900	20.92	33.33	196.54	23.84
1,950	20.94	34.38	197.08	22.62
2,000	20.96	35.43	197.61	21.46
2,050	20.99	36.48	198.13	20.36
2,100	21.01	37.53	198.63	19.31
2,150	21.04	38.58	199.13	18.3
2,200	21.07	39.63	199.61	17.34
2,250	21.1	40.68	200.09	16.42
2,300	21.13	41.74	200.55	15.54
2,350	21.16	42.8	201.01	14.7
2,400	21.19	43.86	201.45	13.89
2,450	21.22	44.92	201.89	13.11
2,500	21.25	45.98	202.32	12.36
2,550	21.29	47.04	202.74	11.64
2,600	21.32	48.11	203.15	10.95

(continued)

C (gas)   (continued)

T (K)	$\hat{c}_p$ (kJ/kmol-K)	$\hat{h} - \hat{h}^0$ (MJ/kmol)	$\hat{s}$ (kJ/kmol-K)	$g^0/\hat{R}_u T$
2,650	21.36	49.17	203.56	10.28
2,700	21.39	50.24	203.96	9.63
2,750	21.43	51.31	204.35	9.01
2,800	21.47	52.39	204.74	8.41
2,850	21.5	53.46	205.12	7.83
2,900	21.54	54.54	205.49	7.27
2,950	21.58	55.61	205.86	6.73
3,000	21.62	56.69	206.22	6.2
3,050	21.66	57.78	206.58	5.7
3,100	21.69	58.86	206.93	5.2
3,150	21.73	59.95	207.28	4.72
3,200	21.77	61.03	207.62	4.26
3,250	21.81	62.12	207.96	3.81
3,300	21.85	63.21	208.3	3.37
3,350	21.89	64.31	208.62	2.95
3,400	21.93	65.4	208.95	2.54
3,450	21.96	66.5	209.27	2.13
3,500	22	67.6	209.59	1.74

C (Solid)

T (K)	$\hat{c}_p$ (kJ/kmol-K)	$\hat{h} - \hat{h}^0$ (MJ/kmol)	$\hat{s}$ (kJ/kmol-K)	$g^0/\hat{R}_u T$
200	4.63	−0.65	3.14	−0.77
250	6.69	−0.37	4.4	−0.7
298	8.51	0	5.74	−0.69
300	8.57	0.02	5.79	−0.69
350	10.3	0.49	7.24	−0.7
400	11.87	1.04	8.72	−0.73
450	13.29	1.67	10.2	−0.78
500	14.58	2.37	11.67	−0.83
550	15.74	3.13	13.12	−0.89
600	16.79	3.94	14.53	−0.96
650	17.72	4.81	15.91	−1.02
700	18.55	5.71	17.26	−1.09
750	19.28	6.66	18.56	−1.16
800	19.92	7.64	19.83	−1.24
850	20.47	8.65	21.05	−1.31
900	20.94	9.68	22.23	−1.38
950	21.33	10.74	23.38	−1.45
1,000	21.65	11.82	24.48	−1.52
1,050	21.92	12.91	25.54	−1.59
1,100	22.18	14.01	26.57	−1.66
1,150	22.43	15.12	27.56	−1.73
1,200	22.66	16.25	28.52	−1.8
1,250	22.89	17.39	29.45	−1.87

(continued)

C (Solid) (continued)

T (K)	$\hat{c}_p$ (kJ/kmol-K)	$\hat{h} - \hat{h}^0$ (MJ/kmol)	$\hat{s}$ (kJ/kmol-K)	$g^0/\hat{R}_u T$
1,300	23.1	18.54	30.35	-1.94
1,350	23.3	19.7	31.23	-2
1,400	23.49	20.87	32.08	-2.07
1,450	23.67	22.05	32.9	-2.13
1,500	23.84	23.24	33.71	-2.19
1,550	24.01	24.43	34.49	-2.25
1,600	24.16	25.64	35.26	-2.31
1,650	24.31	26.85	36	-2.37
1,700	24.45	28.07	36.73	-2.43
1,750	24.58	29.29	37.44	-2.49
1,800	24.71	30.52	38.14	-2.55
1,850	24.82	31.76	38.82	-2.6
1,900	24.94	33.01	39.48	-2.66
1,950	25.05	34.26	40.13	-2.71
2,000	25.15	35.51	40.76	-2.77
2,050	25.25	36.77	41.39	-2.82
2,100	25.34	38.04	42	-2.87
2,150	25.43	39.3	42.59	-2.92
2,200	25.52	40.58	43.18	-2.97
2,250	25.6	41.86	43.75	-3.02
2,300	25.68	43.14	44.32	-3.07
2,350	25.75	44.42	44.87	-3.12
2,400	25.83	45.71	45.41	-3.17
2,450	25.9	47.01	45.95	-3.22
2,500	25.97	48.3	46.47	-3.27
2,550	26.03	49.6	46.98	-3.31
2,600	26.1	50.91	47.49	-3.36
2,650	26.16	52.21	47.99	-3.4
2,700	26.23	53.52	48.48	-3.45
2,750	26.29	54.84	48.96	-3.49
2,800	26.35	56.15	49.43	-3.53
2,850	26.41	57.47	49.9	-3.58
2,900	26.47	58.79	50.36	-3.62
2,950	26.53	60.12	50.81	-3.66
3,000	26.59	61.45	51.26	-3.7
3,050	26.65	62.78	51.7	-3.74
3,100	26.7	64.11	52.13	-3.78
3,150	26.76	65.45	52.56	-3.82
3,200	26.82	66.79	52.98	-3.86
3,250	26.88	68.13	53.4	-3.9
3,300	26.94	69.47	53.81	-3.94
3,350	27	70.82	54.22	-3.98
3,400	27.06	72.17	54.62	-4.02
3,450	27.12	73.53	55.01	-4.05
3,500	27.18	74.89	55.4	-4.09

CH$_4$

T (K)	$\hat{c}_p$ (kJ/kmol-K)	$\hat{h} - \hat{h}^0$ (MJ/kmol)	$\hat{s}$ (kJ/kmol-K)	$g^0/\hat{R}_u T$
200	28.14	−3.12	173.46	−67.78
250	31.9	−1.62	180.15	−58.48
298	35.14	0	186.05	−52.59
300	35.26	0.07	186.27	−52.4
350	38.33	1.91	191.94	−48.17
400	41.19	3.89	197.24	−45.07
450	43.93	6.02	202.25	−42.73
500	46.61	8.29	207.02	−40.92
550	49.26	10.68	211.59	−39.49
600	51.93	13.21	215.99	−38.34
650	54.62	15.88	220.25	−37.41
700	57.33	18.67	224.4	−36.65
750	60.05	21.61	228.45	−36.02
800	62.75	24.68	232.41	−35.5
850	65.38	27.88	236.29	−35.07
900	67.88	31.22	240.1	−34.72
950	70.17	34.67	243.83	−34.42
1,000	72.16	38.23	247.48	−34.18
1,050	73.92	41.88	251.05	−33.98
1,100	75.6	45.62	254.52	−33.81
1,150	77.19	49.44	257.92	−33.68
1,200	78.71	53.34	261.24	−33.58
1,250	80.15	57.31	264.48	−33.5
1,300	81.52	61.35	267.65	−33.45
1,350	82.82	65.46	270.75	−33.41
1,400	84.05	69.63	273.79	−33.38
1,450	85.22	73.86	276.76	−33.37
1,500	86.32	78.15	279.66	−33.38
1,550	87.36	82.49	282.51	−33.39
1,600	88.35	86.89	285.3	−33.41
1,650	89.29	91.33	288.03	−33.45
1,700	90.17	95.81	290.71	−33.49
1,750	91	100.34	293.34	−33.53
1,800	91.78	104.91	295.91	−33.59
1,850	92.52	109.52	298.44	−33.64
1,900	93.22	114.16	300.92	−33.71
1,950	93.87	118.84	303.35	−33.78
2,000	94.49	123.55	305.73	−33.85
2,050	95.07	128.29	308.07	−33.92
2,100	95.62	133.06	310.37	−34
2,150	96.13	137.85	312.62	−34.08
2,200	96.61	142.67	314.84	−34.16
2,250	97.06	147.51	317.02	−34.25
2,300	97.49	152.38	319.15	−34.34
2,350	97.88	157.26	321.25	−34.42
2,400	98.26	162.16	323.32	−34.51

(continued)

CH$_4$ (continued)

T (K)	$\hat{c}_p$ (kJ/kmol-K)	$\hat{h} - \hat{h}^0$ (MJ/kmol)	$\hat{s}$ (kJ/kmol-K)	$g^0/\hat{R}_u T$
2,450	98.61	167.09	325.35	−34.61
2,500	98.94	172.02	327.34	−34.7
2,550	99.25	176.98	329.31	−34.79
2,600	99.54	181.95	331.24	−34.89
2,650	99.82	186.93	333.14	−34.98
2,700	100.08	191.93	335	−35.08
2,750	100.32	196.94	336.84	−35.18
2,800	100.55	201.96	338.65	−35.27
2,850	100.77	207	340.43	−35.37
2,900	100.98	212.04	342.19	−35.47
2,950	101.18	217.09	343.92	−35.57
3,000	101.37	222.16	345.62	−35.67
3,050	101.55	227.23	347.3	−35.76
3,100	101.72	232.31	348.95	−35.86
3,150	101.89	237.4	350.58	−35.96
3,200	102.05	242.5	352.18	−36.06
3,250	102.2	247.61	353.77	−36.16
3,300	102.35	252.72	355.33	−36.26
3,350	102.5	257.84	356.87	−36.35
3,400	102.64	262.97	358.39	−36.45
3,450	102.78	268.11	359.89	−36.55
3,500	102.92	273.25	361.37	−36.65

H$_2$

T (K)	$\hat{c}_p$ (kJ/kmol-K)	$\hat{h} - \hat{h}^0$ (MJ/kmol)	$\hat{s}$ (kJ/kmol-K)	$g^0/\hat{R}_u T$
200	28.52	−2.82	119.13	−16.02
250	28.71	−1.39	125.52	−15.76
298	28.87	0	130.59	−15.71
300	28.88	0.05	130.77	−15.71
350	29.01	1.5	135.23	−15.75
400	29.12	2.95	139.11	−15.84
450	29.21	4.41	142.54	−15.97
500	29.27	5.87	145.63	−16.1
550	29.33	7.34	148.42	−16.25
600	29.37	8.81	150.97	−16.39
650	29.42	10.28	153.32	−16.54
700	29.46	11.75	155.51	−16.69
750	29.51	13.22	157.54	−16.83
800	29.58	14.7	159.45	−16.97
850	29.67	16.18	161.24	−17.1
900	29.79	17.67	162.94	−17.24
950	29.95	19.16	164.56	−17.37

(continued)

H$_2$ (continued)

T (K)	$\hat{c}_p$ (kJ/kmol-K)	$\hat{h} - \hat{h}^0$ (MJ/kmol)	$\hat{s}$ (kJ/kmol-K)	$g^0/\hat{R}_u T$
1,000	30.16	20.66	166.1	−17.49
1,050	30.39	22.18	167.58	−17.62
1,100	30.62	23.7	168.99	−17.73
1,150	30.85	25.24	170.36	−17.85
1,200	31.08	26.79	171.68	−17.96
1,250	31.3	28.35	172.95	−18.07
1,300	31.51	29.92	174.18	−18.18
1,350	31.73	31.5	175.38	−18.29
1,400	31.94	33.09	176.53	−18.39
1,450	32.15	34.69	177.66	−18.49
1,500	32.35	36.3	178.75	−18.59
1,550	32.56	37.93	179.82	−18.68
1,600	32.76	39.56	180.85	−18.78
1,650	32.95	41.2	181.86	−18.87
1,700	33.14	42.86	182.85	−18.96
1,750	33.33	44.52	183.81	−19.05
1,800	33.52	46.19	184.76	−19.14
1,850	33.7	47.87	185.68	−19.22
1,900	33.88	49.56	186.58	−19.3
1,950	34.06	51.26	187.46	−19.39
2,000	34.23	52.97	188.33	−19.47
2,050	34.41	54.68	189.17	−19.55
2,100	34.57	56.41	190	−19.62
2,150	34.74	58.14	190.82	−19.7
2,200	34.9	59.88	191.62	−19.77
2,250	35.06	61.63	192.41	−19.85
2,300	35.21	63.39	193.18	−19.92
2,350	35.37	65.15	193.94	−19.99
2,400	35.52	66.92	194.68	−20.06
2,450	35.66	68.7	195.42	−20.13
2,500	35.81	70.49	196.14	−20.2
2,550	35.95	72.28	196.85	−20.27
2,600	36.09	74.08	197.55	−20.33
2,650	36.23	75.89	198.24	−20.4
2,700	36.36	77.71	198.92	−20.46
2,750	36.49	79.53	199.58	−20.53
2,800	36.62	81.35	200.24	−20.59
2,850	36.75	83.19	200.89	−20.65
2,900	36.87	85.03	201.53	−20.71
2,950	36.99	86.88	202.16	−20.77
3,000	37.11	88.73	202.79	−20.83
3,050	37.23	90.59	203.4	−20.89
3,100	37.34	92.45	204.01	−20.95
3,150	37.45	94.32	204.61	−21.01
3,200	37.56	96.2	205.2	−21.06
3,250	37.67	98.08	205.78	−21.12

(continued)

H_2 (continued)

T (K)	$\hat{c}_p$ (kJ/kmol-K)	$\hat{h} - \hat{h}^0$ (MJ/kmol)	$\hat{s}$ (kJ/kmol-K)	$g^0/\hat{R}_u T$
3,300	37.78	99.96	206.36	−21.18
3,350	37.88	101.86	206.92	−21.23
3,400	37.99	103.75	207.49	−21.29
3,450	38.09	105.65	208.04	−21.34
3,500	38.19	107.56	208.59	−21.39

H

T (K)	$\hat{c}_p$ (kJ/kmol-K)	$\hat{h} - \hat{h}^0$ (MJ/kmol)	$\hat{s}$ (kJ/kmol-K)	$g^0/\hat{R}_u T$
200	20.79	−2.04	106.3	117.07
250	20.79	−1	110.94	91.04
298	20.79	0	114.6	74.15
300	20.79	0.04	114.73	73.61
350	20.79	1.08	117.93	61.09
400	20.79	2.12	120.71	51.66
450	20.79	3.16	123.16	44.29
500	20.79	4.2	125.35	38.37
550	20.79	5.23	127.33	33.5
600	20.79	6.27	129.13	29.42
650	20.79	7.31	130.8	25.95
700	20.79	8.35	132.34	22.97
750	20.79	9.39	133.77	20.37
800	20.79	10.43	135.11	18.09
850	20.79	11.47	136.37	16.06
900	20.79	12.51	137.56	14.26
950	20.79	13.55	138.69	12.63
1,000	20.79	14.59	139.75	11.16
1,050	20.79	15.63	140.77	9.83
1,100	20.79	16.67	141.73	8.61
1,150	20.79	17.71	142.66	7.49
1,200	20.79	18.74	143.54	6.46
1,250	20.79	19.78	144.39	5.51
1,300	20.79	20.82	145.21	4.63
1,350	20.79	21.86	145.99	3.81
1,400	20.79	22.9	146.75	3.04
1,450	20.79	23.94	147.48	2.33
1,500	20.79	24.98	148.18	1.66
1,550	20.79	26.02	148.86	1.03
1,600	20.79	27.06	149.52	0.44
1,650	20.79	28.1	150.16	−0.12
1,700	20.79	29.14	150.78	−0.65
1,750	20.79	30.18	151.38	−1.15
1,800	20.79	31.22	151.97	−1.63

(continued)

H   (continued)

T (K)	$\hat{c}_p$ (kJ/kmol-K)	$\hat{h} - \hat{h}^0$ (MJ/kmol)	$\hat{s}$ (kJ/kmol-K)	$g^0/\hat{R}_u T$
1,850	20.79	32.26	152.54	−2.08
1,900	20.79	33.29	153.09	−2.51
1,950	20.79	34.33	153.63	−2.92
2,000	20.79	35.37	154.16	−3.31
2,050	20.79	36.41	154.67	−3.68
2,100	20.79	37.45	155.17	−4.03
2,150	20.79	38.49	155.66	−4.38
2,200	20.79	39.53	156.14	−4.7
2,250	20.79	40.57	156.61	−5.02
2,300	20.79	41.61	157.06	−5.32
2,350	20.79	42.65	157.51	−5.61
2,400	20.79	43.69	157.95	−5.88
2,450	20.79	44.73	158.38	−6.15
2,500	20.79	45.77	158.8	−6.41
2,550	20.79	46.8	159.21	−6.66
2,600	20.79	47.84	159.61	−6.9
2,650	20.79	48.88	160.01	−7.13
2,700	20.79	49.92	160.4	−7.36
2,750	20.79	50.96	160.78	−7.58
2,800	20.79	52	161.15	−7.79
2,850	20.79	53.04	161.52	−7.99
2,900	20.79	54.08	161.88	−8.19
2,950	20.79	55.12	162.24	−8.38
3,000	20.79	56.16	162.59	−8.57
3,050	20.79	57.2	162.93	−8.75
3,100	20.79	58.24	163.27	−8.92
3,150	20.79	59.28	163.6	−9.09
3,200	20.79	60.31	163.93	−9.26
3,250	20.79	61.35	164.25	−9.42
3,300	20.79	62.39	164.57	−9.58
3,350	20.79	63.43	164.88	−9.73
3,400	20.79	64.47	165.19	−9.88
3,450	20.79	65.51	165.49	−10.02
3,500	20.79	66.55	165.79	−10.16

O

T (K)	$\hat{c}_p$ (kJ/kmol-K)	$\hat{h} - \hat{h}^0$ (MJ/kmol)	$\hat{s}$ (kJ/kmol-K)	$g^0/\hat{R}_u T$
200	22.48	−2.18	152.08	130.26
250	22.15	−1.06	157.06	100.49
298	21.9	0	160.94	81.17
300	21.89	0.04	161.07	80.55
350	21.67	1.13	164.43	66.24

(continued)

O   (continued)

T (K)	$\hat{c}_p$ (kJ/kmol-K)	$\hat{h} - \hat{h}^0$ (MJ/kmol)	$\hat{s}$ (kJ/kmol-K)	$g^0/\hat{R}_u T$
400	21.5	2.21	167.31	55.47
450	21.36	3.28	169.84	47.05
500	21.26	4.35	172.08	40.29
550	21.17	5.41	174.1	34.74
600	21.11	6.46	175.94	30.09
650	21.07	7.52	177.63	26.14
700	21.03	8.57	179.19	22.74
750	21.01	9.62	180.64	19.78
800	20.99	10.67	182	17.18
850	20.97	11.72	183.27	14.88
900	20.95	12.77	184.46	12.82
950	20.93	13.81	185.6	10.97
1,000	20.91	14.86	186.67	9.31
1,050	20.91	15.91	187.69	7.79
1,100	20.9	16.95	188.66	6.41
1,150	20.89	18	189.59	5.14
1,200	20.88	19.04	190.48	3.97
1,250	20.87	20.08	191.33	2.9
1,300	20.87	21.13	192.15	1.9
1,350	20.86	22.17	192.94	0.97
1,400	20.85	23.21	193.7	0.1
1,450	20.85	24.26	194.43	−0.7
1,500	20.84	25.3	195.14	−1.46
1,550	20.84	26.34	195.82	−2.17
1,600	20.83	27.38	196.48	−2.84
1,650	20.83	28.42	197.12	−3.47
1,700	20.83	29.46	197.74	−4.07
1,750	20.82	30.51	198.35	−4.63
1,800	20.82	31.55	198.93	−5.17
1,850	20.82	32.59	199.5	−5.68
1,900	20.82	33.63	200.06	−6.16
1,950	20.82	34.67	200.6	−6.62
2,000	20.82	35.71	201.13	−7.06
2,050	20.82	36.75	201.64	−7.48
2,100	20.82	37.79	202.14	−7.88
2,150	20.82	38.83	202.63	−8.26
2,200	20.82	39.87	203.11	−8.63
2,250	20.83	40.92	203.58	−8.98
2,300	20.83	41.96	204.04	−9.32
2,350	20.83	43	204.48	−9.64
2,400	20.84	44.04	204.92	−9.95
2,450	20.84	45.08	205.35	−10.25
2,500	20.85	46.13	205.77	−10.54
2,550	20.86	47.17	206.19	−10.82
2,600	20.86	48.21	206.59	−11.09
2,650	20.87	49.25	206.99	−11.35

(continued)

O   (continued)

T (K)	$\hat{c}_p$ (kJ/kmol-K)	$\hat{h} - \hat{h}^0$ (MJ/kmol)	$\hat{s}$ (kJ/kmol-K)	$g^0/\hat{R}_u T$
2,700	20.88	50.3	207.38	−11.6
2,750	20.89	51.34	207.76	−11.85
2,800	20.9	52.39	208.14	−12.08
2,850	20.91	53.43	208.51	−12.31
2,900	20.92	54.48	208.87	−12.53
2,950	20.93	55.52	209.23	−12.74
3,000	20.94	56.57	209.58	−12.95
3,050	20.96	57.62	209.93	−13.15
3,100	20.97	58.67	210.27	−13.35
3,150	20.98	59.72	210.61	−13.54
3,200	21	60.76	210.94	−13.72
3,250	21.01	61.81	211.26	−13.9
3,300	21.03	62.87	211.58	−14.08
3,350	21.04	63.92	211.9	−14.25
3,400	21.06	64.97	212.21	−14.41
3,450	21.08	66.02	212.52	−14.57
3,500	21.09	67.08	212.82	−14.73

OH

T (K)	$\hat{c}_p$ (kJ/kmol-K)	$\hat{h} - \hat{h}^0$ (MJ/kmol)	$\hat{s}$ (kJ/kmol-K)	$g^0/\hat{R}_u T$
200	30.14	−2.95	171.6	1.03
250	30.04	−1.44	178.31	−3.39
298	29.93	0	183.6	−6.36
300	29.93	0.06	183.78	−6.45
350	29.82	1.55	188.39	−8.73
400	29.72	3.04	192.36	−10.5
450	29.63	4.52	195.86	−11.93
500	29.57	6	198.97	−13.11
550	29.53	7.48	201.79	−14.11
600	29.53	8.95	204.36	−14.97
650	29.55	10.43	206.72	−15.72
700	29.61	11.91	208.92	−16.38
750	29.71	13.39	210.96	−16.97
800	29.84	14.88	212.88	−17.51
850	30.01	16.38	214.7	−17.99
900	30.21	17.88	216.42	−18.43
950	30.43	19.4	218.06	−18.84
1,000	30.68	20.93	219.62	−19.21
1,050	30.94	22.47	221.13	−19.56
1,100	31.18	24.02	222.57	−19.88
1,150	31.43	25.59	223.96	−20.18
1,200	31.66	27.16	225.31	−20.47
1,250	31.89	28.75	226.6	−20.74

(continued)

OH   (continued)

T (K)	$\hat{c}_p$ (kJ/kmol-K)	$\hat{h} - \hat{h}^0$ (MJ/kmol)	$\hat{s}$ (kJ/kmol-K)	$g^0/\hat{R}_u T$
1,300	32.11	30.35	227.86	−20.99
1,350	32.33	31.96	229.07	−21.23
1,400	32.54	33.58	230.25	−21.46
1,450	32.74	35.22	231.4	−21.68
1,500	32.94	36.86	232.51	−21.88
1,550	33.13	38.51	233.6	−22.08
1,600	33.32	40.17	234.65	−22.27
1,650	33.5	41.84	235.68	−22.46
1,700	33.68	43.52	236.68	−22.63
1,750	33.85	45.21	237.66	−22.8
1,800	34.02	46.91	238.62	−22.96
1,850	34.18	48.61	239.55	−23.12
1,900	34.34	50.33	240.46	−23.27
1,950	34.49	52.05	241.36	−23.42
2,000	34.63	53.77	242.23	−23.56
2,050	34.78	55.51	243.09	−23.69
2,100	34.91	57.25	243.93	−23.83
2,150	35.05	59	244.75	−23.96
2,200	35.18	60.76	245.56	−24.08
2,250	35.3	62.52	246.35	−24.2
2,300	35.42	64.29	247.13	−24.32
2,350	35.54	66.06	247.89	−24.44
2,400	35.65	67.84	248.64	−24.55
2,450	35.76	69.63	249.38	−24.66
2,500	35.87	71.42	250.1	−24.77
2,550	35.97	73.21	250.81	−24.88
2,600	36.07	75.01	251.51	−24.98
2,650	36.17	76.82	252.2	−25.08
2,700	36.26	78.63	252.88	−25.18
2,750	36.35	80.45	253.54	−25.27
2,800	36.44	82.27	254.2	−25.37
2,850	36.52	84.09	254.85	−25.46
2,900	36.6	85.92	255.48	−25.55
2,950	36.68	87.75	256.11	−25.64
3,000	36.76	89.59	256.72	−25.72
3,050	36.83	91.43	257.33	−25.81
3,100	36.9	93.27	257.93	−25.89
3,150	36.97	95.12	258.52	−25.97
3,200	37.04	96.97	259.11	−26.06
3,250	37.1	98.82	259.68	−26.13
3,300	37.16	100.68	260.25	−26.21
3,350	37.22	102.54	260.81	−26.29
3,400	37.28	104.4	261.36	−26.36
3,450	37.34	106.26	261.9	−26.44
3,500	37.4	108.13	262.44	−26.51

N

T (K)	$\hat{c}_p$ (kJ/kmol-K)	$\hat{h} - \hat{h}^0$ (MJ/kmol)	$\hat{s}$ (kJ/kmol-K)	$g^0/\hat{R}_uT$
200	20.79	−2.04	144.88	265.57
250	20.79	−1	149.52	208.91
298	20.79	0	153.18	172.23
300	20.79	0.04	153.31	171.06
350	20.78	1.08	156.51	143.96
400	20.78	2.12	159.29	123.59
450	20.78	3.16	161.74	107.71
500	20.78	4.2	163.93	94.98
550	20.79	5.23	165.91	84.54
600	20.79	6.27	167.72	75.83
650	20.79	7.31	169.38	68.43
700	20.79	8.35	170.92	62.08
750	20.79	9.39	172.35	56.57
800	20.79	10.43	173.7	51.73
850	20.78	11.47	174.96	47.46
900	20.78	12.51	176.14	43.65
950	20.78	13.55	177.27	40.23
1,000	20.78	14.59	178.33	37.15
1,050	20.79	15.63	179.35	34.36
1,100	20.79	16.67	180.32	31.81
1,150	20.79	17.71	181.24	29.48
1,200	20.79	18.75	182.13	27.34
1,250	20.79	19.79	182.97	25.37
1,300	20.79	20.83	183.79	23.55
1,350	20.79	21.87	184.57	21.85
1,400	20.79	22.9	185.33	20.28
1,450	20.79	23.94	186.06	18.81
1,500	20.79	24.98	186.76	17.44
1,550	20.79	26.02	187.45	16.15
1,600	20.79	27.06	188.11	14.94
1,650	20.78	28.1	188.75	13.8
1,700	20.78	29.14	189.37	12.72
1,750	20.78	30.18	189.97	11.71
1,800	20.78	31.22	190.55	10.75
1,850	20.78	32.26	191.12	9.84
1,900	20.78	33.3	191.68	8.97
1,950	20.78	34.34	192.22	8.15
2,000	20.78	35.37	192.74	7.37
2,050	20.78	36.41	193.26	6.62
2,100	20.78	37.45	193.76	5.91
2,150	20.78	38.49	194.25	5.23
2,200	20.78	39.53	194.72	4.58
2,250	20.79	40.57	195.19	3.96
2,300	20.79	41.61	195.65	3.36
2,350	20.79	42.65	196.09	2.79
2,400	20.8	43.69	196.53	2.24

(continued)

N (continued)

T (K)	$\hat{c}_p$ (kJ/kmol-K)	$\hat{h} - \hat{h}^0$ (MJ/kmol)	$\hat{s}$ (kJ/kmol-K)	$g^0/\hat{R}_u T$
2,450	20.81	44.73	196.96	1.71
2,500	20.82	45.77	197.38	1.2
2,550	20.83	46.81	197.79	0.71
2,600	20.84	47.85	198.2	0.24
2,650	20.85	48.89	198.6	−0.22
2,700	20.86	49.94	198.99	−0.66
2,750	20.88	50.98	199.37	−1.08
2,800	20.89	52.02	199.74	−1.49
2,850	20.91	53.07	200.11	−1.88
2,900	20.93	54.11	200.48	−2.27
2,950	20.95	55.16	200.84	−2.64
3,000	20.97	56.21	201.19	−3
3,050	21	57.26	201.54	−3.34
3,100	21.02	58.31	201.88	−3.68
3,150	21.05	59.36	202.21	−4.01
3,200	21.08	60.41	202.55	−4.33
3,250	21.11	61.47	202.87	−4.64
3,300	21.14	62.53	203.2	−4.94
3,350	21.18	63.58	203.51	−5.23
3,400	21.21	64.64	203.83	−5.51
3,450	21.25	65.71	204.14	−5.79
3,500	21.29	66.77	204.44	−6.05

NO

T (K)	$\hat{c}_p$ (kJ/kmol-K)	$\hat{h} - \hat{h}^0$ (MJ/kmol)	$\hat{s}$ (kJ/kmol-K)	$g^0/\hat{R}_u T$
200	29.37	−2.9	198.85	28.64
250	29.56	−1.43	205.42	18.05
298	29.73	0	210.64	11.09
300	29.73	0.06	210.83	10.87
350	29.91	1.55	215.42	5.65
400	30.1	3.05	219.43	1.67
450	30.32	4.56	222.99	−1.47
500	30.57	6.08	226.19	−4.02
550	30.85	7.61	229.12	−6.15
600	31.17	9.16	231.82	−7.95
650	31.53	10.73	234.33	−9.49
700	31.91	12.32	236.68	−10.84
750	32.31	13.92	238.89	−12.02
800	32.71	15.55	240.99	−13.07
850	33.11	17.19	242.99	−14.02
900	33.49	18.86	244.89	−14.87
950	33.82	20.54	246.71	−15.64

(continued)

NO   (continued)

T (K)	$\hat{c}_p$ (kJ/kmol-K)	$\hat{h} - \hat{h}^0$ (MJ/kmol)	$\hat{s}$ (kJ/kmol-K)	$g^0/\hat{R}_u T$
1,000	34.07	22.24	248.45	−16.35
1,050	34.28	23.95	250.12	−17
1,100	34.48	25.67	251.72	−17.6
1,150	34.67	27.4	253.25	−18.15
1,200	34.85	29.13	254.73	−18.67
1,250	35.02	30.88	256.16	−19.15
1,300	35.18	32.64	257.54	−19.6
1,350	35.33	34.4	258.87	−20.03
1,400	35.47	36.17	260.15	−20.43
1,450	35.61	37.95	261.4	−20.8
1,500	35.74	39.73	262.61	−21.16
1,550	35.86	41.52	263.78	−21.5
1,600	35.97	43.32	264.92	−21.82
1,650	36.08	45.12	266.03	−22.13
1,700	36.18	46.92	267.11	−22.42
1,750	36.27	48.73	268.16	−22.7
1,800	36.36	50.55	269.18	−22.97
1,850	36.45	52.37	270.18	−23.22
1,900	36.53	54.19	271.15	−23.47
1,950	36.6	56.02	272.1	−23.7
2,000	36.67	57.85	273.03	−23.93
2,050	36.73	59.69	273.94	−24.15
2,100	36.8	61.53	274.82	−24.36
2,150	36.85	63.37	275.69	−24.56
2,200	36.91	65.21	276.54	−24.76
2,250	36.96	67.06	277.37	−24.95
2,300	37.01	68.91	278.18	−25.13
2,350	37.05	70.76	278.98	−25.31
2,400	37.09	72.61	279.76	−25.48
2,450	37.13	74.47	280.52	−25.65
2,500	37.17	76.33	281.27	−25.82
2,550	37.21	78.19	282.01	−25.97
2,600	37.24	80.05	282.73	−26.13
2,650	37.27	81.91	283.44	−26.28
2,700	37.3	83.78	284.14	−26.42
2,750	37.33	85.64	284.82	−26.56
2,800	37.36	87.51	285.5	−26.7
2,850	37.39	89.38	286.16	−26.84
2,900	37.41	91.25	286.81	−26.97
2,950	37.44	93.12	287.45	−27.1
3,000	37.46	94.99	288.08	−27.22
3,050	37.49	96.86	288.7	−27.34
3,100	37.51	98.74	289.31	−27.46
3,150	37.53	100.62	289.91	−27.58
3,200	37.55	102.49	290.5	−27.69
3,250	37.58	104.37	291.08	−27.81

(continued)

NO   (continued)

T (K)	$\hat{c}_p$ (kJ/kmol-K)	$\hat{h} - \hat{h}^0$ (MJ/kmol)	$\hat{s}$ (kJ/kmol-K)	$g^0/\hat{R}_u T$
3,300	37.6	106.25	291.66	-27.92
3,350	37.62	108.13	292.22	-28.02
3,400	37.64	110.01	292.78	-28.13
3,450	37.66	111.89	293.33	-28.23
3,500	37.68	113.78	293.87	-28.33

NO$_2$

T (K)	$\hat{c}_p$ (kJ/kmol-K)	$\hat{h} - \hat{h}^0$ (MJ/kmol)	$\hat{s}$ (kJ/kmol-K)	$g^0/\hat{R}_u T$
200	32.93	-3.43	226.01	-9.34
250	35.03	-1.73	233.58	-13.01
298	36.88	0	239.91	-15.51
300	36.95	0.07	240.14	-15.59
350	38.71	1.96	245.97	-17.54
400	40.33	3.94	251.25	-19.08
450	41.83	5.99	256.09	-20.35
500	43.23	8.12	260.57	-21.43
550	44.52	10.31	264.75	-22.35
600	45.73	12.57	268.67	-23.16
650	46.86	14.88	272.38	-23.88
700	47.91	17.25	275.89	-24.53
750	48.88	19.67	279.23	-25.12
800	49.76	22.14	282.41	-25.66
850	50.55	24.65	285.45	-26.16
900	51.24	27.19	288.36	-26.63
950	51.82	29.77	291.15	-27.06
1,000	52.27	32.37	293.82	-27.47
1,050	52.64	35	296.38	-27.85
1,100	52.99	37.64	298.84	-28.21
1,150	53.31	40.29	301.2	-28.55
1,200	53.62	42.97	303.47	-28.88
1,250	53.91	45.66	305.67	-29.19
1,300	54.18	48.36	307.79	-29.48
1,350	54.44	51.07	309.84	-29.77
1,400	54.68	53.8	311.82	-30.04
1,450	54.9	56.54	313.75	-30.3
1,500	55.11	59.29	315.61	-30.55
1,550	55.3	62.05	317.42	-30.8
1,600	55.48	64.82	319.18	-31.03
1,650	55.65	67.6	320.89	-31.26
1,700	55.8	70.39	322.55	-31.47
1,750	55.95	73.18	324.17	-31.69
1,800	56.08	75.98	325.75	-31.89

(continued)

NO$_2$ (continued)

T (K)	$\hat{c}_p$ (kJ/kmol-K)	$\hat{h} - \hat{h}^0$ (MJ/kmol)	$\hat{s}$ (kJ/kmol-K)	$g^0/\hat{R}_u T$
1,850	56.2	78.79	327.29	−32.09
1,900	56.31	81.6	328.79	−32.29
1,950	56.42	84.42	330.25	−32.47
2,000	56.51	87.24	331.68	−32.66
2,050	56.6	90.07	333.08	−32.84
2,100	56.68	92.9	334.44	−33.01
2,150	56.76	95.74	335.78	−33.18
2,200	56.82	98.58	337.08	−33.35
2,250	56.88	101.42	338.36	−33.51
2,300	56.94	104.27	339.61	−33.66
2,350	56.99	107.11	340.84	−33.82
2,400	57.04	109.97	342.04	−33.97
2,450	57.08	112.82	343.21	−34.12
2,500	57.12	115.67	344.37	−34.26
2,550	57.15	118.53	345.5	−34.4
2,600	57.18	121.39	346.61	−34.54
2,650	57.21	124.25	347.7	−34.68
2,700	57.24	127.11	348.77	−34.81
2,750	57.27	129.97	349.82	−34.94
2,800	57.29	132.84	350.85	−35.07
2,850	57.31	135.7	351.87	−35.2
2,900	57.33	138.57	352.86	−35.32
2,950	57.35	141.43	353.84	−35.44
3,000	57.37	144.3	354.81	−35.56
3,050	57.39	147.17	355.76	−35.68
3,100	57.4	150.04	356.69	−35.8
3,150	57.42	152.91	357.61	−35.91
3,200	57.44	155.78	358.51	−36.02
3,250	57.45	158.66	359.4	−36.13
3,300	57.47	161.53	360.28	−36.24
3,350	57.49	164.4	361.14	−36.35
3,400	57.51	167.28	362	−36.45
3,450	57.52	170.15	362.84	−36.56
3,500	57.54	173.03	363.66	−36.66

SO$_2$

T (K)	$\hat{c}_p$ (kJ/kmol-K)	$\hat{h} - \hat{h}^0$ (MJ/kmol)	$\hat{s}$ (kJ/kmol-K)	$g^0/\hat{R}_u T$
200	35.59	−3.71	233.08	−208.79
250	37.86	−1.87	241.27	−172.74
298	39.86	0	248.11	−149.6
300	39.94	0.07	248.35	−148.86
350	41.83	2.12	254.66	−131.92
400	43.55	4.25	260.36	−119.3

(continued)

SO$_2$ (continued)

T (K)	$\hat{c}_p$ (kJ/kmol-K)	$\hat{h} - \hat{h}^0$ (MJ/kmol)	$\hat{s}$ (kJ/kmol-K)	$g^0/\hat{R}_u T$
450	45.12	6.47	265.58	-109.56
500	46.54	8.76	270.4	-101.83
550	47.83	11.12	274.9	-95.55
600	48.99	13.54	279.11	-90.37
650	50.03	16.02	283.08	-86.02
700	50.97	18.54	286.82	-82.32
750	51.8	21.11	290.36	-79.15
800	52.53	23.72	293.73	-76.39
850	53.17	26.37	296.94	-73.99
900	53.71	29.04	299.99	-71.87
950	54.17	31.74	302.91	-70
1,000	54.53	34.45	305.7	-68.33
1,050	54.84	37.19	308.36	-66.83
1,100	55.14	39.94	310.92	-65.49
1,150	55.41	42.7	313.38	-64.28
1,200	55.67	45.48	315.74	-63.17
1,250	55.92	48.27	318.02	-62.17
1,300	56.16	51.07	320.22	-61.26
1,350	56.38	53.88	322.34	-60.42
1,400	56.58	56.71	324.4	-59.65
1,450	56.78	59.54	326.39	-58.94
1,500	56.96	62.39	328.31	-58.29
1,550	57.14	65.24	330.18	-57.69
1,600	57.3	68.1	332	-57.13
1,650	57.45	70.97	333.77	-56.61
1,700	57.6	73.85	335.48	-56.13
1,750	57.73	76.73	337.16	-55.68
1,800	57.86	79.62	338.78	-55.26
1,850	57.98	82.51	340.37	-54.87
1,900	58.09	85.42	341.92	-54.51
1,950	58.2	88.32	343.43	-54.17
2,000	58.3	91.24	344.9	-53.85
2,050	58.39	94.15	346.34	-53.55
2,100	58.48	97.07	347.75	-53.27
2,150	58.56	100	349.13	-53.01
2,200	58.63	102.93	350.48	-52.76
2,250	58.71	105.86	351.79	-52.52
2,300	58.78	108.8	353.09	-52.3
2,350	58.84	111.74	354.35	-52.1
2,400	58.9	114.68	355.59	-51.9
2,450	58.96	117.63	356.81	-51.71
2,500	59.02	120.58	358	-51.54
2,550	59.07	123.53	359.17	-51.38
2,600	59.12	126.49	360.31	-51.22
2,650	59.17	129.45	361.44	-51.07
2,700	59.22	132.41	362.55	-50.93

(continued)

SO$_2$ (continued)

T (K)	$\hat{c}_p$ (kJ/kmol-K)	$\hat{h} - \hat{h}^0$ (MJ/kmol)	$\hat{s}$ (kJ/kmol-K)	$g^0/\hat{R}_u T$
2,750	59.27	135.37	363.63	−50.8
2,800	59.31	138.33	364.7	−50.68
2,850	59.36	141.3	365.75	−50.56
2,900	59.4	144.27	366.79	−50.45
2,950	59.44	147.24	367.8	−50.34
3,000	59.49	150.21	368.8	−50.24
3,050	59.53	153.19	369.78	−50.14
3,100	59.57	156.17	370.75	−50.05
3,150	59.61	159.14	371.71	−49.97
3,200	59.66	162.13	372.65	−49.89
3,250	59.7	165.11	373.57	−49.81
3,300	59.74	168.1	374.48	−49.74
3,350	59.78	171.08	375.38	−49.67
3,400	59.83	174.07	376.27	−49.6
3,450	59.87	177.07	377.14	−49.54
3,500	59.92	180.06	378	−49.48

SO$_3$

T (K)	$\hat{c}_p$ (kJ/kmol-K)	$\hat{h} - \hat{h}^0$ (MJ/kmol)	$\hat{s}$ (kJ/kmol-K)	$g^0/\hat{R}_u T$
200	42.79	−4.6	238.07	−269.42
250	47.05	−2.36	248.08	−221.39
298	50.78	0	256.7	−190.54
300	50.91	0.09	257.01	−189.55
350	54.42	2.73	265.13	−166.96
400	57.58	5.53	272.6	−150.13
450	60.42	8.48	279.55	−137.14
500	62.96	11.57	286.05	−126.83
550	65.23	14.77	292.16	−118.46
600	67.23	18.09	297.92	−111.55
650	68.99	21.49	303.38	−105.75
700	70.52	24.98	308.55	−100.82
750	71.86	28.54	313.46	−96.6
800	73	32.16	318.13	−92.93
850	73.97	35.84	322.59	−89.73
900	74.78	39.56	326.84	−86.92
950	75.45	43.31	330.9	−84.43
1,000	76	47.1	334.79	−82.21
1,050	76.47	50.91	338.51	−80.22
1,100	76.91	54.75	342.07	−78.43
1,150	77.33	58.6	345.5	−76.82
1,200	77.72	62.48	348.8	−75.36
1,250	78.08	66.37	351.98	−74.03

(continued)

SO₃ (continued)

T (K)	$\hat{c}_p$ (kJ/kmol-K)	$\hat{h} - \hat{h}^0$ (MJ/kmol)	$\hat{s}$ (kJ/kmol-K)	$g^0/\hat{R}_u T$
1,300	78.42	70.29	355.05	−72.82
1,350	78.74	74.22	358.02	−71.71
1,400	79.04	78.16	360.89	−70.69
1,450	79.31	82.12	363.66	−69.76
1,500	79.57	86.09	366.36	−68.9
1,550	79.81	90.08	368.97	−68.1
1,600	80.03	94.07	371.51	−67.36
1,650	80.23	98.08	373.97	−66.68
1,700	80.42	102.1	376.37	−66.05
1,750	80.59	106.12	378.71	−65.46
1,800	80.75	110.15	380.98	−64.91
1,850	80.9	114.2	383.19	−64.4
1,900	81.03	118.24	385.35	−63.92
1,950	81.16	122.3	387.46	−63.47
2,000	81.27	126.36	389.51	−63.05
2,050	81.37	130.43	391.52	−62.66
2,100	81.46	134.5	393.48	−62.29
2,150	81.54	138.57	395.4	−61.95
2,200	81.62	142.65	397.28	−61.62
2,250	81.69	146.73	399.11	−61.32
2,300	81.75	150.82	400.91	−61.03
2,350	81.8	154.91	402.67	−60.76
2,400	81.85	159	404.39	−60.51
2,450	81.9	163.09	406.08	−60.27
2,500	81.94	167.19	407.73	−60.04
2,550	81.97	171.29	409.36	−59.83
2,600	82	175.39	410.95	−59.62
2,650	82.03	179.49	412.51	−59.43
2,700	82.06	183.59	414.04	−59.25
2,750	82.08	187.69	415.55	−59.08
2,800	82.11	191.8	417.03	−58.92
2,850	82.13	195.9	418.48	−58.77
2,900	82.15	200.01	419.91	−58.63
2,950	82.17	204.12	421.32	−58.49
3,000	82.18	208.23	422.7	−58.36
3,050	82.2	212.34	424.05	−58.24
3,100	82.22	216.45	425.39	−58.12
3,150	82.24	220.56	426.71	−58.01
3,200	82.26	224.67	428	−57.91
3,250	82.28	228.78	429.28	−57.81
3,300	82.29	232.9	430.53	−57.72
3,350	82.31	237.01	431.77	−57.63
3,400	82.34	241.13	432.99	−57.55
3,450	82.36	245.25	434.19	−57.47
3,500	82.38	249.37	435.38	−57.4

C_3H_8

T (K)	$\hat{c}_p$ (kJ/kmol-K)	$\hat{h} - \hat{h}^0$ (MJ/kmol)	$\hat{s}$ (kJ/kmol-K)	$g^0/\hat{R}_uT$
200	52.35	−6.19	245.27	−95.68
250	63.29	−3.3	258.13	−82.6
298	73.55	0	270.16	−74.39
300	73.94	0.14	270.62	−74.13
350	84.23	4.09	282.8	−68.3
400	94.1	8.55	294.69	−64.1
450	103.5	13.49	306.32	−60.99
500	112.4	18.89	317.69	−58.65
550	120.79	24.73	328.8	−56.85
600	128.63	30.96	339.65	−55.46
650	135.95	37.58	350.24	−54.39
700	142.74	44.55	360.57	−53.56
750	149.03	51.85	370.63	−52.92
800	154.86	59.44	380.44	−52.44
850	160.26	67.32	389.99	−52.08
900	165.31	75.47	399.3	−51.82
950	170.06	83.85	408.36	−51.65
1,000	174.61	92.47	417.2	−51.55
1,050	178.22	101.29	425.81	−51.51
1,100	181.68	110.29	434.18	−51.52
1,150	184.99	119.45	442.33	−51.57
1,200	188.16	128.78	450.27	−51.66
1,250	191.18	138.27	458.01	−51.78
1,300	194.07	147.9	465.57	−51.92
1,350	196.82	157.67	472.95	−52.09
1,400	199.45	167.58	480.15	−52.28
1,450	201.95	177.61	487.19	−52.48
1,500	204.33	187.77	494.08	−52.7
1,550	206.59	198.05	500.82	−52.93
1,600	208.74	208.43	507.41	−53.17
1,650	210.79	218.92	513.87	−53.42
1,700	212.72	229.51	520.19	−53.68
1,750	214.56	240.19	526.38	−53.94
1,800	216.3	250.96	532.45	−54.21
1,850	217.94	261.82	538.4	−54.49
1,900	219.5	272.75	544.23	−54.77
1,950	220.96	283.77	549.95	−55.05
2,000	222.35	294.85	555.56	−55.34
2,050	223.65	306	561.07	−55.62
2,100	224.88	317.21	566.47	−55.91
2,150	226.03	328.48	571.78	−56.21
2,200	227.11	339.81	576.99	−56.5
2,250	228.12	351.19	582.1	−56.79
2,300	229.07	362.63	587.13	−57.09
2,350	229.96	374.1	592.06	−57.38
2,400	230.79	385.62	596.91	−57.68

(continued)

C$_3$H$_8$ (continued)

T (K)	$\hat{c}_p$ (kJ/kmol-K)	$\hat{h} - \hat{h}^0$ (MJ/kmol)	$\hat{s}$ (kJ/kmol-K)	$g^0/\hat{R}_u T$
2,450	231.57	397.18	601.68	−57.97
2,500	232.29	408.78	606.37	−58.26
2,550	232.96	420.41	610.97	−58.56
2,600	233.58	432.07	615.5	−58.85
2,650	234.16	443.77	619.96	−59.14
2,700	234.7	455.49	624.34	−59.43
2,750	235.2	467.23	628.65	−59.72
2,800	235.66	479.01	632.89	−60.01
2,850	236.08	490.8	637.07	−60.3
2,900	236.48	502.61	641.18	−60.58
2,950	236.84	514.45	645.22	−60.87
3,000	237.17	526.3	649.21	−61.15
3,050	237.48	538.16	653.13	−61.43
3,100	237.77	550.05	656.99	−61.71
3,150	238.03	561.94	660.8	−61.99
3,200	238.28	573.85	664.55	−62.27
3,250	238.5	585.77	668.25	−62.54
3,300	238.71	597.7	671.89	−62.81
3,350	238.91	609.64	675.48	−63.09
3,400	239.09	621.59	679.02	−63.36
3,450	239.27	633.55	682.51	−63.62
3,500	239.43	645.52	685.96	−63.89

Appendix 4
Elementary Reaction Mechanisms

Table A Elementary reactions for hydrogen combustion $k = A_o T^b \exp\left(-\frac{E_a}{R_u T}\right)$

#	Reaction	A_o	b	E_a (cal/mol)
1	$H + O_2 = O + OH$	$1.92 \cdot 10^{14}$	0.0	16,440.0
2	$O + H_2 = H + OH$	$5.08 \cdot 10^4$	2.7	6,292.0
3	$OH + H_2 = H + H_2O$	$2.16 \cdot 10^8$	1.5	3,430.0
4	$O + H_2O = OH + OH$	$2.97 \cdot 10^6$	2.0	13,400.0
5	$H_2 + M = H + H + M$	$4.58 \cdot 10^{19}$	-1.4	104,400.0
6	$O_2 + M = O + O + M$	$4.52 \cdot 10^{17}$	-0.6	118,900.0
7	$OH + M = O + H + M$	$9.88 \cdot 10^{17}$	-0.7	102,100.0
8	$H_2O + M = H + OH + M$	$1.91 \cdot 10^{23}$	-1.8	118,500.0
9	$H + O_2(+M) = HO_2(+M)$	$1.48 \cdot 10^{12}$	0.6	0.0
	Low pressure limit:	$3.48 \cdot 10^{16}$	-0.41	$-11,150.0$
	Troe centering:	0.5	0	10^{31} 10^{100}
10	$HO_2 + H = H_2 + O_2$	$1.66 \cdot 10^{13}$	0	823.
11	$HO_2 + H = OH + OH$	$7.08 \cdot 10^{13}$	0	295.0
12	$HO_2 + O = OH + O_2$	$3.25 \cdot 10^{13}$	0	0
13	$HO_2 + OH = H_2O + O_2$	$2.89 \cdot 10^{13}$	0	-497.0
14	$H_2O_2 + O_2 = HO_2 + HO_2$	$4.63 \cdot 10^{16}$	-0.3	50,670.0
15	$H_2O_2 + O_2 = HO_2 + HO_2$	$1.43 \cdot 10^{13}$	-0.3	37,060.0
16	$H_2O_2(+M) = OH + OH(+M)$	$2.95 \cdot 10^{14}$	0	48,430.0
	Low pressure limit[a]:	$1.20 \cdot 10^{17}$	0	45,500.0
	Troe centering:	0.5	0	10^{31} 10^{100}
17	$H_2O_2 + H = H_2O + OH$	$2.41 \cdot 10^{13}$	0	3,970.0
18	$H_2O_2 + H = H_2 + HO_2$	$6.02 \cdot 10^{13}$	0	7,950.0
19	$H_2O_2 + O = OH + HO_2$	$9.55 \cdot 10^6$	2.0	3,970.0
20	$H_2O_2 + OH = H_2O + HO_2$	$1.00 \cdot 10^{12}$	0	0
21	$H_2O_2 + OH = H_2O + HO_2$	$5.80 \cdot 10^{14}$	0	9,557.0

Note: E_a units cal/mol, A_o units mol-cm-s-K

[a] At high pressures, the concentration of third body becomes high enough so that the three-body reaction becomes a two-body reaction. A high pressure rate is then modeled through a different model such as that proposed by Troe

Table B Elementary reactions for methane-air combustion $k = A_o T^b \exp\left(-\frac{E_a}{R_u T}\right)$

#	Reaction	A_o	b	E_a (cal/mol)
1 [a]	$CH_4(+M) = CH_3 + H(+M)$	$6.3 \cdot 10^{14}$	0.0	104,000
2	$CH_4 + O_2 = CH_3 + HO_2$	$7.9 \cdot 10^{13}$	0.0	56,000.0
3 [b]	$CH_4 + H = CH_3 + H_2$	$2.2 \cdot 10^4$	3.0	8,750.0
4 [c]	$CH_4 + O = CH_3 + OH$	$1.6 \cdot 10^6$	2.36	7,400.0
5 [a]	$CH_4 + OH = CH_3 + H2O$	$1.6 \cdot 10^6$	2.1	2,460.0
6 [b]	$CH_3 + O = CH_2O + H$	$6.8 \cdot 10^{13}$	0.0	0.0
7 [a]	$CH_3 + OH = CH_2O + H_2$	$1.0 \cdot 10^{12}$	0.0	0.0
8 [a]	$CH_3 + OH = CH_2 + H_2O$	$1.5 \cdot 10^{13}$	0.0	5,000.0
9	$CH_3 + H = CH_2 + H_2$	$9.0 \cdot 10^{13}$	0.0	15,100.0
10	$CH_2 + H = CH + H_2$	$1.4 \cdot 10^{19}$	-2.0	0.0
11 [a]	$CH_2 + OH = CH_2O + H$	$2.5 \cdot 10^{13}$	0.0	0.0
12 [a]	$CH_2 + OH = CH + H_2O$	$4.5 \cdot 10^{13}$	0.0	3,000.0
13 [a]	$CH + O_2 = HCO + O$	$3.3 \cdot 10^{13}$	0.0	0.0
14	$CH + O = CO + H$	$5.7 \cdot 10^{13}$	0.0	0.0
15	$CH + OH = HCO + H$	$3.0 \cdot 10^{13}$	0.0	0.0
16 [c]	$CH + CO_2 = HCO + CO$	$3.4 \cdot 10^{12}$	0.0	690.0
17	$CH_2 + CO_2 = CH_2O + CO$	$1.1 \cdot 10^{11}$	0.0	1,000.0
18	$CH_2 + O = CO + H + H$	$3.0 \cdot 10^{13}$	0.0	0.0
19	$CH_2 + O = CO + H_2$	$5.0 \cdot 10^{13}$	0.0	0.0
20	$CH_2 + O_2 = CO_2 + H + H$	$1.6 \cdot 10^{12}$	0.0	1,000.0
21 [a]	$CH_2 + O_2 = CH_2O + O$	$5.0 \cdot 10^{13}$	0.0	9,000.0
22	$CH_2 + O_2 = CO_2 + H_2$	$6.9 \cdot 10^{11}$	0.0	500.0
23	$CH_2 + O_2 = CO + H_2O$	$1.9 \cdot 10^{10}$	0.0	$-1,000.0$
24	$CH_2 + O_2 = CO + OH + H$	$8.6 \cdot 10^{10}$	0.0	-500.0
25	$CH_2 + O_2 = HCO + OH$	$4.3 \cdot 10^{10}$	0.0	-500.0
26 [a]	$CH_2O + OH = HCO + H_2O$	$3.43 \cdot 10^9$	1.18	-447.0
27 [b]	$CH_2O + H = HCO + H_2$	$2.19 \cdot 10^8$	1.77	3,000.0
28	$CH_2O + M = HCO + H + M$	$3.31 \cdot 10^{16}$	0.0	81,000.0
29	$CH_2O + O = HCO + OH$	$1.81 \cdot 10^{13}$	0.0	3,082.0
30	$HCO + OH = CO + H_2O$	$5.0 \cdot 10^{12}$	0.0	0.0
31 [a]	$HCO + M = H + CO + M$	$1.6 \cdot 10^{14}$	0.0	14,700.0
32	$HCO + H = CO + H_2$	$4.00 \cdot 10^{13}$	0.0	0.0
33	$HCO + O = CO_2 + H$	$1.0 \cdot 10^{13}$	0.0	0.0
34	$HCO + O_2 = HO_2 + CO$	$3.3 \cdot 10^{13}$	-0.4	0.0
35	$CO + O + M = CO_2 + M$	$3.20 \cdot 10^{13}$	0.0	$-4,200.0$
36 [b]	$CO + OH = CO_2 + H$	$1.51 \cdot 10^7$	1.3	-758.0
37	$CO + O_2 = CO_2 + O$	$1.6 \cdot 10^{13}$	0.0	41,000.0
38	$HO_2 + CO = CO_2 + OH$	$5.80 \cdot 10^{13}$	0.0	22,934.0
39	$H_2 + O_2 = 2OH$	$1.7 \cdot 10^{13}$	0.0	47,780.0
40 [b]	$OH + H_2 = H_20 + H$	$1.17 \cdot 10^9$	1.3	3,626.0
41 [b]	$H + O_2 = OH + O$	$5.13 \cdot 10^{16}$	-0.816	16,507.0
42 [a]	$O + H_2 = OH + H$	$1.8 \cdot 10^{10}$	1.0	8,826.0
43 [b]	$H + O_2 + M = HO_2 + M$	$3.61 \cdot 10^{17}$	-0.72	0.0
44 [c]	$OH + HO_2 = H_2O + O_2$	$7.5 \cdot 10^{12}$	0.0	0.0
45 [b]	$H + HO_2 = 2OH$	$1.4 \cdot 10^{14}$	0.0	1,073.0

(continued)

Table B (continued)

#	Reaction	A_o	b	E_a (cal/mol)
46	$O + HO_2 = O_2 + OH$	$1.4 \cdot 10^{13}$	0.0	1,073.0
47[b]	$2OH = O + H_2O$	$6.0 \cdot 10^8$	1.3	0.0
48	$H + H + M = H_2 + M$	$1.0 \cdot 10^{18}$	-1.0	0.0
49	$H + H + H_2 = H_2 + H_2$	$9.2 \cdot 10^{16}$	-0.6	0.0
50	$H + H + H_2O = H_2 + H_2O$	$6.0 \cdot 10^{19}$	-1.25	0.0
51	$H + H + CO_2 = H_2 + CO_2$	$5.49 \cdot 10^{20}$	-2.0	0.0
52	$H + OH + M = H_2O + M$	$1.6 \cdot 10^{22}$	-2.0	0.0
53	$H + O + M = OH + M$	$6.2 \cdot 10^{16}$	-0.6	0.0
54	$H + HO_2 = H_2 + O_2$	$1.25 \cdot 10^{13}$	0.0	0.0
55[c]	$HO_2 + HO_2 = H_2O_2 + O_2$	$2.0 \cdot 10^{12}$	0.0	0.0
56[c]	$H_2O_2 + M = OH + OH + M$	$1.3 \cdot 10^{17}$	0.0	45,500.0
57	$H_2O_2 + H = HO_2 + H_2$	$1.6 \cdot 10^{12}$	0.0	3,800.0
58[c]	$H_2O_2 + OH = H_2O + HO_2$	$1.0 \cdot 10^{13}$	0.0	1,800.0

[a] Subsidiary reaction path
[b] Main reaction path
[c] Additional reaction important for minor species

Table C Elementary reactions for formation of nitrogen oxides $k = A_o T^b \exp\left(-\frac{E_a}{R_u T}\right)$

#	Reaction	A_o	b	E_a (cal/mol)
1	$O + N_2 = N + NO$	$1.82 \cdot 10^{14}$	0.0	76,213.9
2	$O + NO = N + O_2$	$3.80 \cdot 10^9$	1.0	41,356.0
3	$H + NO = N + OH$	$2.63 \cdot 10^{14}$	0.0	50,393.6
4	$NO + M = N + O + M$	$3.98 \cdot 10^{20}$	-1.5	149,945.6
5	$N_2O + M = N_2 + O + M$	$1.60 \cdot 10^{14}$	0.0	51,600.0
6	$N_2O + O = NO + NO$	$6.92 \cdot 10^{13}$	0.0	26,615.8
7	$N_2O + O = N_2 + O_2$	$1.00 \cdot 10^{14}$	0.0	28,006.2
8	$N_2O + N = N_2 + NO$	$1.00 \cdot 10^{13}$	0.0	19,862.1
9	$N + HO_2 = NO + OH$	$1.00 \cdot 10^{13}$	0.0	1,985.3
10	$N_2O + H = N_2 + OH$	$7.60 \cdot 10^{13}$	0.0	15,096.1
11	$HNO + O = NO + OH$	$5.01 \cdot 10^{11}$	0.5	1,985.3
12	$HNO + OH = NO + H_2O$	$1.26 \cdot 10^{12}$	0.5	1,985.3
13	$NO + HO_2 = HNO + O_2$	$2.00 \cdot 10^{11}$	0	1,985.3
14	$HNO + HO_2 = NO + H_2O_2$	$3.16 \cdot 10^{11}$	0.5	1,985.3
15	$HNO + H = NO + H_2$	$1.26 \cdot 10^{13}$	0.0	3,972.9
16	$HNO + M = H + NO + M$	$1.78 \cdot 10^{16}$	0.0	48,663.9
17	$HO_2 + NO = NO_2 + OH$	$2.11 \cdot 10^{12}$	0.0	-479.0
18	$NO_2 + H = NO + OH$	$3.50 \cdot 10^{14}$	0.0	1,500.0
19	$NO_2 + O = NO + O_2$	$1.00 \cdot 10^{13}$	0.0	600.0
20	$NO_2 + M = NO + O + M$	$1.10 \cdot 10^{16}$	0.0	66,000.0

Note: E_a units cal/mol, A_0 units mol-cm-s-K

Appendix 5
Summary of Limits of Flammability[1]

Gas or vapor[a]	In air (%) Lower		In air (%) Higher		In O_2 (%) Lower		In O_2 (%) Higher	
	M1	M2	M1	M2	M1	M2	M1	M2
Inorganic								
Hydrogen	4.0	4.0	75	75	–	4.0	94	–
Deuterium	–	5	–	75	–	5	–	95
Ammonia	–	15	–	28	–	15	–	79
Hydrazine	–	4.7	–	100	–	–	–	–
Hydrogen sulfide	–	4.3	–	45	–	–	–	–
Hydrogen cyanide	–	6	–	41	–	–	–	–
Cyanogen	–	6	–	32	–	–	–	–
Carbon disulfide	1.25	–	44	50	–	–	–	–
Carbon oxysulfide	–	12	–	29	–	–	–	–
Carbon monoxide	12.5	–	74	–	–	15.5	–	94
Chlorine monoxide	–	23.5	–	100	–	–	–	–
Hydrocarbons								
Methane	5.3	5.0	14	15	5.1	–	61	–
Ethane	3.0	–	12.5	15	3.0	–	66	–
Propane	2.2	–	9.5	–	2.3	–	55	–
Butane	1.9	–	8.5	–	1.8	–	49	–
Isobutane	1.8	–	8.4	–	1.8	–	48	–
Pentane	1.5	1.4	7.8	–	–	–	–	–
Isopentane	1.4	–	7.6	–	–	–	–	–
2,2-Dimethyl propane	1.4	–	7.5	–	–	–	–	–
Dimethyl butane	1.2	–	7.0	–	–	–	–	–
2-Methyl pentane	1.2	–	7.0	–	–	–	–	–
Heptane	1.2	1.1	6.7	–	–	–	–	–
2,3-Dimethyl pentane	1.1	–	6.7	–	–	–	–	–
Octane	1.0	–	–	–	–	–	–	–

(continued)

[1]H.F. Coward and G.W. Jones, "Limits of Flammability of Gases and Vapors," Bulletin 503, Bureau of Mines, (1952) US Government of Printing Office, Washington DC.

Gas or vapor[a]	In air (%) Lower		In air (%) Higher		In O$_2$ (%) Lower		In O$_2$ (%) Higher	
	M1	M2	M1	M2	M1	M2	M1	M2
Isooctane	1.1	1.0	–	6.0	–	–	–	–
Nonane	–	0.8	–	–	–	–	–	–
Tetramethyl pentane	0.8	–	4.9	–	–	–	–	–
Diethyl pentane	–	0.7	–	5.7	–	–	–	–
Decane	0.8	–	–	5.4	–	–	–	–
Ethylene	3.1	2.7	32	34	3.0	–	80	–
Propylene	2.4	2.0	10.3	11	2.1	–	53	–
Butylene	2.0	–	9.6	–	–	–	–	–
Butene-1	1.6	–	9.3	–	1.8	–	58	–
Butene-2	1.8	–	9.7	–	1.7	–	55	–
Isobutylene	1.8	–	8.8	–	–	–	–	–
b-n-Amylene	1.5	1.4	8.7	–	–	–	–	–
Butadiene	2.0	–	11.5	–	–	–	–	–
Acetylene	2.5	2.3	–	81	–	–	–	–
Benzene	1.4	–	7.1	–	–	–	–	–
Toluene	1.4	1.3	–	6.7	–	–	–	–
o-Xylene	–	1.0	–	6.0	–	–	–	–
Ethyl benzene	1.0	–	–	–	–	–	–	–
Styrene	–	1.1	–	6.1	–	–	–	–
Butyl benzene	–	0.8	–	5.8	–	–	–	–
Naphthalence	–	0.9	–	5.9	–	–	–	–
Cyclopropane	2.4	–	10.4	–	2.5	–	60	–
Ethyl cyclobutane	1.2	–	7.7	–	–	–	–	–
Ethyl cyclopentane	1.1	–	6.7	–	–	–	–	–
Cyclohexane	1.3	–	8.0	–	–	–	–	–
Methyl cyclohexane	1.2	–	–	–	–	–	–	–
Ethyl cyclohexane	0.9	–	6.6	–	–	–	–	–
Alcohols								
Methyl alcohol	7.3	6.7	–	36	–	–	–	–
Ethyl alcohol	4.3	3.3	–	19	–	–	–	–
n-propyl alcohol	–	2.1	–	13.5	–	–	–	–
Isoproyl alcohol	–	2.0	–	12	–	–	–	–
n-Butyl alcohol	–	1.4	–	11.2	–	–	–	–
Amyl alcohol	–	1.2	–	–	–	–	–	–
Furfuryl alcohol	–	1.8	–	16.3	–	–	–	–
Allyl alcohol	–	2.5	–	18.0	–	–	–	–
Propylene glycol	–	2.6	–	12.5	–	–	–	–
Triethylene glycol	–	0.9	–	9.2	–	–	–	–
Ethers								
Methyl ether	3.4	–	18	27	–	3.9	–	61
Ethyl ether	1.9	1.7	48	–	2.0	2.1	–	82
Ethyl n-propyl ether	1.9	–	24	–	2.0	–	78	–
Isopropyl ether	1.4	1.3	21	–	–	–	69	–
Vinyl ether	–	1.7	27	28	–	1.8	–	85

(continued)

Gas or vapor[a]	In air (%) Lower		In air (%) Higher		In O_2 (%) Lower		In O_2 (%) Higher	
	M1	M2	M1	M2	M1	M2	M1	M2
Ethylene oxide	3.0	3.6	80	100	–	–	–	–
Propylene oxide	2.1	–	21.6	–	–	–	–	–
Dioxane	–	2.0	–	22	–	–	–	–
Trioxane	–	3.6	–	29	–	–	–	–
Acetal	1.6	–	10.4	–	–	–	–	–
Methyl cellosolve	–	2.5	–	19.8	–	–	–	–
Ethyl cellosolve	–	1.8	–	14.0	–	–	–	–
Butyl cellosolve	–	1.1	–	10.6	–	–	–	–
Diethyl peroxide	–	2.3	–	–	–	–	–	–
Aldehydes								
Acetaldehyde	4.1	–	55	–	–	4	–	93
Paraldehyde	–	1.3	–	–	–	–	–	–
Butyraldehyde	2.5	–	–	–	–	–	–	–
Acrolein	2.8	–	31	–	–	–	–	–
Croton aldehyde	–	2.1	–	15.5	–	–	–	–
Furfural	2.1	–	–	–	–	–	–	–
Ketones								
Acetone	3.0	2.5	11	13	–	–	–	–
Methyl ethyl ketone	–	1.8	–	10	–	–	–	–
Methyl propyl ketone	–	1.5	–	8	–	–	–	–
Methyl butyl ketone	–	1.3	–	8	–	–	–	–
Methyl isobutyl ketone	–	1.4	–	7.5	–	–	–	–
Cyclohexanone	–	1.1	–	–	–	–	–	–
Isophorone	–	0.8	–	3.8	–	–	–	–
Acid; anhydrides								
Acetic acid	–	5.4	–	–	–	–	–	–
Acetic anhydride	–	2.7	–	10	–	–	–	–
Phthalic anhydride	–	1.7	–	10.5	–	–	–	–
Esters								
Methyl formate	5.9	5.0	20	23	–	–	–	–
Ethyl formate	2.7	–	13.5	16.4	–	–	–	–
Butyl formate	–	1.7	–	8	–	–	–	–
Methyl acetate	–	3.1	–	16	–	–	–	–
Ethyl acetate	2.5	2.2	9	11	–	–	–	–
Vinyl acetate	2.6	–	13.4	–	–	–	–	–
Propyl acetate	2.0	1.8	–	8	–	–	–	–
Isopropyl acetate	–	1.8	–	8	–	–	–	–
Butyl acetate	1.7	1.4	–	7.6	–	–	–	–
Amyl acetate	–	1.1	–	–	–	–	–	–
Methyl cellosolve acetate	–	1.7	–	8.2	–	–	–	–
Methyl propionate	2.45	–	13	–	–	–	–	–
Ethyl propionate	1.85	–	11	–	–	–	–	–
Methyl lactate	–	2.2	–	–	–	–	–	–
Ethyl lactate	–	1.5	–	–	–	–	–	–

(continued)

Gas or vapor[a]	In air (%) Lower		In air (%) Higher		In O$_2$ (%) Lower		In O$_2$ (%) Higher	
	M1	M2	M1	M2	M1	M2	M1	M2
Ethyl nitrate	4.0	–	–	–	–	–	–	–
Ethyl nitrite	4.1	3.0	–	–	–	–	–	–
Phenols								
Cresol	–	1.1	–	–	–	–	–	–
Amines & imines								
Methylamine	4.9	–	20.7	–	–	–	–	–
Dimethylamine	2.8	–	14.4	–	–	–	–	–
Trimethylamine	2.0	–	11.6	–	–	–	–	–
Ethylamine	3.5	–	14.0	–	–	–	–	–
Diethylamine	1.8	–	10.1	–	–	–	–	–
Triethylamine	1.2	–	–	8.0	–	–	–	–
Propylamine	2.0	–	10.4	–	–	–	–	–
n-Butyl amine	1.7	–	9.8	–	–	–	–	–
Allyamine	2.2	–	22.	–	–	–	–	–
Ethylene imine	3.6	–	46	–	–	–	–	–
Other nitrogen compounds								
Acrylonitrile	3.0	–	–	17	–	–	–	–
Pyridine	–	1.8	–	12.4	–	–	–	–
Nicotine	–	0.7	–	4.0	–	–	–	–
Halogen derivatives								
Methyl chloride	10.7	7.6	17.4	19	–	8	–	66
Methyl bromide	–	13.5	–	14.5	14	–	19	–
Methylene chloride	–	–	–	–	15.5	–	–	66
Ethyl chloride	3.8	–	15.4	–	4.0	–	67	–
Ethyl bromide	–	6.7	–	11.3	6.7	–	44	–
Ethylene dichloride	6.2	–	16	–	–	–	–	–
Vinyl chloride	4.0	–	22	–	4.0	–	70	–
Dichloroethylene	9.7	–	12.8	–	10	–	26	–
Trichloroethylene	–	–	–	–	–	10	–	65
Ethylene chlorohydrin	–	4.9	–	15.9	–	–	–	–
Propyl chloride	–	2.6	–	11.1	–	–	–	–
Propylene dichloride	–	3.4	–	14.5	–	–	–	–
Allyl chloride	3.3	–	11.1	–	–	–	–	–
Allyl bromide	4.4	–	7.3	–	–	–	–	–
2-Chloropopene	4.5	–	16.0	–	4.5	–	54	–
n-Butyl chloride	–	1.8	–	10.1	–	–	–	–
Isobutyl chloride	–	2.0	–	8.8	–	–	–	–
Butyl bromide	–	5.2	–	5.6	–	–	–	–
Chlorobutene	2.2	–	9.3	–	–	–	–	–
Isocrotyl chloride	4.2	–	19.	–	4.2	–	66	–
Isocrotyl bromide	6.4	–	12	–	6.4	–	50	–
n-Amyl chloride	–	1.6	–	8.6	–	–	–	–
tert-Amyl chloride	–	1.5	–	7.4	–	–	–	–
Chlorobenzene	–	1.3	–	7.1	–	–	–	–

(continued)

Gas or vapor [a]	In air (%) Lower		In air (%) Higher		In O$_2$ (%) Lower		In O$_2$ (%) Higher	
	M1	M2	M1	M2	M1	M2	M1	M2
Dichlorobenzene	–	2.2	–	9.2	–	–	–	–
Miscellaneous								
Dimethyl sulfide	2.2	–	19.7	–	–	–	–	–
Ethyl mercaptan	2.8	–	18.0	–	–	–	–	–
Diethyl selenide	–	2.5	–	–	–	–	–	–
Dimethyldichloro-silane	3.4	–	–	–	–	–	–	–
Methyltrichcloro-silane	7.6	–	–	–	–	–	–	–
Tin tetramethyl	–	1.9	–	–	–	–	–	–
Lead tetramethyl	–	1.8	–	–	–	–	–	–
Water gas	7.0	–	72	–	–	–	–	–
Carbureted water gas	5.5	–	36	–	–	–	–	–
Pittsburgh natural gas	4.8	–	13.5	–	–	–	–	–
Other natural gas	3.8—6.5	–	13–17	–	–	–	–	–
Benzine	1.1	–	–	–	–	–	–	–
Gasoline	1.4	–	7.6	–	–	–	–	–
Naphtha	–	0.8	–	5	–	–	–	–
Kerosine	–	0.7	–	5	–	–	–	–
Coal gas	5.3	–	32	–	–	7	–	70
Coke-oven gas	4.4	–	34	–	–	–	–	–
Blast furnace gas	35	–	74	–	–	–	–	–
Producer gas	17	20–35	70	70–80	–	–	–	–
Oil gas	4.7	–	33	–	–	–	–	–

[a] Ordinary temperatures and pressures

M1: upward propagation of flame in large vessels, open at their lower ends.

M2: mixtures contained in closed or small vessels.

Appendix 6
Minimum Ignition Energy

Substance	MIE (mJ)	Ref.	Substance	MIE (mJ)	Ref.
ABS	30	3	Carbon monoxide	<0.3	2
Acetaldehyde	0.37	1	Carbon disulphide	0.009	1
	0.36	3		0.015	3
Acetone	1.15	1	Casein	60	1
Acetyl cellulose	15	1	Celluose	35	1
Acetylene	0.017	1	Celluose acetate	20–50	3
Acrolein	0.13	1	Charcoal	20	3
Acrolonitrile	0.16	1	Chromium	140	3
Adicpic acid	60	1	Cinnamon	30	1
Alfalfa meal	32–5100	3	Coal	40	1
Allyl chloride	0.78	3	Coal pittsburgh	250	3
Aluminum	50	1	Cocoa	100	1
Aluminum stearate	15	1		100–180	3
Ammonia	680	2	Coffee	160	3
Antimony	1920	3	Copal	30	1
Aspirin	25–30	3	Cork powder	45	1
Aziridine	0.48	2		35–100	3
Benzene	0.20	1	Corn meal	40	1
Bisphenol-A	1.8	4	Corn flour	20	1
Black Power	320	3	Corn starch	30–60	3
Boron	60	3	Cotton (filler)	25	1
1,3-butadiene	0.13	1	Cotton linters	1920	3
Butane	0.25	1	Cyclohexane	0.22	1
	0.26	3	Cyclopentane	0.54	1
n-Buyl chloride	0.33	3		0.24	3
Cadimium	4000	3	1,3-Cyclopentadiene	0.67	1

(continued)

Substance	MIE (mJ)	Ref.	Substance	MIE (mJ)	Ref.
Cyclopropane	0.17	1	Grain dust	30	3
	0.18	3	Grass seed	60–260	3
Dextrine	40	1	Hemp	30	1
Dichlorosilane	0.015	3	Heptane	0.24	1
Diethyl ether	0.19	1	Hexamethylene-tetramine	10	1
	0.2	3			
2,3-Dihydopyran	0.36	1	Hexane	0.24	1
Diisobutylene	0.96	1		0.29	3
Diisopropyl ether	1.14	1	Hydrogen	0.011	1
Dimethoxymethane	0.42	3		0.017	3
Dimethyl amine	<0.3	2	Hydrogen sulphide	0.068	1
2,2-dimethyl butane	0.25	1		0.077	3
Dimethyl ether	0.29	1	Isooctane	1.35	1
2,2-Dimethyl propane	1.57	1	Isopentane	0.21	1
Dimethyl sulphide	0.5	2		0.25	3
	0.48	3	Isopropyl alcohol	0.65	1
Dinitrobenzamide	45	3	Isopropyl amine	2.0	1
Dinitrobenzoic acid	45	3	Isopropyl chloride	1.55	1
Dinitro-sym-diphenylurea	60	3		1.08	3
Dinitrotoluamide	15	3	Isopropyl ether	1.14	2
Dioxane	<0.3	2	Isopropyl mercaptan	0.53	1
Di-ter-butylperoxide	0.5	2	Lignin	20	1
	0.41	3	Lycopodium	50	3
Epoxy resin	15	1	Magnesium	80	1
Ethane	0.24	1		40	3
	0.26	3	Manganese	305	3
Ethene	0.07	2	Melamine formaldehyde	50320	3
Ether	0.19	2			
Ethyl acetate	1.42	1	Methane	0.28	1
Ethyl amine	2.4	1		0.3	3
Ethyl cellulose	10	1	Methanol	0.14	1
Ethyl chloride	<0.3	2	Methyl acetylene	0.11	1
Ethylene	0.07	3		0.115	3
Ethylene oxide	0.06	1	Methylal	0.5	2
	0.065	2	Methyl cyclohexane	0.27	1
	0.062	3	Methylene chloride	10000	1
Flour, cake	25–80	3	Methyl ethyl ketone	0.53	1
Furan	0.22	1	Methylformate	0.5	2
Gasoline	0.8	3	Methylmethacrylate	15	1

(continued)

Substance	MIE (mJ)	Ref.	Substance	MIE (mJ)	Ref.
Nitrostarch	40	3	Rubber (hard)	30	1
Nylon	20	1	SAN	30	3
	20–30	3	Shellac	10	1
Paper dust	20–60	3	Silicon	100	3
Paraformaldehyde	20	1	Soap	60	1
Pentaerythritol	10	1		60–120	3
Pentane	0.22	1	Soy flour	100–460	3
2-Pentane	0.18	1	Sugar	30	1
Petroleum ether (benzene)	0.25	3	Sulphur	15	1
			Tantalum	120	3
Phenol formaldehyde	10–6000	3	Tetrafluoroethylene	3.5	3
Phosphorus (red)	0.2	1	Tetrahydrofuran	0.54	1
Phthalic anhydride	15	1	Tetrahydropyran	0.22	1
PMMA	15–20	3	Thiophene	0.39	1
Polyacrylonitrile	20	3	Thorium	5	1
Polycarbonate	25	1	Tin	80	3
Polyethylene	10	1	Titanium	40	1
	70	3		25	3
Polyethylene teraphthalate	35	3	TNT	75	3
			Toluene	0.24	3
Ploypropylene	25–400	3	Trichloroethylene	295	3
Polystyrene	40–120	3	Triethyl amine	0.75	1
Polyvinyl acetate	160	3		1.15	3
			2,2,3-Trimethylbutane	1.0	1
Polyvinyl acetate alcohol	120	3	Uranium	45	1
			Urea formaldehyde	80–1280	3
Polyvinly butyral	10	3	Vanadium	60	3
Potato starch	20	1	Vinyl acetate	0.7	1
	25	3	Vinyl acetylene	0.082	1
Propane	0.25	1	Vinyl chloride	<0.3	2
	0.26	3	Wheat flour	50	1
Propene	0.28	2	Wheat starch	20	1
Propionaldehyde	0.4	2		25–60	3
Propylchloride	1.08	1	Wood bark	40–60	3
Propylene	0.28	1	Wood (filler)	20	1
Propylene oxide	0.13	1	Wood flour	30–40	3
	0.14	3	m-,o-,& p-xylene	0.2	3
Pyrethrum	80	3	Yeast	50	3
Rayon	240	3	Zinc	960	3
Rice	40	1	Zirconium	5	1
	40–120	3		15	3

References

1. Haase, H (1977) Electrostatic Hazards, Their Evaluation and Control, Verlag Chemie, Weinheim
2. Berufsgenossenschaften, Richtlinien Statische Eletrizitat, ZH/200 (1980), Bonn. Buschman, C.H. (1962) De Veiligheid 38: 20–28
3. Babrauskak, V. (2003) Ignition Handbook, Fire Science Publishers, Issaquah WA
4. Bisphenol-A: Safety & Handling Guide, Publication Number AE-154, Bisphenol-A Global Industry Group

Appendix 7
Antoine Equation

$$\ln P = A - \frac{B}{T + C}$$

where P is in kPa and T is in K. The calculations are based on the Antoine correlation and the Peng-Robinson equation of state. For convenience the appropriate Antoine parameters are given for the calculation.

Species	A	B	C
Ammonia	15.494	2363.24	−22.6207
Aniline	15.0205	4103.52	−62.7983
Acetone	14.7171	2975.95	−34.5228
Ammonia	15.494	2363.24	−22.6207
Acetonitrile	14.8766	3366.49	−26.6513
Acetylene	14.8321	1836.66	−8.4521
Benzene	14.1603	2948.78	−44.5633
Biphenyl	14.4481	4415.36	−79.1919
Carbon disulfide	15.2388	3549.9	15.1796
Carbon tet	14.6247	3394.46	−10.2163
Chlorobenzene	14.305	3457.17	−48.5524
cis-2-Butene	13.8005	2209.76	−36.08
Cyclohexane	13.7865	2794.58	−49.1081
Cyclopentane	13.844	2590.03	−41.6716
Diethyl ether	14.1675	2563.73	−39.3707
Diethyl ketone	14.3864	3128.36	−54.4122
Dimethyl ether	14.3448	2176.84	−24.6733
Ethane	13.8797	1582.18	−13.7622
Ethanol	16.1952	3423.53	−55.7152
Ethyl benzene	13.9698	3257.17	−61.0096
Ethyl formate	14.4017	2758.61	−45.7813
Ethyl amine	14.4758	2407.6	−45.7539
Ethylene	13.8182	1427.22	−14.308
Formaldehyde	14.3483	2161.33	−31.9756
Hydrogen chloride	14.7081	1802.24	−9.6678
Hydrogen cyanide	15.4856	3151.53	−8.8383
Hydrogen sulfide	14.5513	1964.37	−15.2417

(continued)

Species	A	B	C
iso-Butane	13.8137	2150.23	−27.6228
iso-Butene	13.9102	2196.49	−29.863
iso-Butanol	15.4994	3246.51	−826,994
iso-Pentane	13.6106	2345.09	−40.2128
Iso-Propanol	15.6491	3109.34	−73.5459
n-Butane	13.9836	2292.44	−27.8623
n-Butanol	14.6961	2902.96	−102.912
n-Butylbenzene	14.0579	3630.48	−71.8524
n-Butylcyclohexane	13.8938	3538.87	−72.5651
nHexane	14.0568	2825.42	−42.7089
n-Decane	13.9899	3452.22	−78.8993
n-Octane	14.2368	3304.16	−55.2278
n-Propylbenzene	13.9908	3433.51	−66.0278
n-Pentane	13.9778	2554.6	−36.2529
n-Propanol	15.2175	3008.31	−86.4909
Nitric oxide	16.9196	1319.11	−14.1427
Nitrogen	13.4477	658.22	−2.854
Nitrogen dioxide	21.9837	6615.36	86.878
m-Xylene	14.1146	3360.81	−58.3463
Methyl ethyl ketone	14.2173	2831.82	−57.3831
Methane	13.584	968.13	−3.72
Methyl acetate	14.7074	2917.7	−41.3724
Methanol	16.4948	3593.39	−35.2249
Methyl amine	14.8909	2342.65	−38.7081
Methylcyclohexane	13.763	2965.76	−49.7775
Methylcyclopentane	13.8064	2742.47	−46.5148
o-Dichlorobenzene	14.3011	3776.97	−63.6069
o-Xylene	14.1257	3412.02	−58.6824
Oxygen	13.6835	780.26	−4.1758
p-Xylene	14.0891	3351.69	−57.6
Phenol	15.2767	4027.98	−76.7014
Propane	13.7097	1872.82	−25.1011
Propionic acid	15.4276	3761.14	−66.0009
Propylene	13.8782	1875.25	−22.9101
Styrene	14.3284	3516.43	−56.1529
Toluene	14.2515	3242.38	−67.1806
Trimethylamine	13.865	2239.1	−33.8347
Vinyl chloride	13.6163	2027.8	−33.5344
1-Butene	13.8817	2189.45	−30.5161
1-Heptene	13.8747	2895.9	−53.9388
1-Hexene	13.7987	2657.34	−47.1749
1-Pentene	13.7564	2409.11	−39.4834
1,1-Dichloroethane	13.8796	2607.81	−48.9442
1,2-Butadiene	14.4754	2580.48	−22.2012
1,2-Dichloroethane	14.3572	3069.08	−42.3468
1,3-Butadiene	14.0719	2280.96	−27.5956
Water	16.5362	3985.44	−38.9974

Appendix 8
Flash Points for Common Fuels

Fuel	Formula	T_L (K) Closed	T_L (K) Open	T_b (K)	T_a (K)	$T_{f,ad}$[a] (K)	LFL (%)	h_{fg} (kJ/g)	Q_c[b] (kJ/g)
Methane	CH_4	–	–	111	910	2,226	5.3	0.59	50.2
Propane	C_3H_8	–	169	231	723	2,334	2.2	0.43	46.4
n-Butane	C_4H_{10}	–	213	273	561	2,270	1.9	0.39	45.9
n-Hexane	C_6H_{14}	251	247	342	498	2,273	1.2	0.35	45.1
n-Heptane	C_7H_{16}	269	–	371	533	2,274	1.2	0.32	44.9
n-Octane	C_8H_{18}	286	–	398	479	2,275	0.8	0.3	44.8
n-Decane	$C_{10}H_{22}$	317	–	447	474	2,277	0.6	0.28	44.6
Kerosene	$\approx C_{14}H_{30}$	322	–	505	533	–	0.6	0.29	44.0
Benzene	C_6H_6	262	–	353	771	2,342	1.2	0.39	40.6
Toluene	C_7H_8	277	280	383	753	2,344	1.3	0.36	41.0
Naphthalene	$C_{10}H_8$	352	361	491	799	–	0.9	0.32	40.3
Methanol	CH_3OH	285	289	337	658	2,183	6.7	1.10	20.8
Ethanol	C_2H_5OH	286	295	351	636	2,144	3.3	0.84	27.8
n-Butanol	C_4H_9OH	302	316	390	616	2,262	11.3	0.62	36.1
Formaldehyde	CH_2O	366	–	370	703	2,334	7.0	0.83	18.7
Acetone	C_3H_6O	255	264	329	738	2,121	2.6	0.52	29.1
Gasoline	$\approx C_{8.26} H_{15.5}$	228	–	306	644	–	1.4	0.34	44.1

T_L = flash point; T_a = autoignition; T_b = boiling point; $T_{f,ad}$ = adiabatic flame temperature; LFL = lean flammability limit; h_{fg} = heat of vaporization; Q_c = heat of combustion
[a] Based on stoichiometric combustion with air
[b] Water and fuel in gaseous state

Appendix 9
Properties of Some Alcohol Fuels and Ammonia

Methanol

Critical temperature: 513.15 K
Chemical formula: CH_3OH
Critical pressure: 7,950 kPa
Molecular weight: 32.0
Critical density: 275 kg/m^3

T_{sat} (K)	337.85	353.2	373.2	393.2	413.2	433.2	453.2	473.2	493.2	511.7
P_{sat} (kPa)	101.3	178.4	349.4	633.3	1,076	1,736	2,678	3,970	5,675	7,775
ρ_l (kg/m^3)	751.0	735.5	714.0	690.0	664.0	634.0	598.0	533.0	490.0	363.5
ρ_v (kg/m^3)	1.222	2.084	3.984	7.142	12.16	19.94	31.86	50.75	86.35	178.9
h_{lv} (kJ/kg)	1,101	1,070	1,022	968	922	843	756	645	482	
c_{pl} (kJ/kg-K)	2.88	3.03	3.26	3.52	3.80	4.11	4.45	4.81		
c_{pv} (kJ/kg-K)	1.55	1.61	1.69	1.83	1.99	2.20	2.56	3.65	5.40	
μ_l (μNs/m^2)	326	271	214	170	136	109	88.3	71.6	58.3	41.6
μ_v (μNs/m^2)	11.1	11.6	12.4	13.1	14.0	14.9	16.0	17.4	20.1	26.0
k_l (mW/m-K)	191.4	187.0	181.3	178.5	170.0	164.0	158.7	153.0	147.3	142.0
k_v (mW/m-K)	18.3	20.6	23.2	26.2	29.7	33.8	39.4	46.9	60.0	98.7
Pr_l	5.13	4.67	4.15	3.61	3.34	2.82	2.56	2.42		
Pr_v	0.94	0.91	0.90	0.92	0.94	0.97	1.04	1.35	1.81	
σ (mN/m)	18.75	17.5	15.7	13.6	11.5	9.3	6.9	4.5	2.1	0.09

Ethanol

Critical temperature: 516.25K
Chemical formula: CH_3CH_2OH
Critical pressure: 6,390 kPa
Molecular weight: 46.1
Critical density: 280 kg/m^3

T_{sat} (K)	351.45	373	393	413	433	453	473	483	503	513
P_{sat} (kPa)	101.3	226	429	753	1,256	1,960	2,940	3,560	5,100	6,020
ρ_l (kg/m^3)	757.0	733.7	709.0	680.3	648.5	610.5	564.0	537.6	466.2	420.3
ρ_v (kg/m^3)	1.435	3.175	5.841	10.25	17.15	27.65	44.40	56.85	101.1	160.2
h_{lv} (kJ/kg)	963.0	927.0	885.5	834.0	772.9	698.9	598.3	536.7	387.3	280.5
c_{pl} (kJ/kg-K)	3.00	3.30	3.61	3.96	4.65	5.51	6.16	6.61		
c_{pv} (kJ/kg-K)	1.83	1.92	2.02	2.11	2.31	2.80	3.18	3.78	6.55	
μ_l (μNs/m^2)	428.7	314.3	240.0	185.5	144.6	113.6	89.6	79.7	63.2	56.3
μ_v (μNs/m^2)	10.4	11.1	11.7	12.3	12.9	13.7	14.5	15.1	16.7	18.5
k_l (mW/m-K)	153.6	150.7	146.5	141.9	137.2	134.8	129.1	125.6	108.0	79.11
k_v (mW/m-K)	19.9	22.4	24.5	26.8	29.3	32.1	35.3	37.8	43.9	50.7
Pr_l	8.37	6.88	5.91	5.18	4.90	4.64	4.28	4.19		
Pr_v	0.96	0.95	0.96	0.97	1.02	1.20	1.31	1.51	2.49	
σ (mN/m)	17.7	15.7	13.6	11.5	9.3	6.9	4.5	3.3	0.9	0.34

1-Propanol

Critical temperature: 536.85K
Chemical formula: $CH_3CH_2\ CH_2OH$
Critical pressure: 5,050 kPa
Molecular weight: 60.1
Critical density: 273 kg/m^3

T_{sat} (K)	373.2	393.2	413.2	433.2	453.2	473.2	493.2	513.2	523.2	533.1
P_{sat} (kPa)	109.4	218.5	399.2	683.6	1,089	1,662	2,426	3,402	3,998	4,689
ρ_l (kg/m^3)	732.5	711	687.5	660	628.5	592.0	548.5	492.0	452.5	390.5
ρ_v (kg/m^3)	2.26	4.43	8.05	13.8	22.5	35.3	55.6	90.4	118.0	161.0
h_{lv} (kJ/kg)	687	645	594	544	486	427	356	264	209	138
c_{pl} (kJ/kg-K)	3.21	3.47	3.86	4.36	5.02	5.90	6.78	7.79		
c_{pv} (kJ/kg-K)	1.65	1.82	1.93	2.05	2.20	2.36	2.97	3.94		
μ_l (μNs/m^2)	447	337	250	188	148	119	90.6	70.0	61.4	53.9
μ_v (μNs/m^2)	9.61	10.3	10.9	11.5	12.2	12.9	14.2	15.7	17.0	19.3
k_l (mW/m-K)	142.4	139.2	138.4	133.5	127.9	120.7	111.8	100.6	94.1	89.3
k_v (mW/m-K)	20.9	23.	26.2	28.9	31.4	34.7	38.0	43.9	47.5	53.5
Pr_l	10.1	8.40	6.97	5.14	5.81	5.82	5.50	5.42		
Pr_v	0.76	0.82	0.80	0.82	0.85	0.88	1.11	1.41		
σ (mN/m)	17.6	16.15	14.42	12.7	10.77	8.85	6.35	4.04	2.6	0.96

Ammonia

Critical temperature: 405.55K
Chemical formula: NH_3
Critical pressure: 11,290 kPa
Molecular weight: 17.03
Critical density: 235 kg/m^3

T_{sat} (K)	239.75	250	270	290	310	330	350	370	390	400
P_{sat} (kPa)	101.3	165.4	381.9	775.3	1,425	2,422	3,870	5,891	8,606	10,280
ρ_l (kg/m^3)	682	669	643	615	584	551	512	466	400	344
ρ_v (kg/m^3)	0.86	1.41	3.09	6.08	11.0	18.9	31.5	52.6	93.3	137
h_{lv} (kJ/kg)	1,368	1,338	1,273	1,200	1,115	1,019	899	744	508	307
c_{pl} (kJ/kg-K)	4.472	4.513	4.585	4.649	4.857	5.066	5.401	5.861	7.74	
c_{pv} (kJ/kg-K)	2.12	2.32	2.69	3.04	3.44	3.90	4.62	6.21	8.07	
μ_l (μNs/m^2)	285	246	190	152	125	105	88.5	70.2	50.7	39.5
μ_v (μNs/m^2)	9.25	9.59	10.30	11.05	11.86	12.74	13.75	15.06	17.15	19.5
k_l (mW/m-K)	614	592	569	501	456	411	365	320	275	252
k_v (mW/m-K)	18.8	19.8	22.7	25.2	28.9	34.3	39.5	50.4	69.2	79.4
Pr_l	2.06	1.88	1.58	1.39	1.36	1.32	1.34	1.41	1.43	
Pr_v	1.04	1.11	1.17	1.25	1.31	1.34	1.49	1.70	1.86	
σ (mN/m)	33.9	31.5	26.9	22.4	18.0	13.7	9.60	5.74	2.21	0.68

Index